CHEMICAL ANALYSIS for Antibiotics Used in Agriculture

Edited by Hisao Oka, Hiroyuki Nakazawa,
Ken-ichi Harada, and James D. MacNeil

AOAC INTERNATIONAL
The Scientific Association Dedicated to Analytical Excellence*

by AOAC INTERNATIONAL
2200 Wilson Boulevard, Suite 400
Arlington, VA 22201-3301

Printed in the United States of America

Printed on acid-free paper

ISBN 0-935584-57-9

CONTENTS

Preface

Antibiotics are used frequently in modern agricultural practice, both on a herd basis as additives to feed and water to promote weight gain, improve feed conversion efficiency and to prevent disease, and, in higher dosage forms, for the treatment of individual animals for specific disease conditions. Such usages may lead to problems with residues in agricultural products and to environmental contamination. The inspection of agricultural products for unacceptable residues is therefore one of the most important duties of a public health agency or regulatory authority and it is from this perspective that the material in this book has been prepared. This is not meant to minimize or ignore environmental concerns that may relate to the use of these compounds. The editors hope that those with responsibilities for the analysis of environmental samples for these compounds will find ideas that they can use or adapt in their work.

Although microbiological assays have been more commonly used for the detection of antibiotic residues because of their wide applicability, they can also be complicated, time-consuming, nonspecific, and in some cases, lacking in adequate sensitivity. Recent developments in chemical residue techniques such as thin-layer chromatography, gas chromatography, high performance liquid chromatography and mass spectrometry have allowed analysts in this field to develop simple, rapid chemical analytical methods which feature the required sensitivity and specificity.

The objective of the editors is that the content of this book will prove useful for analytical chemists in regulatory government and university laboratories, as well as for newcomers to the field. For those whose training is not primarily in analytical chemistry, but who are faced with a need to analyze for theses chemicals, we hope the book will also serve as a helpful introduction to this field.

In chapters 1 through 3, current practices in various countries are introduced with respect to both the regulation and control of antibiotic residues in foods. Chapter 4 introduces the variety of test kit technologies that are presently available to be used both in the laboratories by technically trained staff and also by inspectors under field conditions, as well as initiatives to provide third-party validation of these kits. In Chapter 5 throughout Chapter 11, chemical residue methods available for the seven major categories of antibiotics used in agriculture, classified according to their chemical structures, are reviewed by scientists who are actively involved in the analysis of these compounds within a regulatory framework. In each of these chapters the author describes the structure of the antibiotics, the recent developments in the chemical analysis, and the limitations of the methods described as well as provide ideas for approaches to analytical methods and theoretical considerations to why particular approaches have been or could have been taken.

The editors hope that the ideas described in this book will inspire further development and application of chemical methods of analysis for antibiotics in foods and related products, and will prove useful also to those who may in the future be responsible for developing methodology for environmental samples. We express our sincere thanks to all contributors who have helped us to compile this book and to the scientific reviewers who have added their expertise to the final product. We also acknowledge the support of our various organizations for the time and facilities required to bring this idea to reality.

Finally, the enthusiasm, support, and assistance of the AOAC publications staff for this project has been deeply appreciated.

Hisao Oka
Hiroyuki Nakazawa
Ken-ichi Harada
James D. MacNeil
May 1995

Chapter 1

Regulatory Overview of Antibiotic Use in Food-Producing Animals in North America and Current Methods of Detection and Analysis

James D. MacNeil

Health of Animals Laboratory, Agriculture & Agri-Food Canada, 116 Veterinary Road, Saskatoon, Saskatchewan S7N 5E3, Canada

AND

Richard L. Ellis

Chemistry Division, Science & Technology, Food Safety & Inspection Service, United States Department of Agriculture 300 12th Street SW, Room 405-Annex, Washington, DC 20250, U.S.A.

Contents

Regulatory Overview

Historically, the legislative authority for food safety in general can be traced to the preamble of the United States Constitution's, "providing for the general welfare." In the United States, the regulatory authority for veterinary drug products in food-producing animals is delegated to the United States Food and Drug Administration's Center for Veterinary Medicine, Department of Health and Human Services. Legislative authority for the United States Food and Drug Administration (FDA) to carry out this responsibility lies within the broad framework of the 1938 United States Federal Food, Drug, and Cosmetic Act (FFDCA), even though this Act has been amended several times over the past 50 years to provide more comprehensive authority for food safety, medical devices and other public health matters. The FDA's general safety provision for new animal drugs is specifically referenced in section 512 of the FFDCA. Comparable sections for food additives and color additives are referenced in sections 409 and 706, respectively, where these types of substances are proposed for use in food-producing animals.

Substances intended for use in food-producing animals must be safe and effective in animals, and the edible products derived from these treated animals must be safe for human consumption. The Code of Federal Regulations (CFR), subpart E: "Regulation of Carcinogenic Compounds Used in Food-Producing Animals", relates directly to the safety of foods consumed by humans. Pertinent regulations that implement the statutory provisions of the FFDCA are found in 21 CFR part 514.1 for new animal drugs, part 70 for food additives and part 570 for color additives.

In Canada, the registration of veterinary drug products, including those for use in food-producing animals, is the responsibility of the Bureau of Veterinary Drugs, Health Protection Branch, Federal Department of Health, acting under the Food and Drugs Act and Regulations. This process closely parallels the registration process in the United States as provided for under a harmonization initiative of the United States-Canada Free Trade Agreement, although both countries retain their individual processes, responsibilities, and terminologies. The registration of a new veterinary drug product in Canada begins with an application by the Sponsor, known as a New Drug Submission (NDS). An addition to an existing drug label, such as the addition of a species, is applied for through a Supplementary New Drug Submission (S/NDS). Permits for the investigative use of a product are obtained through an application for an Investigational Drug (IND) or an Experimental Studies Certificate (ESC), which allows limited investigative use of the drug. Generally, the issuance of an IND or an ESC precedes the application for an NDS.

While it is not the intent here to describe in detail all the requirements for filing an NDS, a S/NDS, or an IND, there are general requirements that must be met to demonstrate that the product may be safely used for the intended purpose. For example, the Sponsor must provide information on the clinical trials for the product in the target species, human safety assessment data, and procedures for the analysis of the active ingredient in a medicated feed (if that is the route of administration). Also, a suitable method should be provided for the analysis of the residues in edible tissue at the maximum residue limit (MRL) that has been established by the regulatory authorities.

Regulatory requirements in the United States are similar to those in Canada. It should be noted, however, that procedures in the United States establish an MRL for nonedible tissues when this tissue is the best target for a marker residue. Based on the analysis of such tissue, a determination may be made about the residue safety of the entire carcass. In practical terms, the target tissue is typically liver or kidney.

To assist the animal health industry, the FDA has developed a set of general principles for evaluating the safety of compounds used in food-producing animals (*1*). Briefly, the Sponsor is required to furnish the FDA with the necessary scientific data for demonstrating the safety and efficacy of the product and proof that the residues of the substance in the edible products of treated animals are safe. These data must be provided from studies and procedures approved by the FDA, and must be consistent with the regulatory requirements in force at the time of registration. The FDA believes that complying with the following seven guidelines will provide an acceptable basis for determining the safety of the compound.

I. Guideline for metabolism studies and for the selection of residues for toxicological testing
II. Guideline for toxicological testing
III. Guideline for threshold assessment
IV. Guideline for establishing a tolerance
V. Guideline for the approval of a method of analysis for residues
VI. Guideline for establishing a withdrawal period
VII. Guideline for new animal drugs and food additives derived from fermentation

Sponsors may use these guidelines with the assurance that they describe FDA approved procedures. If a Sponsor believes that an alternative approach may be more applicable to a new substance, the guidelines do not prevent the Sponsor from proposing that alternative. In these instances, the Sponsor is encouraged to review the appropriateness of their approach with the FDA before proceeding with their studies.

The guidelines describe studies that the Sponsor needs to conduct in order to meet the statutory provisions of the FFDCA. Guidelines I and IV describe the appropriate scientific studies for gathering information on the probable consumption of the substance and its residues. The FDA describes the approved studies for obtaining information on the toxicity (including cumulative effects) of the substance and its residues as well as on the safety factors normally used in guidelines

II and III. Guidelines V and VI describe the appropriate studies for obtaining residue depletion information and analytical methods required to ensure that animal-derived foods are safe for human consumption. The final guideline, VII, describes the appropriate studies for showing the safety of "biomass" products. The more pertinent guidelines for this chapter are I, IV, V, and VII.

The data required of the Sponsor must include metabolism studies in the species for which the drug use approval is being sought. This will establish the fate of the residues and their distribution within animal tissues, and it will identify a target tissue and a marker residue that may be used for residue control programs. The target tissue is the tissue where the residues occur with the most persistence and the highest concentration. The marker residue is the individual substance (parent molecule or metabolite) for which residue analysis is carried out. Data must be presented demonstrating the quantitative relationship of the marker residue to the total residues of the drug in the tissues of the animal. This establishes an MRL that indicates when the toxicologically significant residues are at a safe concentration level in the animal.

A battery of toxicity tests are required to characterize the toxicological profile of the substance, to determine a no-effect level (NOEL), and to establish an acceptable daily intake of residues based on the consumption of the residues in edible food products. Using this basic information, FDA can establish a tolerance (equivalent to an MRL) for the marker residue. Current regulations require that the Sponsor must submit an analytical method that will successfully pass a multilaboratory validation study suitable for determining the marker residue at the residue tolerance. Ideally, the analytical method should be able to quantitate residues at one-half the residue tolerance. To determine adequate method performance, residues from treated animals are often analyzed. The Sponsors must also provide residue depletion data for the target species to establish an appropriate withdrawal period for the approved use of the substance.

The six-step procedure for conducting a human safety assessment to meet Canadian requirements has been described in Korsrud et al. (*2*) and has been harmonized with the FDA process under the terms of the United States-Canada Free Trade Agreement. To establish a target tissue and a marker residue for monitoring programs, the data package from the Sponsor must include metabolism studies in the target species to determine the fate of the residues, the form in which the residues are found, and their distribution between tissues. Comparative metabolism studies must also be carried out in laboratory test species. Toxicity and carcinogenicity testing must be conducted to establish what constitutes an acceptable residue risk level for those consuming the foods derived from the animals treated with the product. From this information and the information on the metabolism of the compound in the treated species, regulators can set an MRL for the marker residue based on the total residues found. The Sponsor must provide a method of analysis for the marker residue that is sensitive enough to detect residues quantitatively at the MRL. Finally, depletion studies must be provided for the target species, enabling the Bureau of Veterinary Drugs to set a withdrawal period for the approved use of the drug. In Canada, these methods usually are reviewed by a desk audit, not through actual testing in a regulatory laboratory prior to approval.

Although the regulatory authority for approving veterinary substances in the United States resides with the FDA's Center for Veterinary Medicine (CVM), the regulatory authority for determining compliance of animal-derived foods with established tolerances (MRLs) rests with the U. S. Department of Agriculture, Food Safety and Inspection Service (FSIS). The residue control programs established by the FSIS include the National Residue Program (NRP), exploratory residue testing programs, and inspector-generated in-plant residue test samples. Selection of these samples is based on available information (e.g., injection sites in edible tissue) that may indicate the animal carcass is adulterated and unsuitable for human consumption. To support the NRP inspection programs, a combination of in-plant tests and laboratory tests are employed. The NRP monitoring program contains several classes of veterinary drugs and is developed and published annually by FSIS (*3*). A major component of the NRP includes antibiotic residue testing.

In Canada, once a drug has been approved for use by the Bureau of Veterinary Drugs, residue monitoring of the drug on a routine basis in edible animal tissues is an integral part of the Canadian federal meat inspection system of the Food Production and Inspection Branch, Federal Department of Agriculture and Agri-Food (Agriculture and Agri-Food Canada). Antibiotic monitoring is conducted at the slaughter plants by veterinary meat inspectors who test both a random selection of slaughter animals, according to a national survey plan, and also any suspect animals. Monitoring for antibiotic residues in cultured finfish is the responsibility of the Inspection Service Branch, Federal Department of Fisheries and Oceans.

The marketing of milk and dairy products in Canada is controlled by the provincial governments. Dairy products intended for export, however, are subject to federal inspection, and additional testing of retail meat and dairy products for drug residues may be conducted by the Canadian Federal Department of Health (Health Canada).

There are less than 30 antimicrobial drugs approved by FDA for use in food-producing animals in the United States. The tolerance of antimicrobial residues in food permitted by FDA typically ranges from approximately 0–10 ppm. The tolerances vary based on the toxicity of the drug as determined by its toxicology, residue, and metabolism studies. In some instances other specialized studies may be required to assess drug-specific toxicological concerns. For example, hypersensitivity tests may be required for the β-lactam antibiotics. Recently, FDA has been concerned with how this standard human food safety assessment process accurately determines the safe concentration of antibiotic residues based on these traditional toxicological endpoints. Of particular concern has been the impact of low levels of antibiotics on the intestinal microflora. This has not been directly examined in these contemporary toxicology studies. In June 1992, CVM and the Animal Health Institute sponsored a symposium on the Microbiological Significance of Drug Residues in Food to assess whether or not the current toxicology models are appropriate in determining the impact of low concentrations of antibiotics on the human intestinal microflora (*4*).

It is not the intent here to give a comprehensive summary of this CVM document, but there are some significant points from the proposed guideline "Microbiological Testing of Antimicrobial Drug Residues in Food" worth highlighting.

1. After considering data from a large number of substances, including animal drugs and food additives, CVM has concluded that, based on the results of the standard toxicology tests, noncarcinogens will receive a maximum safe concentration of 1 ppm in the total diet of 1.5 kg. Therefore, the total residue limit for new animal antimicrobial products with no carcinogenic potential will be 1 ppm in the total diet unless additional microbiological testing is performed.

2. In instances where there are antimicrobial residues with only limited antimicrobial activity, CVM notes that all new animal antimicrobial drugs consumed as residues will have to be exposed to the metabolic activities of the target animal. Again, based on their review of the data, in most instances the metabolic action of the microflora- and drug-metabolizing enzymes of the target species will inactivate the antibiotic. When the total microbiologically active residue is below 1 ppm in the total diet, no further microbiological testing will be required.

One conclusion in the proposed guidelines is that it is the Sponsor's responsibility to comprehensively identify the microbiological activity of their product, to determine the appropriate microbiological endpoints to be measured, and to establish the antimicrobial no-effect level in an appropriate model system. These new guidelines may have a significant impact on future registration of antimicrobial animal drug products.

Table 1 lists the antimicrobial animal drugs and tolerances approved by the FDA, while Table 2 provides this information for those compounds registered for use in Canada.

Screening Methods

In both Canada and the United States, testing programs based on evaluating animals at slaughter for the presence of antibiotic residues utilize one of two commercially available microbial growth inhibition tests. The programs use either the Swab Test on Premises (STOP) (7), or the Calf Antibiotic and Sulfa Test (CAST) (*8*). Both tests were originally developed by scientists at the USDA–FSIS. These tests, while similar in principle, use different growth media and different test organisms, providing a different range of test sensitivities to various antibiotic compounds that may be present in the sample (*9*). Both tests are designed to be used on a test sample collected by inserting a swab into the kidney tissue, causing

Table 1 Antibiotic Maximum Residue Tolerances in Food-Producing Animals in the United States

Compound	Use	Cattle	Sheep and goats	Swine	Poultry
Amoxicillin	Bacterial enteritis	0.01 ET	—	—	—
Ampicillin	Bacterial enteritis	0.01 ET	—	0.01 ET	—
Apramycin	Bacterial enteritis	—	—	0.4 F[a]	—
				0.4 K	
				0.3 L	
				0.1 M	
Bacitracin	Bacterial enteritis	0.5 ET	—	0.5 ET	0.5 ET
Carbomycin		—	—	—	0 (0.5) ET
Ceftiofur	Respiratory disease	—[b]	—	—	—
Cephapirin	Mastitis	0.1 ET	—	—	—
Chlortetra-cycline	Bacterial enteritis	0 F[c]	—	0.2 F	1 F
		0.1 K	1 K[d]	4 K	4 K
		0.1 L	0.5 L	2 L	1 L
		0.1 M	0.1 M	1 M	1 M, 1 S
Cloxacillin	Mastitis	0.01 ET	—	—	—
Dihydro-streptomycin	Bacterial scours	0(0.5) ET[e,f]	—	—	—
Erythromycin	Respiratory disease, Growth promotion	0(0.3) ET	—	0.1 ET	0.125 ET

Table 1 Antibiotic Maximum Residue Tolerances in Food-Producing Animals in the United States-*continued*

Compound	Use	Cattle	Sheep and goats	Swine	Poultry
Gentamycin	Bacterial enteritis	—	—	0.4 F, 0.4 K	0.1 ET[g]
	Swine dysentery	—	—	0.3 L, 0.1 M	—
Lincomycin	Respiratory disease, swine dysentery	—	—	0.1 ET	—[b]
Neomycin	Bacterial enteritis	0.25 ET[e]	—	—	—
		1.00 F[b]	1.25 F	1.00 F	0.50 F
		0.75 K	1.25 K	1.00 K	1.00 K
		0.50 L	1.25 L	0.75 L	0.75 L
		0.25 M	0.25 M	0.25 M	0.25 M
Novobiocin	Mastitis, Synovitis	1ET	—	—	1ET
Oxytetracycline	Respiratory disease, Bacterial enteritis, Pneumonia	0.1 ET	0.1 ET	0.1 ET	1 F, 3 K, 1 L, 1 M
Penicillin	Respiratory disease, pneumonia	0.05 ET	0(0.04)ET	0(0.04)ET	0(0.04) ET[i]
Spectinomycin	Bacterial enteritis, Synovitis, Fowl cholera	—	—	—[j]	0.1 ET[k]
Streptomycin	Diarrhoea, Respiratory disease, Bacterial enteritis	—[l]	—	0 (0.5) ET	0 (0.5)ET
Tetracycline	Diarrhoea, Bacterial pneumonia, Bacterial enteritis	0.25 ET[e]	0.25 ET	0.25 ET	0.25 ET

Table 1 Antibiotic Maximum Residue Tolerances in Food-Producing Animals in the United States-*continued*

Compound	Use	Cattle	Sheep and goats	Swine	Poultry
Tiamulin	Dysentery	—	—	0.4 L[m]	—
Tilmicosin	Respiratory disease	1.2 L	—	—	—
Tylosin	Respiratory disease	0.2 F	—	0.2 F	0.2 F
		0.2 K	—	0.2 K	0.2 K
	Dysentery(s)	0.2 L	—	0.2 L	0.2 L
		0.2 M	—	0.2 M	0.2 M
	Sinusitis				
Virginiamycin	Dysentery(S)	—	—	0.4 F	0.2 F[n]
				0.4 K	0.5 K
	Growth promotion (P)	—	—	0.3 L	0.3 L
	Necrotic	—	—	0.1 M	0.1 M
	Enteritis(P)	—	—	0.4 S	0.2 S

ET: edible tissue; F: fat; K: kidney; L: liver; M: muscle; S, skin.
[a]Total residues.
[b]The tolerance for marker residue is not needed.
[c]Cattle only; calves 1 F, 4 K, 4 L, 1 M.
[d]Sheep only.
[e]Calves only.
[f]Administrative tolerance in calves and cattle, 2.0 K.
[g]Turkeys only.
[h]Administrative tolerance for all tissues and species except calves. Letter from J. Taylor, FDA, to L. Crawford, FSIS, January 26, 1988.
[i]Chickens, pheasants, and quail; turkeys 0.01 ET; ducks and geese 0.01 ET (action level.).
[j]For pigs less than 4 weeks of age and less than 15 pounds.
[k]Chickens only.
[l]Administrative tolerance, 2.0 K.
[m]Tolerance for 8-α-hydroxy-mutilin; corresponds to 10.8 ppm total residues in liver.
[n]Broiler chickens only.

tissue disruption and soaking up the released fluids with the swab. The swab is then placed in the assay dish, with a disc containing an antibiotic standard used to monitor the viability of the organism and its inhibition, and is then incubated overnight. The presence of a zone of inhibition around the swab indicates a microbial inhibitor is in the sample. Although the tests are intended for kidney tissue, inspectors have also successfully used these tests on liver and normal

Table 2 Recommended Usages, Withdrawal Periods, and Maximum Residue Limits for Antibiotics Registered for Treatment of Food-Producing Animals in Canada[a]

				Maximum withdrawal period[b]		Maximum residue limit[c]
Antibiotic	Form	Usage	Species	milk (hours)	meat (days)	(mg/kg)
Ampicillin	Injectable	Various	Cattle	48	6	—
			Swine	—	4	0.01 ET, Mk
(with sulbactam)	Injectable	Pneumonia, pasteurellosis	Cattle	—[d]	14	—
Bacitracin	Feed additive	Enteritis	Swine	—	0	—
		Growth promotion	Turkey	—	0	—
Ceftiofur	Injectable	Respiratory disease	Cattle	—[d]	0	—
Cephapirin	Mammary infusion	Mastitis (dry)	Dairy cattle	—[d]	42	0.1 ET
		Mastitis (lactating)	Dairy cattle	96	4	0.02 Mk
Chlortetra-cycline (CTC)	Various	Various	Calf	—[d]	18	4 L, K; 1 M, F
			Cattle	—[d]	10	0.1 ET
			Lamb	—[d]	4	0.1 M, 1 K
			Poultry	—[d]	10	1 M, L, F, 4 K
			Swine	—[d]	10	1 M; 2 L; 4 K, 0.2 F
Cloxacillin	Mammary infusion	Mastitis (dry)	Dairy cattle	—[d]	30	—
		Mastitis (lactating)		60	—	—
Dihydro-streptomycin	Injectable	Various	cattle swine	96	30 30	0.125 Mk
Erythromycin	Various	Various	Poultry	—	1	0.125 ET

Table 2 Recommended Usages, Withdrawal Periods, and Maximum Residue Limits for Antibiotics Registered for Treatment of Food-Producing Animals in Canada[a]—*continued*

Antibiotic	Form	Usage	Species	Maximum withdrawal period[b] milk (hours)	Maximum withdrawal period[b] meat (days)	Maximum residue limit[c] (mg/kg)
			Cattle	72	14	0.05 Mk
			Sheep	—	3	—
			Swine	—	7	0.1 ET
Gentamycin	Various	Various	Cattle	—	30	—
			Chicken	—	35	—
			Swine	—	42	0.4 K
			Turkey	—	63	0.1 ET
Hygromycin B	Feed additive	Anthelmintic	Chicken	—	2	—
			Swine	—	2	—
Lasalocid	Feed additive	Growth promotion	Cattle	—[c]	0	—
		Coccidiosis	Chicken	—	0	—
Lincomycin	Various	Growth promotion	Chicken	—	0	—
		Various	Swine	—	2	—
(with specti-nomycin)	Water	Respiratory disease	Poultry	—	3	—
		Dysentery	Swine	—	5	—
Monensin	Feed additive	Feed efficiency	Cattle	—	0	0.05 ET
		Coccidiosis	Chicken	—	3	—
Neomycin (alone or in combination)	Oral (water or feed)	Enteritis	Cattle	—[d]	30	calves: 0.25 ET
			Poultry	—	14	—
			Swine	—	14	—
			Sheep	—	14	—
	Bolus		Calves	—	30	—
			Swine	—	30	—

Table 2 Recommended Usages, Withdrawal Periods, and Maximum Residue Limits for Antibiotics Registered for Treatment of Food-Producing Animals in Canada[a] —*continued*

				Maximum withdrawal period[b]		Maximum residue limit[c]
Antibiotic	Form	Usage	Species	milk (hours)	meat (days)	(mg/kg)
Neomycin (plus OTC)	Feed additive	Stress	Cattle	—[d]	7	—
Novobiocin	Feed additive	Synovitis	Turkey	—	4	1.0 ET
(in combination)	Mammary Infusion	Mastitis	Cattle	72	35	1.0 ET
Oxytetracycline (OTC)	Various	Various	Cattle	—[c]	28	—
			Poultry	—	7	—
			Sheep	—	18	—
			Swine	—	28	—
Penicillin G (procaine)	Injectable	Various	Cattle	72	5	0.05 ET, 0.01 IU/mL Mk
			Horse	—	5	—
			Sheep	—	5	—
			Swine	—	5	—
			Turkey	—	—	0.01 ET
Penicillin G (benzathine)	Injectable	Various	Cattle	—[c]	14	—
			Horse	—	14	—
			Sheep	—	14	—
			Swine	—	14	—
Spectinomycin	Oral	Various	Chicken	—	5	0.1 ET
		Enteritis	Swine	—	21	—
	Injectable	Fowl cholera	Turkey	—	5	—
Streptomycin (in combinatons)	Bolus	Diarrhoea	Calves	—	30	—
	Oral (water or feed)	Various	Calves	—	7	0.125 Mk
			Poultry	—	7	—
			Swine	—	7	—

Table 2 Recommended Usages, Withdrawal Periods, and Maximum Residue Limits for Antibiotics Registered for Treatment of Food-Producing Animals in Canada[a] —*continued*

Antibiotic	Form	Usage	Species	Maximum withdrawal period[b] milk (hours)	meat (days)	Maximum residue limit[c] (mg/kg)
Tetracycline	Oral (water or feed)	Various	Calves	96	5	0.25 ET
			Poultry	—	5	0.25 ET
			Sheep	—	5	0.25 ET
			Swine	—	5	0.25 ET
	Uterine bolus	Various	Cattle	72	18	—
			Horse	—	18	—
Tiamulin	Feed additive	Dysentery	Swine	—	2	0.4 L
Tilmicosin	Injectable	Respiratory diseases	Cattle	—[c]	28	—
Tylosin	Various	Various	Cattle	—[c]	21	0.2 ET
	Oral	Respiratory diseases	Chicken	—	1	0.2 ET
	Various	Dysentery	Swine	—	14	0.2 ET
	Oral	Infectious sinusitis	Turkey	—	3	0.2 ET
Virginiamycin	Feed additive	Growth promotion	Chicken	—	0	—
			Swine	—	0	—

ET: edible tissue; Mk: milk; L: liver; M: muscle; F: fat; K: kidney.

[a]See references (5), (6).

[b]Represents maximum withdrawal period for a registered product containing this antibiotic as the primary or most persistent component of the formulation. Other approved formulations containing this antibiotic may have shorter withdrawal periods.

[c]Where no official MRL has been established for a compound or species, either an administrative action level or a zero tolerance may exist.

[d]Not approved for use in lactating dairy cattle. In cases where a withdrawal period for milk is given, there may be some approved formulations containing this compound not approved for use in lactating dairy cattle.

muscle tissue. Attempts to assay other tissues, such as heart muscle, can lead to a high rate of false positives due to interfering substances that are naturally present in the tissue. Because most antibiotics concentrate in the kidney, it remains the tissue of choice for testing.

In Canada, the STOP and CAST test kits are used in random surveys of slaughter animals, primarily beef and pork, to test for the prevalence of violative antibiotic residues. The kits are also used on tissues collected from animals that are identified as suspect during the inspection process (suspect animals may have visible injection sites, signs of illness, positively tested herdmates, etc.). When samples from an animal test positive, or when an inspector has reason to suspect the presence of antibiotic residues even if the test kit result is negative, the carcass is detained pending laboratory testing of kidney and muscle tissues. A maximum turnaround time of 4 days from receipt to report has been established for these held carcass samples. Shipment of sample to the laboratory can add another 1–3 days to this time. The average laboratory turnaround time for these samples, using the current testing scheme, is under 2 working days.

In the United States, the FSIS residue control programs primarily support animal inspection operations in Federal establishments. These programs integrate Federal or FSIS screening tests with available laboratory-based quantitative and confirmatory methods. The single greatest use of screening methods in the FSIS is for antibiotic residues in red meat animals.

CAST was developed to provide a simple, rapid, and reliable test for detecting violative concentrations of sulfonamide and antibiotic residues in meat and meat products from fancy veal calves, and has been in use since mid-1984. Within two years of its introduction, a greater diversity of antibiotic therapy was being used to replace the sulfonamide or antibiotic treatment practices prevalent when the CAST program was initiated. These changes in treatment made it very important for the scientist to adhere to specific procedures to ensure that a CAST response from a nonviolative antibiotic residue was not reported as a CAST-positive subject to regulatory action. For this reason, the FSIS standardized the zone of inhibition for CAST to be at least 18 mm when determining if violative concentrations of residues are present.

The FSIS Microbiology Division has recently developed a new antimicrobial screening test designed to detect the presence of antibiotic and sulfonamide residues in animal tissues. The new test, the Fast Antimicrobial Screen Test (FAST) intended for use in fancy veal calves, was designed to read an in-plant test result in a shorter time than either the STOP or CAST. The main difference with the FAST compared to the STOP and CAST is in the agar plate composition where two specific changes were made. A carbohydrate nutrient as well as a pH indicator, bromocresol purple were incorporated to provide a characteristic color change to the agar plate for positive results. With this test, a preliminary determination of an antibiotic positive is available in as little as 6 hours, and final test results are read at about 18 hours. A report on the field trials conducted on fancy veal calves for residues in liver and kidney in the western United States was recently published by FSIS (*10*). The significant findings of the field study comparing STOP and CAST with FAST, are noted below.

- Kidney is a better choice than liver as a test target tissue for detecting positive residue concentrations in muscle tissue.
- FAST in-plant test results at 6 and 18 hours are essentially equivalent for fancy veal calves.
- FAST is equivalent to CAST for screening fancy veal calves based on kidney test results at two calf slaughtering plants.
- Analytical sensitivities for the 18 hour FAST and CAST are equivalent for both antibiotics and sulfonamides in the CAST plants tested.
- Statistically, FAST is significantly better than STOP for screening cows, based on its ability to detect positive residues in kidney.
- Sensitivities for the STOP and 18 hour FAST are similar in kidney tissue, but may vary for individual compounds.

FAST is currently being implemented in calf slaughtering plants starting in the western United States.

In Canada, federally inspected eggs are also randomly tested for antibiotic residues using the Brilliant Black Reduction Test (BBR) (Laboratorium Enterotox, Germany). This test, originally developed for testing milk and other fluid samples, was adapted by Agriculture and Agri-Food Canada researchers for use on fluid egg samples. In collaboration with the manufacturer, a kit was developed for egg testing with a protocol specific to eggs. Following laboratory testing, inspectors were trained in the use of the test kit, which was introduced into field use in 1990. Any samples testing positive with the kit are submitted for laboratory confirmation.

The Canadian dairy industry has developed a system in collaboration with regulatory authorities whereby the lead role in product testing is taken by the industry itself, with regulatory monitoring conducted primarily by provincial authorities. Testing is usually conducted on bulk tank milk received at the dairy prior to processing for human consumption. A variety of test kits are used for residue monitoring, including the Brilliant Black Reduction Test (Laboratorium Enterotox), the Charm II tests (Charm Sciences, U.S.A.), Delvo tests (Gist-Brocades, Inc., the Netherlands) and LacTek tests (Idetek, U.S.A.). Industry and provincial dairy laboratories provide confirmatory back-up to the test kit results.

Laboratory Methods

The historical procedures for determining antibiotics are microbial inhibition bioassays. In most instances, these bioassays are the current primary methods for residue screening and quantitative analysis in the United States. Before proceeding with other microbiological or chemical bioassays for identifying antibiotic residues, the results of STOP in-plant tests need to be verified through the laboratory analysis of liver, kidney, and muscle tissue.

The FDA is encouraging scientists to develop chemical methods for supplementary quantitation and confirmation for old and new antibiotics. However, there are regulatory questions that must be addressed before chemical methods can be considered as alternatives or replacements for the microbial inhibition assays. Recently, the FSIS laboratories have been exploring the correlation between a liquid chromatography procedure and the official microbial inhibition assay for tetracyclines. We have prepared a preliminary report for the FSIS and the FDA to consider (*11*). There is some correlation between the two procedures, but more data are needed from the chemical methods before they can be submitted for consideration by the FDA as an alternative methodology.

The basic procedures in the FSIS residue control programs for determining antibiotic residues in animal tissue require a suitable test organism that can be incorporated into an agar layer for the microbial inhibition bioassay. Current laboratory procedures use a variety of test organisms for these multiresidue assays. Table 3 summarizes the analyte class and test organisms used for identification purposes (*12*). Through the use of standards, fingerprint profiles for the classes of antimicrobials have been developed.

A strength of the generalized response of microbial inhibition assays is that it detects almost all members of several antibiotic classes simultaneously at a relatively low cost. However, drawbacks may occur because some analytes have a reduced detection sensitivity to the microorganism or an inhibition by possible coextractives (such as lipids from animal tissues or lysozyme). Shepherd (*13*) describes several schemes used to minimize the influence of coextractive materials.

The use of chemical antibiotic methods is limited, particularly for older antibiotics, because microbial inhibition methods are commonly used for regulatory purposes. In many instances chemical methods are only used for chemical identification. More than 15 years ago, thin-layer chromatography methods were developed for identifying tetracyclines. Applications of high-performance liquid chromatography (LC) methods for antimicrobials have been described by Moats (*14–16*), but there are no published reports that any of these methods have been successfully subjected to interlaboratory validation studies.

In some instances, the FSIS has used a combination of thin-layer chromatography (TLC) and microbial inhibition for laboratory analysis. An example of this approach is in the analysis of apramycin. The analyte is extracted from tissues following treatment with aqueous potassium hydroxide. Potentially interfering substances are removed by precipitation with trichloroacetic acid and centrifugation. The supernatant is then passed through an ion-exchange column and the analyte is eluted with ammonium hydroxide. An aliquot of the eluate is spotted on

Table 3 Analyte Class and Test Organisms for Antimcrobials

Analyte class	Test organism
Penicillin (susceptible)	*Sarcina lutea* (ATCC 9341a)
Tetracycline (susceptible)	*Bacillus cereus var. mycoides* (ATCC 11778)
Streptomycin (susceptible)	*Bacillus subtilis* (ATCC 6633)
Tetracycline (resistant)	*Staphylococcus epidermidis* (ATCC 12228)
Streptomycin (resistant)	*Sarcina lutea* (ATCC 9341a)
Neomycin (susceptible)	*Staphylococcus epidermidis* (ATCC 12228)
Erythromycin (susceptible)	*Sarcina lutea* (ATCC 9341a)
Neomycin (resistant)	*Sarcina lutea* (ATCC 9341)
Erythromycin (resistant)	*Sarcina lutea* (ATCC 15957)

a silica gel plate, developed, and covered with an agar containing *Bacillus subtilis* as the test organism. The zone of inhibition is used to estimate the analyte concentration (*17*).

Aminoglycosides such as neomycin, gentamicin, and streptomycin, lack validated interlaboratory analytical chemistry methods, as is typical for antibiotics. These analytes are very polar compounds and have poor to nonexistent chromophores suitable for determinative analytical methods. Chemical analysis for these compounds in animal tissue is difficult, and the FSIS currently has no validated analytical methods for these analytes. The limited number of published methods for kidney tissue residue analysis (the target tissue) are high-performance liquid chromatography (LC) methods. These methods use fluorescence detection following post-column derivatization with ortho-phthalaldehyde or its derivatives. Most published nonvalidated methods for tissue analysis have had detection limits of approximately 100 µg/kg (*13*). The most sensitive published method for aminoglycosides is a fluorescence polarization immunoassay system for the quantitation of gentamicin residues in milk as well as in bovine and ovine species using kidney and muscle (*18*) where the reported detection limits are approximately 20 µg/kg.

A specific and reproducible liquid chromatographic procedure has been developed for the quantitative determination of gentamicin in porcine tissue (*17*). Concentrations of 0.4 mg/kg can be determined without interference from other chemical substances. Tissues are extracted with dilute sulfuric acid and purified using column chromatography, liquid-liquid extraction, and solid-phase extraction chromatography. Detection of gentamicin requires post-column derivatization using netilmicin as an internal standard to check the reproducibility of the retention time.

Effective chemical methods for detecting macrolide antibiotics are also limited and are usually only used to confirm their identity following identification by microbial inhibition assays. Macrolides may be complex mixtures of closely related compounds whose composition can depend on their source. Some members of this class, however, are generally pure substances. In this class of antibiotics sensitivity is often lacking because many of its members have marginal functional groups or chromophores suitable for chemical detection systems. Fortunately, macrolides such as erythromycin, novobiocin, and tylosin are not as broadly used as penicillins because they are generally less potent. Drugs such as oxacillin that are penicillinase resistant have been developed. Some determinative methods are available, though no collaboratively studied methods have been reported (*13*).

A liquid chromatographic method is available for the analysis of novobiocin and virginiamycin residues (*17*) that provides quantitative supporting data for the official microbiological method. Residues are extracted with methanol, filtered, and an aliquot is submitted to liquid chromatography directly for novobiocin. A second aliquot of the original extract is partitioned with dichloromethane for virginiamycin. Both compounds are determined in a single-gradient liquid chromatographic system. The method is suitable for bovine, porcine, and avian species using liver, muscle, or kidney tissue. The method has been automated for high sample throughput using a commercial robotic system.

Penicillins in animal tissue, however, may be detected using biochemical and physicochemical methods, which are more specific than microbial inhibition assays. The biochemical assays use competitive binding or similar principles and they may use penicillinase to block the action of the penicillin in the sample. Physicochemical methods usually employ high-performance liquid chromatography. These physicochemical methods tend to be more specific, more precise, and have lower limits of detection. However, none of these physicochemical methods have been subjected to a successful interlaboratory study.

A sensitive, determinative method is available for amoxicillin from bovine and swine tissues that uses aqueous extraction, removes potential interfering substances with trichloroacetic acid, and uses liquid chromatography following heat treatment to generate a fluorescent compound is available (*17*). The detection limit is approximately 10 μg/kg. Very little developmental work has been done on the cephalosporins. One reason for this is their relatively high safe concentration residue limits.

Tetracyclines have been extensively used as veterinary drugs. Although tetracyclines have been analyzed by chemical and microbiological methods, good analytical methods are rare. The classical method for residues in milk and animal tissues is a microbiological procedure (*12*) where the assay is an agar diffusion method using *Bacillus cereus* (ATCC 11778) as the test organism. The assay has a detection limit of 2–200 μg/kg depending on the tissue matrix being analyzed, and requires good quality control of the assay pH if acceptable results are to be achieved.

The FSIS has not found a satisfactory regulatory chemical method for the tetracyclines to date. However, the FSIS is currently seeking a suitable procedure. Presently, method development is focused on liquid chromatography, and to a lesser extent on thin-layer chromatography methods because of the polar characteristics of the tetracyclines.

A creative approach to an acceptable method has been developed. This method turns the chelating capability of tetracycline into an asset by using a chelated Sepharose adsorbent in a copper (II) form as part of the chromatographic cleanup procedure (*19*). Using this approach, acceptable recoveries have been reported with detection limits approaching 10 μg/kg. Another promising approach uses a solid-phase extraction procedure (*20*). Recoveries for seven tetracyclines are 20–50 μg/kg in honey and 10 μg/kg in bovine and swine tissue. Collaborative studies of methods based on both approaches are currently in progress.

Few veterinary drugs have been targeted for method development in North America and Europe like chloramphenicol. Because of the toxicological properties of chloramphenicol, the FSIS focused method development in the early 1980s at or below 10 μg/kg, which is consistent with the proposed Codex MRL of 10 μg/kg in food-producing animals and birds. Muscle tissue is the FSIS matrix of choice for determining chloramphenicol residues in red meat animals. One particular benefit of this is that these methods are applicable to both domestic and import residue testing programs. Approaches include various immunochemical and virtually every chromatographic and instrumental procedure and, in some instances, combinations of them. Some methods have been successfully subjected to interlaboratory studies (*21*, *22*).

For example, the method of Epstein et al. (*23*), which uses gas–liquid chromatography following derivatization and negative ion chemical ionization mass spectrometry, was successfully evaluated in an interlaboratory study. This latter procedure is the basis of the FSIS tissue residue method, which is suitable for urine and muscle tissue, includes a glucuronidase digestion step to convert any glucuronide metabolites to the parent drug, and permits free chloramphenicol and at least one metabolite of chloramphenicol to be determined. This is relatively important because the presence of this enzyme in the human digestive system would readily cause hydrolysis of the metabolite to the parent drug. Modifications to the method (*17*) include using meta-chloramphenicol as an internal standard and recovery index, and using a solid-phase extraction step to provide a cleaner extract. As a result of these changes, the procedure can readily determine chloramphenicol residues at concentrations of 1 μg/kg in tissue and in urine.

Although the demonstrated presence of drug residues in an edible tissue is a requirement for regulatory action in most cases, the FDA will take action on positive urine results for chloramphenicol. In both Canada and the United States, regulatory authorities will usually act on any proof that a food-producing animal has been exposed to a banned drug, such as chloramphenicol. In such situations, the qualitative confirmatory identification of the banned drug is the primary requirement. Quantitative analysis is not mandatory, but would usually be performed.

The FSIS has included approximately 20 of these antibiotic compounds in its National Residue Program (NRP) over the last 22 years. Table 4 lists the compounds included in the monitoring phase of the NRP (*24*). This phase of the NRP is based on a statistically designed random test sample basis. In addition to the discussion on the availability of validated analytical methods, there are several reasons that carbomycin, ceftiofur, cephapirin, spectinomycin, tiamulin, and tilmicosin have not been specifically tested in the NRP. These reasons include, but may not be limited to, the fact that the safe concentrations of residues are higher than the total residues of toxicological concern, that these compounds may not have been ranked by the FSIS Compound Evaluation System (CES), or if they have been, that they may have lower CES rankings than other compounds selected for the NRP, or they may not have sufficient veterinary use to warrant being included in the NRP. One of the strengths of the NRP, as Table 4 indicates, is the dynamic nature of the FSIS residue test programs. There are some analytes that have been in the program on a continuous basis while others have been cycled in and out of the NRP.

In Canada, testing to confirm positive rapid test results from slaughter plant meat samples is conducted within laboratories of the Food Production and Inspection Branch, Agriculture and Agri-Food Canada. Prior to 1984, confirmations were conducted primarily using microbial inhibition tests at laboratories within the various regions of the country. Research on chemical methods of analysis was begun in the late 1970s using such techniques as thin-layer chromatography (one- and two-dimensional) in combination with various detection systems, particularly bioautography and LC. In 1984, departmental scientists completed the development of an assay method that combined the separating power of thin-layer chromatography with microbiological detection to identify the antibiotics commonly used in food animals in Canada (*25*). The introduction of the thin-layer chromatography–bioautography (TLCBA) method resulted in centralizing all antibiotic confirmatory testing for Canadian federal meat inspection programs at the Health of Animals Laboratory, Saskatoon. While the use of TLCBA was a significant step forward in identifying those antibiotics that had produced positive test kit results at the slaughter plant, the methodology did have its limitations. Separations achieved by TLC were generally insufficient to do more than identify the grouping to which an antibiotic belonged, and would not, for example, distinguish between individual penicillins or tetracyclines. Test sensitivity was also variable, but because the assay organism *B. subtilis* and the growth medium were the same ones used in the STOP, it was believed that the TLCBA method matched the sensitivity of the STOP. In controlled experiments conducted with several antibiotics including chloramphenicol (*26*), chlortetracycline (*27*), oxytetracycline (*28*), and penicillin G (*29*), this was found to be the case.

However, the sensitivities of the STOP and TLCBA detection methods were considered inadequate to detect chloramphenicol residues. The addition of a gas chromatographic assay with a sensitivity of 5 ppb (*30*) was required, and this was

Table 4 Compounds Included in the NRP from 1972 to 1994

Antibiotic	Years
Apramycin	1983–1985, 1987–1990
Bacitracin	1987–1990
Chloramphenicol	1974–1975, 1989–1990
Chlortetracycline	1972–1994
Erythromycin	1972–1994
Gentamicin	1984–1994
Lincomycin	1985–1987, 1990
Neomycin	1972–1994
Novobiocin	1985–1990
Oxytetracycline	1972–1994
Penicillins[a]	1972–1994
Streptomycins[b]	1972–1994
Tetracycline	1972–1994
Tylosin	1984–1994
Virginiamycin	1985–1990

[a]Penicillins include all β-lactams such as amoxicillin, ampicillin, cloxicillin, and hetacillin.
[b]Streptomycins includes dihydrostreptomycin and streptomycin.

applied to samples from target populations beginning in 1984. The need for more selective identification and for quantitative results led to further work in developing additional chemical assays. Based on the chemical structures of the various antibiotics, research on suitable methods focused primarily on liquid chromatography. Advances in the preparation of solid-phase extraction cartridge technology made this a complementary approach to the initial extraction and the clean-up steps of the STOP and TLCBA methods.

In 1986, after testing several published methods while developing a liquid chromatographic assay for tetracyclines, a published method by Oka et al. *(20)* was investigated and found to give good results in the hands of Agriculture Canada analysts. This method was then added to the confirmatory scheme for STOP-positive samples. Samples testing positive for tetracyclines by TLCBA were subsequently analyzed by LC to determine which tetracycline was present and in what concentration. The method has been routinely used in Agriculture Canada's meat residue testing program since that time, and has been subjected to a collaborative study that is nearing completion.

A regulatory change in Canada to replace a zero tolerance (no detectable residue) with specific numerical tolerances, in keeping with Codex MRLs, provided an impetus for further research on quantitative chemical methods. In 1990, a liquid chromatographic assay for penicillin residues was developed and successfully implemented by the Health of Animals Laboratory (*31*), followed in 1991 by liquid chromatographic assays for streptomycin and dihydrostreptomycin (*32*) and tylosin and tilmicosin (*33*). A project was started in 1994 to investigate the stabilization of incurred tissue residues of these compounds to facilitate future interlaboratory validation trials of these methods.

In addition to the work done on liquid chromatographic assays, research had simultaneously been conducted on various test kits to screen samples for specific compounds in order to shorten the sample turnaround time in the laboratory. In 1992, it was realized that the initial screening of samples with TLCBA, which required 2.5 days to obtain a result, followed by the application of quantitative liquid chromatographic assays to obtain more specific information on compound identity and quantity present, was expensive both in terms of materials and analyst time and was not meeting laboratory needs to reduce costs and provide even faster service (*34*). As a result, TLCBA was dropped as an initial screening method. Instead, all samples are analyzed immediately upon receipt for penicillin (day 1), and samples above the tolerance for penicillin G are reported immediately without further analysis. Samples that test negative for penicillin are then screened for streptomycins and macrolides using Charm II tests, and any positives are then assayed by liquid chromatography (day 2). Simultaneous with the Charm tests, the samples are analyzed by liquid chromatography for tetracyclines and the results are reviewed and reported (day 3). Under the previous approach, only the initial TLCBA assay would be completed at this point. The additional selective assays that follow the TLCBA results would still need to be conducted before the sample could be reported. Because a large number of the samples tested are found to be in violation of tolerances for penicillin, the net result of this change has been to reduce sample turnaround time by about 2 working days. The average monthly turnaround times for these carcass-held samples is now under 2 working days from sample receipt. TLCBA is now reserved only for use in special situations where adequate confirmatory results cannot be obtained using alternative methods.

The current approach is based on the premise that the microbiological assay conducted at the slaughter plant provides initial screening information that indicates the probable presence of a microbial inhibitor. The purpose of the laboratory tests, therefore, is to provide confirmation of the presence of a specific antibiotic and to develop quantitative information to determine if the residues are above a violative level. The strategy is to combine a screening method (the test at the slaughter plant) with quantitative and confirmatory methods (the tests at the laboratory).

Results of Antibiotic Residue Test Programs

Table 5 summarizes the results of the U.S. domestic antibiotic residue testing programs. The results from the import testing program are not included because the residue testing on imported products is not based on the same principles as the domestic program. The domestic NRP monitoring program generally selects a number of specimens from normal-appearing carcasses to provide a 95% probability of detecting at least one violation when 1% of the animal population being tested is violative. In addition to profile information, the results are also used to identify producers or other entities marketing animals with violative concentrations of residues. The import residue testing is more of a quality assurance based sampling program to assure the United States that the foreign country residue control system, including residue testing, is operating in a state of control.

In reviewing Table 5, it is important to note that the CAST in-plant test is used only with veal calves under 150 pounds and less than three weeks old. Calves found violative with CAST are condemned without laboratory confirmation. The STOP in-plant test was originally designed and used only with culled dairy cows, but beginning in 1987 its use was expanded to other species suspected of having antibiotic or sulfonamide residues. Laboratory confirmation is needed to condemn the animal carcass. Variations in the number of samples tested in Table 5 have occurred for different reasons, for example special CAST sampling programs for plants that have unusually high numbers of violative animals (including some instances of 100% testing).

The results in Table 5 record only violative results and do not reflect that a significant number of other laboratory confirmation tests may have yielded positive but nonviolative samples. These confirmatory tests, as noted earlier, initially include laboratory analysis of liver, kidney, and muscle tissue with the STOP test procedure. Primary sources of violative antibiotic residues routinely occur in bovine animals, and secondary sources occur in other red meat species. Antibiotic violations in poultry species are low.

Subsequent testing of samples includes a series of seven laboratory tests, as noted in Table 3, to provide a fingerprint of microbiological activity used to characterize the antibiotic class of drugs. Other confirmatory tests are used to specifically identify as many of these specific antibiotics as possible. In some instances, the antibiotic results may only be identified as unidentified microbial inhibitors (UMIs), which is one reason that additional antibiotic residue methods are being developed to quantify and confirm drug residues.

Table 6 shows that the frequency of positive results on samples in Canadian federal meat inspection programs is low, with most of the positive findings and laboratory confirmed violative results being found in suspect animals. While there is some variability from species to species and year to year between random and suspect samples, generally about 3000 each of random beef and pork are tested

Table 5 U.S. Domestic Residue Program Results on Antibiotics

Year	NRP monitoring sample units (samples tested/violative)	Surveillance sample units (sample tested violative) In-plant	Followup
1988	4,849/21	CAST: 168,210/3,034	
		STOP: 56,565/2,108	6,005/151,
1989	8,038/52	CAST: 175,427/4,599	
		STOP: 85,083/2,937	1,733/74,
1990	7,299/78	CAST: 115,403/2,070	
		STOP: 118,533/3,202	430/35,
1991	5,909/51	CAST: 79,666/196	
		STOP: 117,850/2,701	196/24,
1992	4,044/17	CAST: 168,210/3,034	
		STOP: 124,461/2,206	1,371/37

annually, producing positive findings in only a few of the samples. Smaller numbers of random samples are tested for other species, including horses, sheep, and poultry.

Most test kit results are based on kidney testing, although other tissues are occasionally used by inspectors. For carcasses that have tested positive, samples of kidney, muscle, and injection site tissue (if apparent) are sent for laboratory confirmation. Approximately 80% of these submissions contain an identifiable antibiotic residue in at least one of the submitted tissues, but not all of these findings cause carcass condemnation. A positive result on kidney tissue currently leads to condemnation of the organ tissue only. A violative result in muscle tissue is required if the total carcass is to be condemned. A similar situation exists in the United States, where the Meat Inspection Act specifies that a violative residue must be found in muscle tissue if the whole carcass is to be condemned. It is also standard practice to trim out any visible injection sites from the carcass, even if the remaining muscle tissue contains no identified residues in violation of the tolerances.

It is somewhat difficult to draw comparisons between the data of the violations found or the carcasses condemned for the various fiscal years in Table 6 because of changes in both the analytical methodology and in the tolerances (MRLs) in Canada during the period discussed. Through much of this period, any detectable residue of any antibiotic was classed as a violation. This began to change as MRLs were set for the tetracyclines later in the 1980s, followed by penicillin and a number of other antibiotics in the 1990s. As a result, we may identify, for example, penicillin in tissue submitted from a STOP-positive sample, but it may

Table 6 Results of Antibiotic Testing in Federally Inspected Slaughter Houses in Canada[a]

Fiscal Year (April 1–March 31)	Number of animals tested (positive kidney results)	Muscle samples confirmed positive and violations[b]
1983–1984	4,904 (55)	28
1984–1985	6,016 (139)	28
1985–1986	9,031 (379)	56
1986–1987	13,335 (669)	149
1987–1988	22,509 (971)	547
1988–1989	26,327 (819)	186
1989–1990	42,425 (787)	416
1990–1991	40,677 (802)	324
1991–1992	44,935 (827)	296
1992–1993	47,143 (728)	276

[a]Based on the information contained in the "Annual Report on Chemical and Biological Testing of Agri-Food Commodities During the Fiscal Year," for the fiscal years listed above, published annually by the Agri-Food Safety and Strategies Division, Agriculture and Agri-Food Canada, Ottawa, Ontario, Canada.
[b]The number of carcasses in violation is based on the number of positive findings in muscle samples.

be at a level below the MRL. Prior to the 1990 Canadian establishment of 50 ppb MRL for penicillin in beef, any detectable residue using our LC method that had a quantitation limit of 5 ppb would have been considered a violation. These changes have contributed to the reduced number of muscle positives or violations reported in Table 6, even though the number of tests performed has increased since the mid-1980s and the number of positive tests conducted at slaughter plants has remained fairly constant for the last four years.

In the 1989–1990 fiscal year, Agriculture and Agri-Food Canada collected a random survey of 386 hen shell egg samples which were tested in our laboratory for antibiotic residues using TLCBA. No antibiotic residues were detected. In 1990–1991, random testing using the BBR was begun at the plant level. A total of 3,075 random samples and 38 monitoring samples were collected and tested, with no antibiotic residues being confirmed. In 1991–1992, 3,686 in-plant tests were conducted with the BBR, again with no antibiotic residues being confirmed. While the total numbers of samples tested since 1992 had not been published by Agriculture and Agri-Food Canada at the time this chapter was in preparation, the testing program has continued and it is our understanding that several samples have been confirmed to contain detectable residues of tetracyclines. The violation rate, based on these findings, is considered to be well below 1%.

Products such as evaporated milk, milk powders, whey powders, butter, and ice cream have also been tested for antibiotics by Agriculture & Agri-Food Canada laboratories using a variety of test-kit technologies, supplemented by TLCBA and LC methods. No violations were found in the 412 samples tested in 1989–1990 or the 589 samples tested in 1990–1991. Results of provincial and industry testing programs have not been published.

The Future

Looking to the future, we can expect to see a continued evolution of test methods for antibiotics in food animals in both the field tests used in slaughter plants and confirmatory tests applied in laboratories. The third party test kit evaluation programs reviewed in Chapter 4, New Test Kit Technology, for example, will have a significant impact on the manner in which test kits are evaluated and approved for regulatory use.

With the signing of the new General Agreement on Tariffs and Trade (GATT) agreement, increased recognition of validated laboratory methods that will withstand international scrutiny can be expected. The sanitary and phytosanitary provisions in the GATT will focus on the Codex Alimentarius Commission and its subsidiaries for scientific support in the resolution of disputes. The Codex Committee on Veterinary Drug Residues in Foods (CCVDRF) meets annually to evaluate recommendations from the Joint Expert Committee on Food Additives (JECFA) for MRLs. Through CCVDRFs subsidiary *ad hoc* Working Group on Methods of Sampling and Analysis, a review of current methods used in regulatory laboratories is also done to identify those methods that perform sufficiently well enough to be used on a routine basis for analyzing a given compound at the suggested MRL. The Working Group reviews methods that have been collaboratively studied under the criteria established by bodies such as AOAC INTERNATIONAL and the International Union of Pure and Applied Chemistry (IUPAC), as well as methods validated through other documented procedures such as those used by the U.S. Food and Drug Administration (FDA). Methods must be shown to be acceptable in enforcing the recommended MRL, they must be rugged and reliable, and they must be suitable for use in a normally well-equipped residue laboratory. Methods requiring unusual equipment or reagents that are not commercially available are not generally regarded as suitable.

The past decade has brought a shift from a total reliance on microbiological assays for the detection and identification of antibiotic residues to an increased use of chemical separation techniques. New techniques, such as supercritical fluid extraction, capillary zone electrophoresis, and the various combinations of separation techniques with mass spectrometry and tandem mass spectrometry, will bring further evolution to the North American regulatory program tests. Other developments in immunochemical and biochemical analysis, such as bioluminescence, offer opportunities for fast, sensitive, and selective screening tests. The recent conclusion of the North American Free Trade Agreement (NAFTA) linking Mexico more closely to the markets of the United States and Canada will, we hope, also lead to increased collaboration among our respective regulatory laboratories.

The evolution of methodologies is not, however, solely at the discretion of the laboratory analyst or manager. Changes in methodology will be undertaken with careful consideration of international implications, while fully respecting national requirements. Currently in the United States, microbial assays have official status for the determination of residues of many of the antibiotics. Thus, considerable time and effort will be required to produce the data that will enable chemical methods for these compounds to achieve official recognition for regulatory use.

References

1. Food and Drug Administration, *General Principles for Evaluating the Safety of Compounds in Food-Producing Animals* (Rockville, MD: Food and Drug Administration, 1986).
2. G. O. Korsrud, J. O. Boison, M. G. Papich, W. D. G. Yates, J. D. MacNeil, E. D. Janzen, R. D. H. Cohen, D. A. Landry, G. Lambert, and M. S. Yong, *Can. J. Vet. Res.* **57**: 223, 1993.
3. United States Department of Agriculture, *Compound Evaluation and Analytical Capability National Residue Program Plan* (Washington, D.C.: United States Department of Agriculture Food Safety and Inspection Service, 1993.
4. United States Food and Drug Administration, *Proposed Guideline. Microbiological Testing of Antimicrobial Drug Residues in Food* (Rockville, MD: United States Food and Drug Administration, 1993).
5. K. Bennett, ed., *Compendium of Veterinary Products,* 3rd ed., (Hensall, ON: Canadian Animal Health Institute, North American Compendiums Ltd., 1993).
6. *Canada Gazette, Part I,* Schedule No. 753, (Sept. 29, 1990), 3544.
7. R. W. Johnston, R. H. Reamer, E. W. Harris, H. G. Fugate, and B. Schwab *J. Food Prot.* **44**: 828, 1981.
8. United States Department of Agriculture, *Performing the Calf Antibiotic and Sulfa Test. A Self-Instructional Guide* (Washington, D.C.: United States Department of Agriculture, Food Safety and Inspection Service, Meat and Poultry Inspection Program, 1984).
9. G. O. Korsrud and J. D. MacNeil *J. Food Prot.* **51**: 43, 1988.
10. N. H. Thaker, W. F. Kelly, P. D. Saunders, N. R. Clich, and S. A. Bright, *Fast Antimicrobial Screen Test (FAST) In-Plant Study* (Washington, D.C.: United States Department of Agriculture, Food Safety and Inspection Service, 1993).

11. H. M. Marks and A. M. Yoffe *Evaluation of Different Methods for the Detection and Determination of Tetracyclines in Bovine Tissue* (Washington, D.C.: United States Department of Agriculture, Food Safety and Inspection Service, 1993).
12. United States Department of Agriculture, *Microbiology Laboratory Guidebook* (Washington, D.C.: United States Department of Agriculture, Food Safety and Inspection Service, 1974).
13. M. J. Shepherd, "Chapter 8," *Food Contaminants—Sources and Surveillance*, ed. C. Creaser and R. Purchase (Cambridge, U.K.: Thomas Graham House, 1991), 109.
14. W. A. Moats, *J. Chromatogr.*, **317**: 311, 1984.
15. W. A. Moats, "Agricultural Uses of Antibiotics," *ACS Symposium Series*, No. 320, ed. W. A. Moats (Washington, D.C.: American Chemical Society, 1985), 154.
16. W. A. Moats, *J. Assoc. Off. Anal. Chem.*, **73**: 343, 1990.
17. United States Department of Agriculture, *Analytical Chemistry Laboratory Guidebook* (Washington, D.C.: United States Department of Agriculture, Food Safety and Inspection Service, 1991).
18. S. A. Brown, D. R. Newkirk, R. P. Hunter, G.G. Smith, and K. Sugimoto, *J. Assoc. Off. Anal. Chem.*, **73**: 479, 1990.
19. W. H. H. Farrington, J. Tarbin, S. Bygrave, and G. Shearer, *Food Addit. Contam.*, **8**: 55, 1991.
20. H. Oka, K. Uno, K. -I. Harada, K. Yasaka, and M. Suzuki, *J. Chromatogr.*, **325**: 265, 1985.
21. R. M. L. Aerts, H. J. Keukens, and G. A. Werdmuller, *J. Assoc. Off. Anal. Chem.*, **72**: 570, 1989.
22. Codex Alimentarius Commission, *Seventh Report, Codex Committee on Residues of Veterinary Drugs in Food* (Washington, D.C.: 1992).
23. R. A. Epstein, R. A. Ashworth, and R. M. Simpson *Am. J. Vet. Res.*, **49**: 2075, 1986.
24. United States Department of Agriculture, *Compound Evaluation and Residue Information* (Washington, D.C.: United States Department of Agriculture, Food Safety and Inspection Service, 1994).
25. E. Neidert, P. W. Saschenbrecker, and F. Tittiger, *J. Assoc. Off. Anal. Chem.*, **70**: 197, 1987.
26. G. O. Korsrud, J. M. Naylor, C. D. Salisbury, and J. D. MacNeil, *J. Agric. Food Chem.*, **35**: 556, 1987
27. G. O. Korsrud and J. D. MacNeil, *Food Addit. Contam.*, **5**: 149, 1988.
28. J. D. MacNeil, G. O. Korsrud, J. M. Naylor, and W. D. G. Yates, *Am. J. Vet. Res.* **50**: 72, 1989.
29. J. D. MacNeil, G. O. Korsrud, J. O. Boison, M. G. Papich, and W. D. G. Yates, *J. Food Prot.* **54**: 37, 1991.

30. United States Department of Agriculture, "Chemistry Laboratory Guidebook, Science" (Washington, D.C.: United States Department of Agriculture, Food Safety and Inspection Service, 1983).
31. J. O. Boison, C. D. C. Salisbury, W. Chan, and J. D. MacNeil, *J. Assoc. Offic. Anal. Chem.*, **74**: 497, 1991.
32. G. C. Gerhardt, C. D. C. Salisbury, and J. D. MacNeil, *J. AOAC Int.*, **77**: 334, 1994.
33. W. Chan, G. C. Gerhardt, and C. D. C. Salisbury, *J. AOAC Int.*, **77**: 331, 1994.
34. J. D. MacNeil, J. R. Patterson, C. D. C. Salisbury, G. O. Korsrud, and J. O. K. Boison, *Proceedings of the Euroresidue II Conference on Residues of Veterinary Drugs in Foods*, ed. N. Haagsma, A. Ruiter and P.B. Czedik-Eysenberg (Veldhoven, The Netherlands, 1993), 469.

Chapter 2

Overview of Antibiotics Used for Agriculture and Residual Analysis in Japan

Hiroyuki Nakazawa

Department of Analytical Chemistry, Faculty of Pharmaceutical Science, Hoshi University, Ebara 2-4-41, Shinagawa-ku, Tokyo 146, Japan

Contents

Introduction

The Japanese life expectancy has increased every year, with both male and female Japanese enjoying the longest life spans in the world. In 1991, the average life expectancy for Japanese men was 76.11 years and for Japanese women was 82.11 years. The primary reason for this increase is the diversification of the Japanese diet. Also, there has been a significant change since World War II when Western foods were introduced and quickly became part of the Japanese diet. As a result, the consumption of rice has declined and dishes based on meat and dairy products have become popular. Young people, in particular, prefer western style meals, in which meat is a major ingredient as compared to the traditional Japanese style foods. As a result, the livestock industry in Japan has prospered.

Outlook of Livestock and Fishery Industries

Livestock production accounted for 27% of Japan's gross agricultural output in 1991 and outperformed the 25% figure set by rice production. In 1992 there were approximately 4.9 million cattle (including both dairy and beef), 10.9 million pigs, 336.2 million fowl (including broilers and layers), 29 thousand sheep, 35 thousand lambs, and 26 thousand horses in Japan as shown Figure 1 (*1*). The breeding scale in the livestock industry has been enlarged and has become more intense on reducing operating expenses from year to year. While the number of livestock breeders has generally been increasing, the number of farming households has been decreasing annually. The following statistics for the type and number of farms in 1992 (*1*) are followed by the rate of decline relative to 1991 and are indicated in parentheses: dairy farms, 55,000 (8%); beef cattle farms, 210,000 (5%); pig farms, 30,000 (17%); egg farms, 9,200 (9%); and chicken farms, 4,700 (7%). This reduction in the total number of each type of livestock farms is a result of the increasing age of farmers and their lack of successors for these farms and is especially prevalent among small- to medium-sized farms. In contrast, the

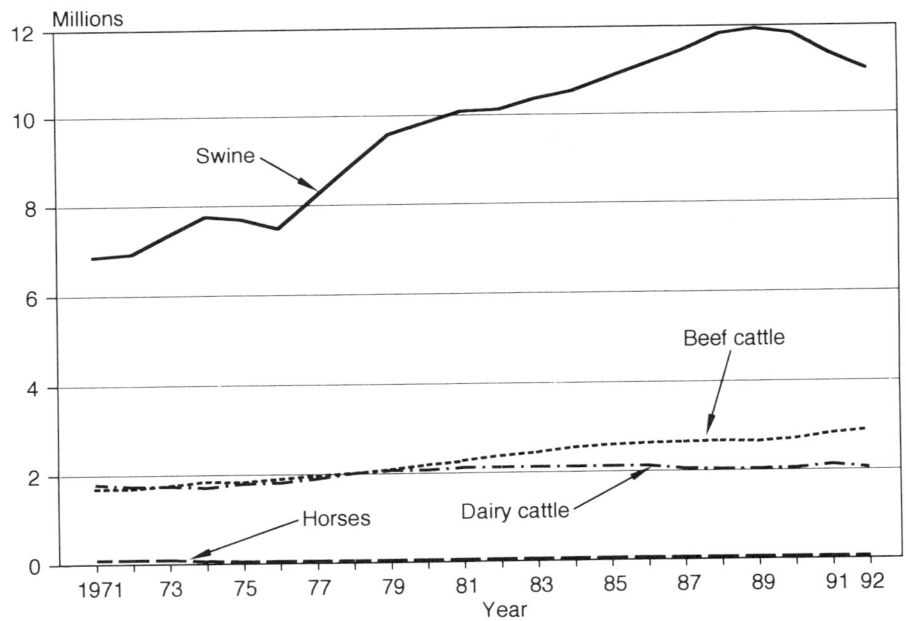

Figure 1 Number of livestock in Japan

number of large-scale livestock producers has increased, and their output per facility has been growing. For example, Figure 2 illustrates the trend towards large scale broiler breeding. On a typical chicken farm, birds are reared on a conveyor system, continuous feeding, and with no sunshine or exercise.

Although the average Japanese intake of meat has increased, the consumption of fish remains high when compared to the consumption of fish in other countries, as shown in Figure 3 (*1*). Marine products provided a daily supply of 18.4 g of animal protein per capita in 1991, accounting for 40% of the total protein supply (45.6 g) (*2*). Japan, which had been the world leader in fisheries production for many years, slipped to second place in 1989. In 1990 the Japanese accounted for 10.9% of the world's production of 101.58 million tons of fish. The total fish catch in 1991 was 9.98 million tons, 1.07 million tons less than the previous year. In per capita terms, the supply of fish and shellfish in 1991 decreased by 4.5 kg to 66.7 kg, compared to 1990. Factors contributing to this decline from 1989 to 1990 included a decrease in domestic catch, the scale of supply–demand for fish and shellfish, intensified restrictions on international fishing operations, and increased fishery imports, all of which contributed to a decrease for three consectuive years from 1989. In addition, the source of fisheries production in Japan has also changed, with the production of the marine culture, or aquaculture, industry increasing over the last 20 years. Concern over these issues in Japan has recently focused public attention on the aquaculture industry.

The livestock and fisheries industries which feature intensive practices, maintain a large population in a small area where the incidence of disease can result in a heavy economic loss. To decrease the cost of production and to improve the quality of livestock and fishery products while raising productivity, various compounds have proven effective against disease, including antibiotics used as feed additives and veterinary drugs. Figure 4 shows that the production of veterinary drugs, such as antibiotics, has increased in proportion to the increase in livestock production. Because antibiotics are officially inspected before marketing, the quantities produced for livestock use can be estimated. In 1991, approximately 58.9 tons (*3*) of antibiotic assays for animal use and veterinary pharmaceuticals sold, including antibiotics, had a market value of 75 billion Yen (*4*).

Outline of Feed Additives and Veterinary Drugs

Table 1 lists the feed additives used in Japan, which are divided into three categories according to their usage. The first group, comprised of antioxidants and antifungal agents, is used to prevent the quality of feed from deteriorating. The second group, which includes 7 amino acids, 28 vitamins, and 32 minerals, is used as dietary supplements. To improve the efficiency of feed utilization, a third group consisting of 7 synthetic antibacterials (such as amprolium and sulfa drugs) 7 and

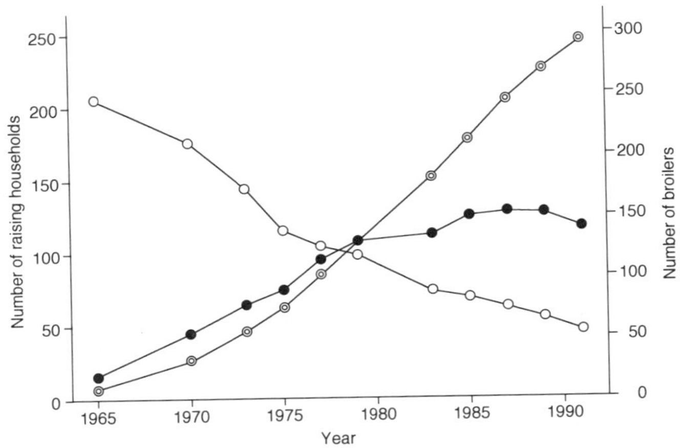

○: Number of raising households (× 100)
●: Total number of broilers (× 1,000,000)
◎: Number of broilers per one farm (× 100)

Figure 2 Comparison of annual population of broilers and number of raising households

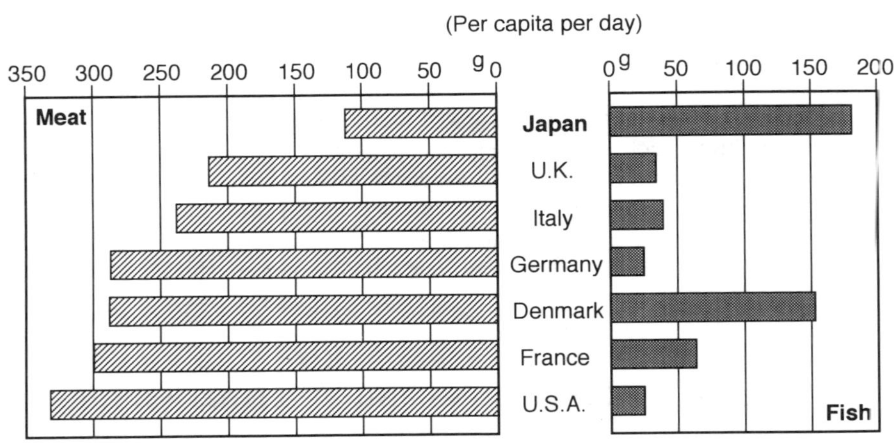

Figure 3 The amount of consumption of meat and fish of principal countries
SOURCE: Ministry of Agriculture, Forestry and Fisheries; "Food Comsumption Statistics," OECD.
Note: Japan 1991, others 1988.

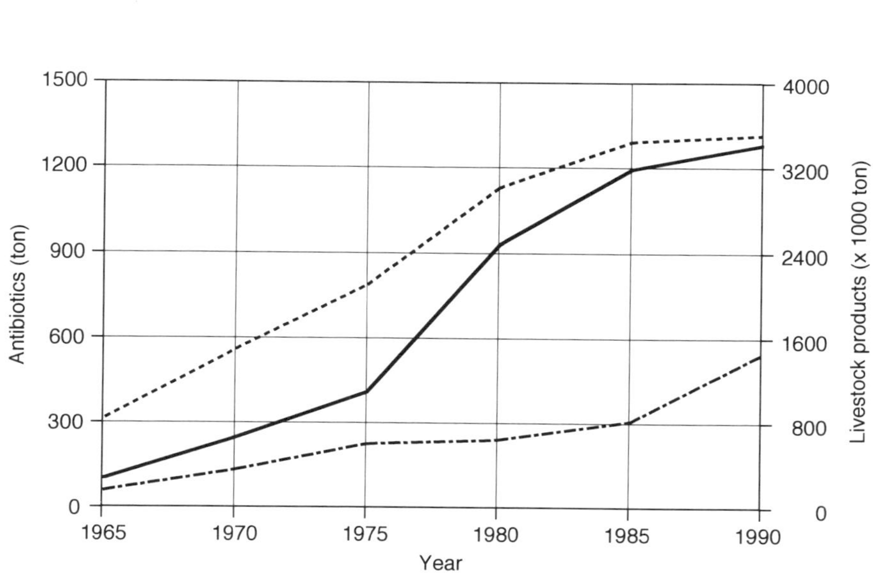

Figure 4 Production amount of antibiotics as veterinary drugs (——) and livestock products (domestic: - - - -, imported: — - — -)

21 antibiotics (avoparcin, tetracycline, etc.) has been used as feed additives. These substances are specified in the Law Concerning Safety Assurance and Quality Improvement of Feed in Animal Husbandry and Aquaculture (Government Ordinance No. 68 of 1976).

The principal legal regulation dealing with feed additives is the Pharmaceutical Affairs Law (Law No. 145) established in 1960. Veterinary drugs are intended exclusively for use in animals. Table 2 summarizes the antibiotics currently regulated as feed additives and veterinary drugs by the Pharmaceutical Affairs Law and the Law Concerning Safety Assurance and Quality Improvement of Feed. Synthetic antibacterials used as veterinary drugs and feed additives have been the subject of a recent review (*5*). Table 3 gives a brief comparison of feed additives and veterinary drugs. It indicates that feed additives are used at trace levels for a relatively long period of time for promoting growth, while veterinary drugs are prescribed at relatively large dosages over a short period of time for preventing and treating infectious diseases.

Legal Restrictions on Drug Residues in Livestock Products

Table 4 indicates the approved dosage of the antibiotics used in the treatment of fish diseases. In the case of spiramycin, for example, the drug is given in feed at 40 mg/kg of fish weight per day to treat streptococcists in yellowtail. Table 4

Table 1 Outlook of Feed Additives Used in Japan

Use	Classification	Designated feed additives
Prevention of quality deterioration	Antioxidants (3)[a]	BHA, BHT, Ethoxyquin
	Antifungal agents (3)	Propionic acid and its Na and Ca salt
	Emulsifiers (4)	Propyleneglycol
Dietary supplements	Amino acids (7)	DL-Alanine, L-tryptophan
	Vitamins (28)	L-Ascorbic acid, thiamine hydrochloride
	Minerals (32)	
Promotion of effective utilization of nutritional ingredients in feed	Synthetic antibacterials (7)	Amprolium, ethopabate, sulfaquinoxaline
	Antibiotics (21)	Avoparcin, tylosin, tetracycline

[a]The numbers in parenthesis indicate the feed additive numbers.

Table 2 Antibiotics Regulated as Veterinary Drugs, Feed Additives, or Both

Class	Antibiotics	Veterinary Drugs	Feed Additives
β-lactams	Amoxicillin	*	
	Ampicillin	*	
	Cloxacillin	*	
	Dicloxacillin	*	
	Mecillinam	*	
	Nafcillin	*	
	Penicillin G	*	
	Cephalonium	*	
	Cephazolin	*	
Aminoglycosides	Apramycin	*	
	Destomycin A	*	*
	Dihydrostreptomycin	*	

Table 2 Antibiotics Regulated as Veterinary Drugs, Feed Additives, or Both–*continued*

Class	Antibiotics	Veterinary Drugs	Feed Additives
	Fradiomycin	*	
	Gentamicin	*	
	Hygromycin B	*	*
	Kanamycin	*	
	Spectinomycin	*	
	Streptomycin	*	
Tetracyclines	Chlortetracycline	*	*
	Doxycycline	*	
	Oxytetracycline	*	*
	Tetracycline	*	
Macrolides	Erythromycin	*	
	Josamycin	*	
	Kitasamycin	*	*
	Mirosamycin	*	
	Oleandmycin	*	
	Sedecamycin	*	
	Spiramycin	*	
	Terdecamycin	*	
	Tylosin	*	*
Polypeptides	Avoparcin		*
	Bacitracin	*	*
	Colistin	*	*
	Enramycin	*	*
	Noshiheptide		*
	Thiopeptin	*	*
	Virginiamycin	*	*
Polysaccharides	Flavophospholipol	*	*
Polyethers	Lasalosid		*
	Monensin	*	*
	Salinomycin	*	*
	Avilamycin		*
Others	Bicozamycin	*	*

Table 2 Antibiotics Regulated as Veterinary Drugs, Feed Additives, or Both–*continued*

Class	Antibiotics	Veterinary Drugs	Feed Additives
	Chloramphenicol	*	
	Fosfomycin	*	
	Lincomycin	*	
	Novobiocin	*	
	Polynactin		*
	Tiamulin	*	

[a]Veterinary drugs: regulated by the Pharmaceutical Affairs Law.
[b]Feed additives: regulated as feed additives in the Law Concerning Safety Assurance and Quality Improvement of Feed.

also shows that a period of 30 days is required by law between the time of final treatment and the time of harvesting the fish for human consumption.

In recent years, public concern over the presence of drug residues in meat products has grown rapidly in Japan. To prevent the occurrence of veterinary drug residues, the law prescribes that animals are not to be slaughtered shortly after the drugs are administered or while the concentration of the drugs remains at therapeutically effective levels. However, instances of illegal or extra-label uses of these drugs are occasionally found.

Table 3 Comparison of Feed Additives and Veterinary Drugs

	Feed Additives	Veterinary drugs
Application	Growth promotion	Prevention and treatment of infectious diseases
Period	Long (2–3 months)	short
Dose	Trace	Relatively large amount
Regulation	SAQIF	Pharmaceutical Affairs Law
Synthetic antibacterials	Sulfaquinoxaline, clopidol, nicarbazin, olaquindox (7)	Sulfadimidine, carbadox, oxolinic acid etc (ca. 25)
Antibiotics	Chlortetracycline, lasalosid etc (21)	Ampicillin, kanamycin etc (ca. 50)

Residual Analysis of Feed Additives and Veterinary Drugs in Livestock Products

Another essential law is the Japanese Food Sanitation Law (Law No. 233) established in 1947, which describes standards and criteria of foods. According to this law, no food should contain antibiotics and meat, poultry, eggs, fish, shellfish, milk, and dairy products should not contain any synthetic antibacterial substances. Based on these legal regulations, it is necessary to analyze foods for residues of synthetic antibacterials and antibiotics used as feed additives or veterinary drugs. For this purpose, the screening techniques used should be relatively simple, rapid, and inexpensive and permit a large number of samples to be analyzed with high reliability. It is also necessary to use a different analytical technique to confirm the presence of a detected residue. The analytical methods approved by the Japanese Government are published as two volumes. Volume I contains the microbial inhibition tests organized by the Veterinary Sanitation Division, the Ministry of Health and Welfare, using antibiotic-sensitive strains of bacteria as test organisms (see Table 5). The microbial inhibition tests use a disk placed on an agar plate that contains an antibiotic-sensitive strain of bacteria that is incubated for a specified period of time (e.g., 17 hrs), depending on the specific assay. Although these bioassays are effective in detecting antibiotic residues, they suffer certain disadvantages including a requirement of having some technical skills and equipment, delayed results, and susceptibility to interfering substances which may yield false positives.

Table 4 Application and Dose of Antibiotics Used in the Treatment of Fish Diseases

Active drugs	Fish	Disease	Application and dose	Day[a]
Spiramycin	Yellowtail	Streptococcisis	40 mg/kg of fish weight per day in feed	30
Oxytetracycline hydrochloride	Red sea bream	Vibriosis	50 mg/kg of fish weight per day in feed	3
Oxytetracycline hydrochloride	Kuruma shrimp	Vibriosis	50 mg/kg of shrimp weight per day in feed	25

[a]Period required by law for stopping treatments before harvesting for human consumption.

Table 5 Individual Methods by Microbiological Assay for Residual Antibiotics in Livestock Products in Japan[a]

Test organisms	Antibiotics
Bacillus stearothermophilus var. calidolactis C-953	Ampicillin, cloxacillin, lasalosid, dicloxacillin, salinomycin
Bacillus cereus var. mycoides ATCC 11778	Oxytetracycline, chlortetracycline
Micrococcus flavus ATCC 10240	Bacitracin
Micrococcus luteus ATCC 9341	Tylosin, spiramycin, kitasamycin, oleandomycin, erythromycin, penicillin G
Bacillus subtilis ATCC 6633	Kanamycin, streptomycin, enramycin, monensin
Bacillus cereus ATCC 19637	Flavophospholipol
Pseudomonas syringae X 205	Hygromycin B
Corynebacterium xerosis NCTC 9755	Thiopeptin, virginiamycin
Bordetella bronchiseptica ATCC 4617	Colistin
Bacillus brevis ATCC 8185	Destomycin A
Staphylococcus epidermidis ATCC 12228	Fradiomycin, novobiocin
Escherichia coli NIHG	Chloramphenicol

[a]Announced Analytical Methods for Residual Substances in Livestock Products, Vol. 1, Veterinary Sanitation Division, Environmental Health Bureau, Ministry of Health and Welfare, Japan.

The simplified classification method for antibacterial residues, based on microbiological assay, was published in 1991 (*6*). Scheme I shows that this method applies a pulp disc technique using *Bacillus subtilis* (ATCC 6633), *Micrococcus luteus* (ATCC 9341), and *Bacillus cereus* var. *mycoides* (ATCC11778), in identifying antibacterial residues in fractions obtained by chemical separation. Fraction A contained macrolide antibiotics (MLs) and sulfa drugs (SAs), fraction B contained tetracyclines (TCs) and penicillins (PCs), and fraction C contained aminoglycoside antibiotics (AGs). Antibacterial agents detected in each fraction were confirmed by the different patterns of growth inhibition of *B. subtilis, M. luteus,* and *B. cereus* (see Table 6). The purpose of a screening test is to quickly determine whether or not an analyte is present at or near the level of concern in the target sample. While the development of analytical methods has made it possible to detect residual drugs at trace levels, the detection limit required for each drug in livestock products should be defined by a safety evaluation based on the toxicological aspects. So far, a level of 50 parts per billion (ppb) has been considered suitable for a provisional detection limit for the analysis of most drugs used in Japan. Chromatographic methods of analysis for antibiotics in animal tissues and foods have been the subject of several recent reviews (*7–9*) and are dealt with in detail in chapters 5–11.

Various chemical techniques have been developed for the analysis of residual antibacterials using spectrophotometric procedures, thin-layer chromatography (TLC), enzyme immunoassay (EIA), gas chromatography (GC), high-performance liquid chromatography (LC), and mass spectrometry (MS) (*10*, *11*). Volume II contains approved chemical methods based on TLC, GC and spectrophotometry for synthetic antibacterials (*12*). When using the methods in Volume II, some drugs cannot be detected with sufficient sensitivity to measure trace amounts and to simultaneously determine each individual drug, so the government has proposed a multiresidue analytical method using LC (*13*).

After reading recent reviews (*5*, *10–12*), it is clear that using LC with a UV detector has been widely applied for the routine analysis of residuals of veterinary

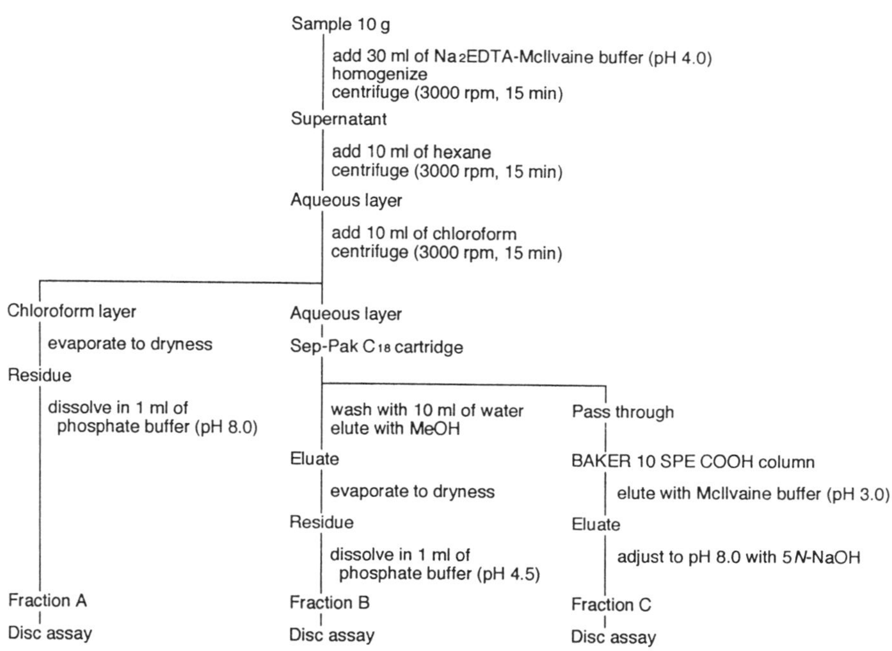

Scheme 1 Microbiological assay for antibiotics in livestock products

Table 6 Classification of Antibacterial Agents in Fractions by Inhibition Patterns of Test Organism

	Test organism			
Fraction	*B. subtilis*	*M. luteus*	*B. cereus*	Determination of antibacterial agent
A	+	#	–	MLs
A	#	–	–	SAs
B	+	–	#	TCs
B	+	#	–	PCs
C	#	–	+	AGs

SOURCE: See reference (*7*).

drugs, and that it is more accepted than GC techniques for these analytes. However, when the peak of a target drug has appeared on the LC chromatogram, conventional LC methods lack qualitative information. It is necessary to ensure the identification of the observed peak before taking regulatory action for food hygiene. Although a few GC–MS methods have been developed to confirm target drugs, these methods are complicated and time consuming, requiring suitable volatile derivatives to be prepared for the large-scale screening tests. For this purpose, alternative instrumental analyses using a photodiode-array detector or LC–MS (*14–19*) provide several powerful approaches in identifying the drugs detected in regulatory samples. The combination of LC with photodiode-array detection and with MS holds tremendous potential for the analysis of residual synthetic antibacterials, and there is considerable urgency for developing reliable instrumental methods of analysis for antibiotics in the edible tissues of farmed livestock and fish.

Surveillance for Residues in Livestock Products

Surveillance of food for chemical contaminants is carried out in many countries in order to ensure that the human diet is safe (*20, 21*). Table 7 shows the results of a national survey conducted by the Veterinary Sanitation Division, Environmental Health Bureau, Ministry of Health and Welfare, on residual antibiotics in domestic meat and fish in Japan, for the fiscal years 1990–1992. The monitoring samples collected at the urban and rural prefectures are analyzed by the laboratories of government authorities such as Meat Inspection Offices, Market Food

Table 7 Antibiotics Residues in Domestic Meat and Fish in Japan

Fiscal year	Sample number tested	Meat			Fish											Egg	Honey	Tot.
		Beef	Pork	Ch.	Y	Bream	Salmon	Carp	Trout	Ayu[a]	Eel	Mack-erel	Tilapia	Flat-fish				
1990	Total	594	910	301	40	72	1	53	38	12	67	—	—	—		—	137	2,224
	Penicillins	(1)	(3)	—	—	—	—	—	—	—	—	—	—	—		—	(2)	(6)
	Tetra-cyclines	—	(13)	—	—	—	—	—	—	—	—	—	—	—		—	(13)	(26)
1991	Total	1,866	4,974	612	88	83	10	61	73	51	93	1	13	10		341	210	8,486
	Penicillins	(5)	(7)	—	—	—	—	—	—	—	—	—	—	—		—	—	(12)
	Tetra-cyclines	—	(5)	(1)	—	—	—	—	—	—	—	—	—	—		(1)	(8)	(15)
	Aminogly-cosides	—	(1)	—	—	—	—	—	—	—	—	—	—	—		—	—	(1)
1992	Total	1,939	3,063	3,116	103	134	20	57	86	118	128	4	14	20		608	354	9,764
	Penicillins	(1)	(1)	—	—	—	—	—	—	—	—	—	—	—		—	—	(2)
	Tetra-cyclines	—	(9)	(1)	—	—	—	—	—	—	—	—	—	—		—	(7)	(17)

Ch.: Chicken; Y: Yellowtail, Tot.: Total. The numbers in parenthesis indicate the detected number.
[a]Ayu means sweetfish in Japanese.

Inspection Offices, and Institutes of Public Health. As was seen for the three year surveillance program using the approved biological assay, the livestock products available to the public contain a relatively low incidence of antibiotic residues.

Recent Developments

According to the Japanese Food Sanitation Law, food should not contain antibiotics and synthetic antibacterial substances. In order to monitor the drug residue levels in livestock products, simple and reliable methods are required. In addition to developing improved methods for residual analysis, evaluating veterinary drug residues should include the parent compounds and their metabolites in any edible portion of the animal product. The development of new veterinary drugs must also be evaluated based on the drugs' efficacy, their safety to the intended animal species, and their safety to the humans consuming animal products. Moreover, although animal drug residues in livestock do not appear to be a problem, it is necessary to monitor the various products by appropriate methods to ensure continued adherence to food hygiene standards.

In January 1994, the Japanese Health and Welfare Ministry requested the Food Hygiene Investigative Committee to determine maximum residue limits for those antibiotics, antibacterial agents, and hormones that have not been banned. According to the Japanese Food Sanitation Law, there is no legal regulation on anthelmintic residues in various livestock. However, in addition to antibacterial agents and hormones, tolerances are to be set for residual anthelmintics.

In Japan, approximately 30 anthelmintics are used to prevent and cure parasitosis from trematodes, nematodes, cestodes, and mites in fowl, swine, beef and dairy cattle, sheep, goats, horses, and honeybees. The anthelmintics used are synthetic antibacterials, benzimidazoles, organophosphorates, phenols, salicylanilides, and three antibiotics including destomycin A, hygromycin B, and ivermectin. Tolerances established for these compounds must be based on a consideration of the toxicological data and the establishment of a no-effect level (NOEL) for that substance in an experimental animal. As lower acceptable residue concentrations are recommended, it will be necessary to ensure the availability of a practical analytical method suitable for routine residue analysis at the limit established for the target drug in edible tissues.

The different methodological approaches, which include both microbiological assays and the current range of chemical analysis techniques such as LC with photodiode-array detection, GC–MS, and LC–MS, are powerful and useful for residual analysis. However, complete facilities for residue analysis using all such methods are not available in all regulatory laboratories. Although LC analyses are currently only used for official residual analysis of synthetic antibacterials, the development of reliable LC methods for antibiotics will help overcome these problems. The recruitment and promotion of talented analysts is an essential element in developing a successful program.

Conclusion

As the scale of production expands, the production cost declines and the time required for upkeep diminishes. There is a movement away from family-run farms to more efficient, large-scale operations managed as a company. Although the livestock industry has grown to meet the increased demand, the moderation of consumer demand and deregulation of beef imports have created a need for livestock producers to improve their business foundation.

Even though the incidence of antibiotic residues in foods of animal origin remains low and the human health risks associated with these residues are small compared with other food safety problems, the public's attention has focused on the residue issue. The establishment of international standards that ensure human food safety and the harmonization of international trade in animal products is necessary. Studies for the safety evaluation of veterinary drugs in food must be continued with respect to pharmacological, toxicological, and chemical aspects. In particular, the effects of residual antibiotics in foods on consumers, e.g. allergic reactions, or the induction of resistant bacteria must be investigated.

References

1. H. Arima, *NIPPON, A Charted Survey of Japan 1993–1994*, (Tokyo, Japan: the Kokusei-sha Corporation, 1993).
2. J. Ueda, *Asahi Shimbun Japan Almanac 1994* (Tokyo, Japan: Asahi Shimbun, 1993).
3. Ministry of Agriculture, Forestry, and Fisheries, *Statistics on Animal Hygiene 1992, Kachiku Eisei Tokei 1992*, (Livestock Industry Bureau, Ministry of Agriculture, Forestry & Fisheries, Japan: 1992).
4. Management and Coordination Agency, *Administrative Inspection Report on Veterinary Drugs 1991, Dobutsuyo Iyakuhin Ni Kansuru Gyosei Kansatsu Hokoku 1991*, (Management and Coordination Agency, Japan: 1991).
5. H. Nakazawa and M. Fujita, *Eisei Kagaku*, **36**: 163, 1990.
6. K. Jinbo, C. Monma, T. Maruyama, and M. Matsumoto, *J. Food. Hyg. Soc. Jpn.*, **32**: 86, 1991.
7. D. R. Bobbitt and K. W. Ng, *J. Chromatogr.*, **624**: 153, 1992.
8. J. O. Boison, *J. Chromatogr.*, **624**: 171, 1992.
9. S. A. Barker and C. C. Walker, *J. Chromatogr.*, **624**: 195, 1992.
10. H. Nakazawa, M. Fujita, M. Horie, and K. Takeba, *Anim. Husbandry*, **46**: 135, 1992.

11. H. Nakazawa, M. Fujita, M. Horie, and K. Takeba, *Anim. Husbandry*, **46**: 287, 1992.
12. H. Nakazawa, M. Fujita, and M. Horie, *Analysis of Antibiotic Drug Residues in Food Products of Animal Origin*, ed. V. K. Agarwal (New York: Plenum Press, 1992), 187.
13. M. Murayama, S. Uchiyama, and Y. Saito, *J. Food Hyg. Soc. Jpn.*, **32**: 155, 1991.
14. M. Horie, K. Saito, Y. Hoshino, N. Nose, N. Hamada, and H. Nakazawa, *J. Food Hyg. Soc. Jpn.*, **31**: 171, 1990.
15. M. Horie, K. Saito, Y. Hoshino, N. Nose, N. Hamada, and H. Nakazawa, *J. Chromatogr.*, **502**: 371, 1990.
16. F. A. Mellon, J. R. Chapman, and J. A. E. Pratt, *J. Chromatogr.*, **394**: 209, 1987.
17. M. Horie, K. Saito, Y. Hoshino, N. Nose, M. Tera, T. Kitsuwa, H. Nakazawa, and Y. Yamane, *Eisei Kagaku*, **36**: 283, 1990.
18. S. Pleasance, P. Blay, M. A. Quilliam, and G. O'Hara, *J. Chromatogr.*, **558**: 155, 1991.
19. Y. Ikai, H. Oka, and M. Kadota, *Tech. J. Food Chemistry & Chemicals*, **9**: 102, 1993.
20. M. S. Brady and S. E. Katz, *Analysis of Antibiotic Drug Residues in Food Products of Animal Origin*, ed. V. K. Agarwal (New York: Plenum Press, 1992), 5.
21. S. N. Dixon, D. R. Tennant, and J. F. Kay, *Safety of Chemicals in Food*, ed. D. Watson (New York: Ellis Horwood Limited, 1993), 8.

Chapter 3

Antibiotic Use in Animal Production in the European Union—Regulation and Current Methods for Residue Detection

Kevin N. Woodward
Veterinary Medicines Directorate, Woodham Lane, New Haw, Addlestone, Surrey KT15 3NB, Great Britain

AND

George Shearer
CSL Food Science Laboratory, Norwich Research Park, Colney, Norwich NR4 7UQ, Great Britain

Contents

Introduction

In the 1930s, prontosil, a dyestuff, was found to have antibacterial activity in animals but no similar effects in in vitro experiments. Further research demonstrated that the activity was due to the in vivo release of sulfanilamide, a sulfonamide and the forerunner of this group of drugs that are now widely used as antibacterial agents in human and veterinary medicine (*1*).

Sulfonamides are not strictly antibiotics but rather antibacterial compounds. The term antibiotics is reserved for agents derived from living organisms, or for synthetic or semisynthetic analogues of such compounds. The first true antibiotic to be isolated was penicillin, which was identified by Alexander Fleming and introduced into human clinical practice following the investigative work by Florey in the 1940s (*2*). Clinical use in animals soon followed and a wide variety of antibacterial compounds and antibiotics is now available for therapeutic and prophylactic purposes in animal medicine, both in companion animals and in food-producing species (*1*). This chapter will concentrate on the latter use only.

Antibacterial compounds and antibiotics usually belong to well-defined classes of chemical substances, although there are a number that can best be categorized into a "miscellaneous" group since their structures do not fit comfortably along side those of other molecules. As indicated earlier, the first major class of antibacterial compounds to be discovered and to be used extensively, was the sulfonamides.

Because antibiotics and antibacterials are regulated in the European Union (EU) by the same legislation, in the interest of completeness for the Regulatory

Section, the general classes of both groups have been briefly discussed. Structures of typical members of each class are shown in Figure 1.

Drug Classes

Sulfonamides

For reasons already explained, sulfonamides are not true antibiotics, but instead are synthetic chemicals originally derived from the dyestuffs industry. Nevertheless, no account of antibiotics would be complete without some reference to the sulfonamides (*1–3*). Most sulfonamide formulations are supplied as combination products having two main components, sulfonamide and the synthetic diaminopyrimidine, trimethoprim. These combinations are believed to act synergistically on specific targets in bacterial DNA synthesis with the sulfonamide blocking the conversion of *p*-aminobenzoic acid to dihydrofolic acid, and trimethoprim inhibiting the conversion of dihydrofolic acid to tetrahydrofolic acid in the folic acid pathway (*1*, *4*). However, doubts have been expressed in recent years about the veracity of this argument, but nevertheless trimethoprim, in practice, does potentiate the antibacterial effects of the sulfonamide.

The sulfonamides are generally wide-spectrum drugs active against a range of bacterial species, both Gram-positive and Gram-negative. Perhaps one of the most widely used is sulfamethazine (sulfadimidine), a drug employed largely in the mass medication of pigs to control atrophic rhinitis and other infections, although it is also used in other species such as cattle. Other commonly used orally absorbed sulfonamides include sulfadimethoxine, sulfapyridazine, sulfamerazine, sulfadoxine, and sulfadiazine. However, many are given orally and parenterally as potentiated formulations with trimethoprim. These include sulfadoxine, sulfaquinoxaline, and sulfatroxazole, which are used to treat diseases caused by a range of Gram-positive and Gram-negative organisms in horses, cattle, sheep, goats, pigs, and poultry, and particularly diseases of the respiratory tract (*1*, *4*). Sulfadiazine and trimethoprim are also used to treat bacterial diseases of salmon and rainbow trout (*6*).

A new generation of diaminopyrimidines is being developed, and several, like baquiloprim, have already been introduced in the markets of some EU Member States.

β-Lactam Antibiotics

The β-Lactam group of antibiotics is comprised of penicillin and semisynthetic penicillins, including cephalosporin antibiotics. They are widely used in veterinary medicine with products intended for intramammary use, particularly in treating mastitis in cattle (*1*). Many of the penicillins and related compounds are unstable when heated. Moreover, many β-lactams are unstable in water unless

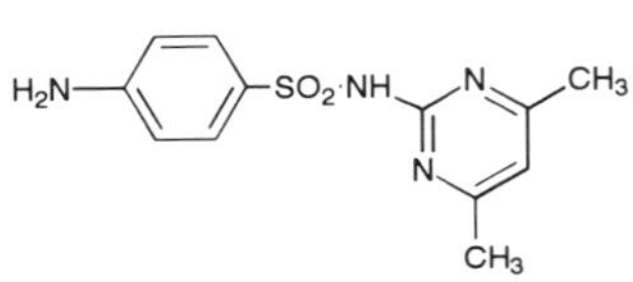

(a) Sulphadimidine (a sulphonamide)

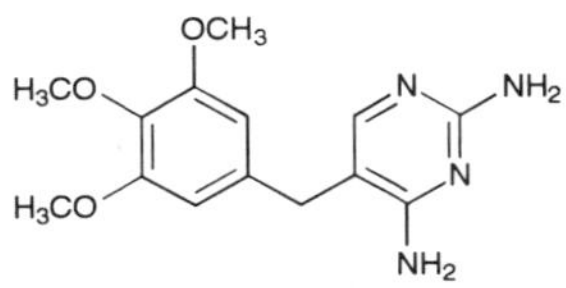

(a) Trimethoprim

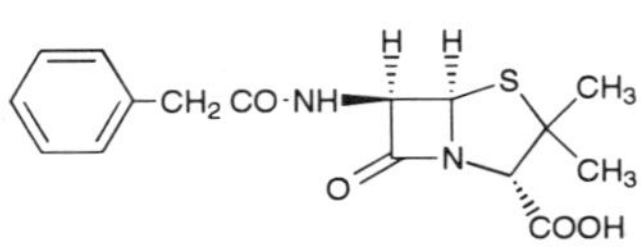

(b) Benzylpenicillin (a penicillin)

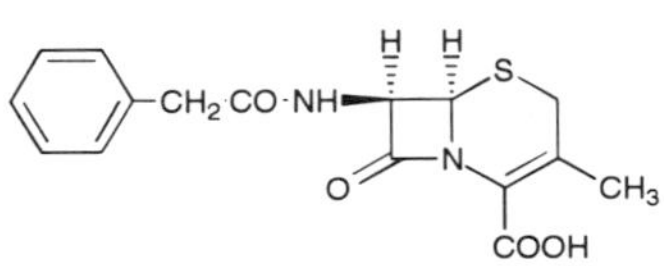

(b) Cephalexin (a cephalosporin)

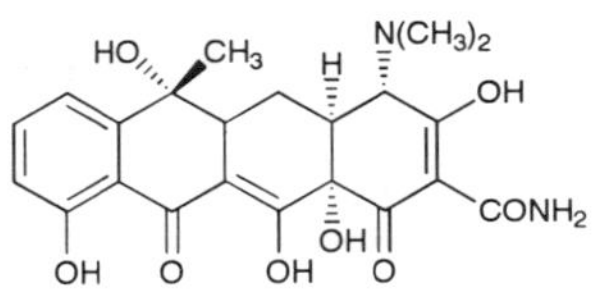

(c) Tetracycline

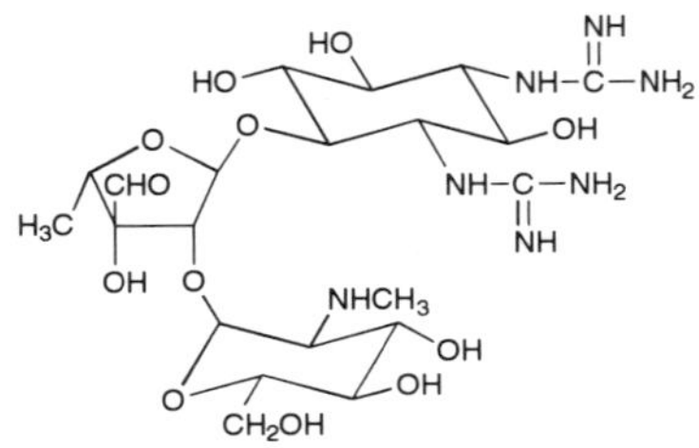

(d) Streptomycin (an aminoglycoside)

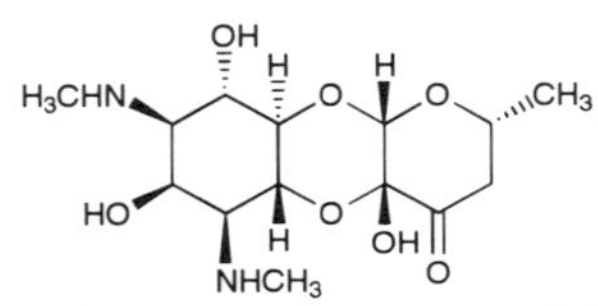

(d) Spectinomycin (an aminocyclitol)

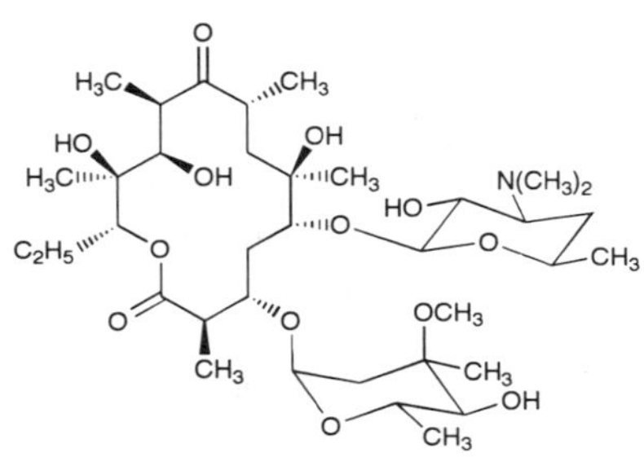

(e) Erythromycin (a macrolide)

Figure 1 Typical Antibiotics from each Class of Drug. (a) sulfadimidine (a sulfonamide); (a) Trimethoprim; (b) Benzylpenicillin (a penicillin); (b) Cephalexin (a cephalosporin); (c) Tetracycline; (d) Streptomycin (an aminoglycoside); (d) Spectinomycin (an aminocyclitol); (e) Erythromycin (a macrolide)

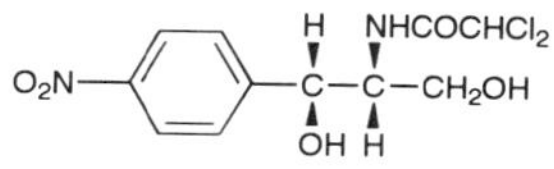

(f) Chloramphenicol

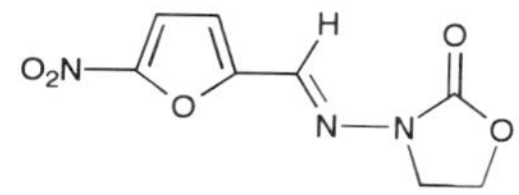

(j) Furazolidone

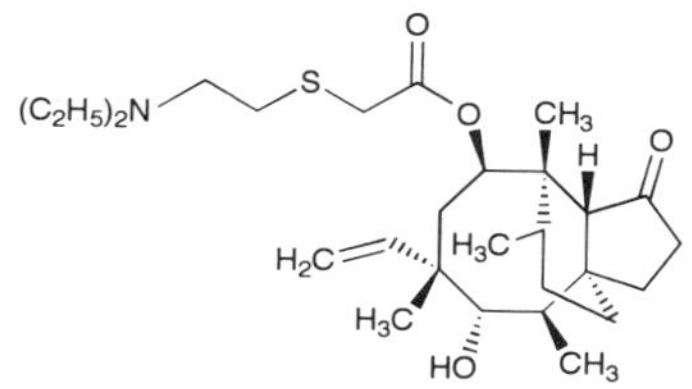

(g) Tiamulin

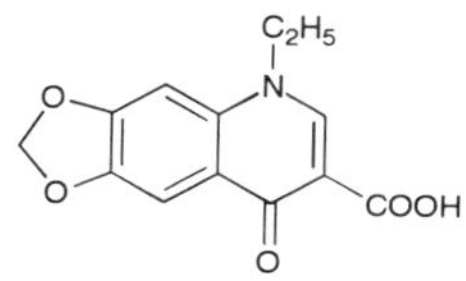

(j) Oxolinic acid

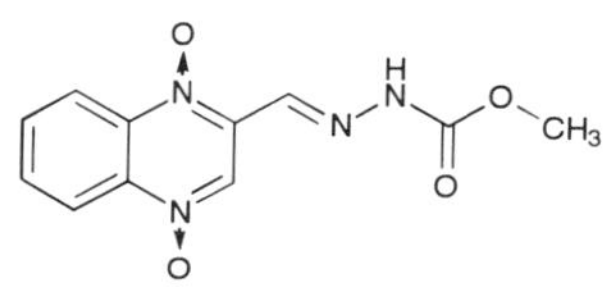

(j) Carbadox

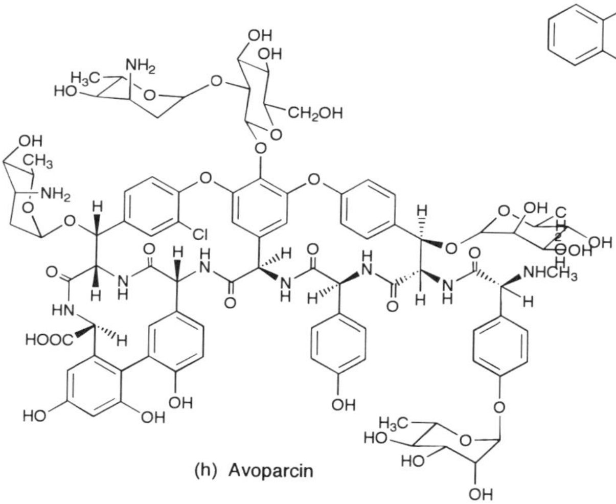

(h) Avoparcin

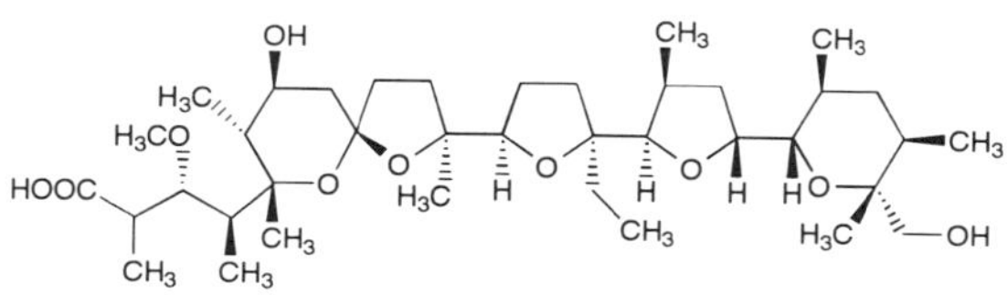

(i) Monensin

Figure 1—*continued* Typical Antibiotics from each Class of Drug. (f) Chloramphenicol; (g) Tiamulin; (h) Avoparcin; (i) Monensin; (j) Furazolidone; (j) Oxolinic acid; (j) Carbadox

present in buffered formulations. The more common ones, including benzyl penicillin, procaine penicillin, oxacillin, cloxacillin, ampicillin, and amoxycillin are wide spectrum antibiotics active against a range of Gram-positive and Gram-negative organisms.

The cephalosporins, like the penicillins, are β-lactams based on a molecule originally isolated from a natural source. However, they are more resistant to breakdown than many other β-lactam compounds. The so-called first generation cephalosporins, such as cephacetrile, are somewhat more stable than the older β-lactam drugs, exemplified by cefuroxine. New, highly stable third generation cephalosporins like ceftiofur and cefquinone are now available. All three generations of these compounds are available on the European market for use in food-producing animals and all have wide spectrum activity (*1*, *3*).

Tetracyclines

The major representatives of the tetracycline group available in the EU for treating food-producing animals are tetracycline, oxytetracycline and chlortetracycline, and to a lesser extent doxycycline (*1*). In the EU, the tetracyclines, which are broad spectrum antibiotics, are used to treat respiratory disease in cattle, sheep, pigs, and chickens, and they may be given in feed or drinking water. Tetracycline itself is active against *Clostridium perfringens*, while oxytetracycline is used to treat enteric disease in a number of species. Oxytetracycline is used to treat various bacterial diseases of trout and salmon (*1*, *3*, *4*).

Aminoglycosides and Aminocyclitols

Streptomycin and dihydrostreptomycin are the two compounds that typify the aminoglycoside group of antibiotics. Streptomycin was isolated from a mold in 1944 and since that time, some 10–15 compounds from the group have been used in veterinary medicine in the EU. In addition to the two compounds mentioned above, neomycin, framycetin, tobramycin, kanamycin, and gentamicin are also used in animals (*1*, *7*). Like the β-lactams, these compounds are wide-spectrum antibiotics and are active against numerous Gram-positive and Gram-negative bacteria. Because most aminoglycosides are poorly absorbed after oral dosing they are useful in treating enteric infections (*1*).

Aminocyclitols are closely related to aminoglycosides. They differ in that aminoglycosides contain amino sugars joined by a glycosidic link while aminocyclitols have the amino group on the cyclitol ring (*8*). The two major aminocyclitols are spectinomycin, which is active against mycoplasmas and organisms implicated in chronic respiratory disease in poultry (*1*), and apramycin.

Many aminoglycosides are toxic in humans .It has been reported that streptomycin damages the VIII cranial nerve in humans causing vertigo and balance impairment while dihydrostreptomycin causes damage to the auditory element of the VIII cranial nerve resulting in deafness (9, 10). It is likely that both compounds are toxic. Several other aminoglycosides have similar effects whereas the aminocyclitols are inactive, although apramycin is toxic to cats (*1*, *3*).

Macrolides

The macrolides are characterized by a large lactone ring structure with sugars attached to it. Erythromycin, a typical macrolide antibiotic, was isolated in 1952 is active against Gram-positive bacteria and mycoplasmas (*1*), and is widely available in animal medicine throughout the EU. Other examples of macrolides include oleandomycin, tylosin, and spiramycin. Like erythromycin, these drugs are active against Gram-positive organisms and mycoplasmas with some activity against Gram-negative organisms. Most are widely available in animal medicines in Europe (*1*, *3*).

Chloramphenicol

Chloramphenicol has been used in antibiotics therapy for over 40 years. However, it induces blood dyscrasias in humans including (rarely) potentially fatal aplastic anaemia (*11*). As a result of the blood dyscrasias and because there are fears that resistance to *Salmonella typhi* (the causative organism of typhoid in humans) is emerging, the use of chloramphenicol is very restricted in the EU, and it is banned for use in food-producing animals. Thiamphenicol, a drug similar to chloramphenicol but seemingly without chloramphenicol's potential to induce blood dyscrasias, is frequently used in some parts of the EU. Currently in the EU, there is a high degree of interest in using thiamphenicol and another congener, fluorphenicol, against bacterial diseases in fish (*1*, *3*).

Tiamulin

Tiamulin is a diterpene antibiotic with an eight-membered carbocyclic ring at the center of its structure (*1*). Its antibacterial profile is poor, but it is active against Gram-positive organisms and mycoplasmas.

Peptides

Peptide antibiotics include avoparcin, a polypeptide antibiotic used to improve the rate of weight gain in chickens, turkeys, pigs, and calves, and to enhance milk production in lactating cattle (*1*, *7*). Other polypeptide antibiotics including zinc bacitracin, virginiamycin, and bambermycin (flavomycin, flavophospholopitol) are used as growth enhancers in the EU. They are regulated under a separate legislation within the EU (Directive 70/524/EEC).

Ionophores

The ionophores are polyether antibiotics that are used not against bacteria, but rather against coccidial (protozoa) infections in poultry. They include monensin, narasin, lasolocid, salinomycin, and maduramycin (*12*). In the EU, they are regulated by the same legislation that regulates the growth-promoting antibiotics (*1*).

Other Antibacterials

For completeness the following three classes of compound are mentioned. Like the sulfonamides, because the nitrofurans, the quinolones, carbadox and olaquindox are synthetic they are not strictly antibiotics but instead are antibacterials.

Nitrofurans. Nitrofurans are synthetic antibacterial compounds used to treat salmonellosis and *E. coli* septicaemia in poultry, and salmonella infections in pigs and calves. They are very effective drugs and do not appreciably contribute to the development of resistance (*1*). However, several have been shown to be mutagenic and cause tumors in laboratory animals, and as a result (*13*), these have been prohibited for use in food-producing animals in a number of countries. In the EU, only furazolidone is currently permitted for use in food-producing species.

Quinolones. Like the sulfonamides, the quinolones are synthetic chemicals with antibacterial activity. Typical first generation compounds are nalidixic acid and oxolinic acid. Nalidixic acid has been widely used in human medicine but less so in veterinary medicine in the EU. Oxolinic acid is restricted to treating diseases in fish such as *Aeromonas salmonicida*, the causative agent of furunculosis in the Atlantic salmon, and against other pathogens such as *Yersinia ruckeri*, and *Flexibacter* species (*1*, *3*, *6*).

Representatives of second generation compounds are the fluoroquinolones, including ciprofloxacin and enrofloxacin. These antibacterials are more potent than their first generation counterparts.

Carbadox and olaquindox. There are two widely available antibacterial synthetic *N,N′-di-N*-oxide quinoxaline compounds used as growth promoters. Neither compound has been without some controversy because carbadox is both mutagenic and carcinogenic in animals, while olaquindox is strongly mutagenic but seemingly devoid of carcinogenic activity. Concern has been expressed about the safety of any residues to the consumer, but evidence suggests that these residues, when present, are devoid of carcinogenic and mutagenic activity and any risk is likely to be to the workers handling the drugs (*1*, *3*, *4*).

Regulation in the European Union

In the EU, veterinary medicinal products are regulated as two distinct groups. The conventional therapeutic drugs, including those for prophylactic treatment, vaccines, and immunological products, are regulated under the Veterinary Medicines Directives 81/851/EEC and 81/852/EEC. The growth-promoting antibiotics (h), (i), and j(iii), the peptides, the coccidiostats, carbadox, and olaquindox are regulated under Directive 70/524/EEC. As if this situation were not confusing enough, Directives 81/851/EEC and 81/852/EEC are dealt with by one Directorate General (DG) in the European Commission, DG III, which is responsible for human and veterinary pharmaceuticals, while Directive 70/524/EEC is handled by DG VI, the DG in the European Commission responsible for agriculture. Hence, there exists the anomaly that these two types of veterinary medicinal products are regulated in two different ways in the EU (*4, 14–16*).

Drugs to be authorized under 81/851/EEC and 81/852/EEC must meet strict criteria of safety, quality, and efficacy. Here, "safety" means safety to the target species, the environment, the consumer, and the workers administering the medicines (*16*). At the present time, EU Member States may, with some exceptions, make their own authorizations at the national level. However, Directive 81/851/EEC offers alternative approaches in the multistate procedure and the concertation procedure. In essence, these are centralized procedures, and advice is provided by the Committee for Veterinary Medicinal Products (CVMP) which regularly meets in Brussels (*17*).

These centralized procedures, along with the concept of mutual recognition where products authorized in one Member State of the EU will be recognized for authorization in another, will be taken further now that the European Medicines Evaluation Agency (EMEA) has taken over much of this work. The EMEA, which is based in the UK, will work within a new regulatory framework. It will continue to draw on CVMP advice, but many of the decisions will be binding on Member States.

One very important piece of EU legislation concerns the establishment of maximum residue limits (MRLs). These MRLs are established by the CVMP through its Working Group on the Safety of Residues. In January 1992, a new piece of legislation, Regulation 2377/90, took effect. The Regulation requires that no new veterinary medicinal product is to be authorized in EU Member States until a Union-wide MRL has been set, and that MRLs for all existing drugs must be established before 1997 (*4, 15*). These MRLs use the concept of an acceptable daily intake (ADI). A package of toxicological tests is provided to Member States, as detailed below (*18*):

- single dose toxicity;
- repeated dose toxicity;
- reproductive effects;
- reproduction;
- embryo/foetotoxicity and teratogenicity;
- mutagenicity data;
- carcinogenicity data;
- pharmacodynamics in laboratory species;

- pharmacokinetics in laboratory species;
- microbiological effects; and
- observations in humans.

Suitable no-observed effect levels are determined from these data and an ADI is calculated. This ADI is used in conjunction with residues depletion data and a knowledge of target tissues and marker residues to establish an MRL or MRLs for veterinary drugs or their metabolites in food of animal origin, including muscle, fat, liver, kidney, milk, and honey (*19*).

For antibiotic and antimicrobial compounds, it is essential that potential effects on the human gut flora are also considered. There is concern that residues of antibiotics and antibacterial substances may give rise to resistant pathogens in the human gastrointestinal tract, or that adventitious organisms may gain a hold. This is a particularly controversial area of hazard and risk assessment, and there are no agreed guidelines for the tests involved (*4*). Nevertheless, tests do exist and these comprise three types:

1. studies in human volunteers;
2. studies in germ-free rats, i.e., holoxenic rats, the intestines of which harbor human gut bacteria; and
3. in vitro studies with bacterial populations.

The studies investigate minimum inhibitory concentrations (MICs) that can be used in ADI calculations, which in turn can be used to elaborate MRLs. However, such MRLs tend to be very conservative and may lead to long withdrawal periods (the time from the last administration of a drug until slaughter is permitted, so that the MRL is not violated) for antibacterial substances.

Under Regulation 2377/90, in addition to residues depletion data, it is required that a method of analysis for routine monitoring is supplied. This must be a practical proposal and so esoteric methods are unlikely to be considered. To ensure this practicality, analytical standards must be made available by the drug sponsor (*19*).

Veterinary medicinal products authorized as growth promoters and coccidiostats under Directive 70/524/EEC are not currently subject to these authorization procedures or to the requirements for establishing MRLs. However, they do follow a centralized procedure through the EU systems, and the type of toxicity data and residues data mentioned above are required, although no formal MRLs are established (*14*, *19*).

Veterinary medicinal products authorized under Directives 81/851/EEC and 81/852/EEC, including antibiotic products, are subject to residues monitoring requirements under other EU legislation (*see* section Control of Residues in Europe). However, this does not apply to the growth promoters and coccidiostats authorized under Directive 70/524/EEC (*19*).

Control of Residues in Europe

Regulations

Under various Directives, notably Directive 86/469/EEC (*20*), EU countries are required to test for the presence of antibiotics and other veterinary residues under a program known as the National Plan. Under this plan 0.1% of cattle, swine, sheep, goats, and horse slaughtered must be sampled at the abattoir for testing for antibiotic residues. This small percentage represents massive sample numbers for the authorities. The slaughter of the main farm animals in the EU is approximately 250 million per annum and in the United Kingdom about 35 million per annum. There is no official sampling requirement for poultry or farmed fish yet, although a pilot scale study of farmed fish has been conducted within those countries with a significant aquaculture industry.

To oversee this surveillance, the EU has appointed four Community Reference Laboratories (Decision 91/664/EEC) and a further 36 National Laboratories (Decision 93/257/EEC) . The National Laboratories may undertake work on all or limited classes of the veterinary drug areas listed in Directive 86/469. The EU Community Reference Laboratories have been assigned responsibility for specific groups of drugs. The Rijksinstituut voor de Volksgezondheid en Milieuhygiene (RIVM) in Holland is responsible for hormones, the Bundesgesundheitsamt (BGA) in Germany covers the β-agonists, chloramphenicol, and sulfonamides, the Laboratoire des Medicaments Veterinaires (CNEVA) in France is interested in antibiotics, and the Istituto Superiore di Sanita (ISS) in Italy is responsible for trace metals and pesticides.

The responsibility of the Community Reference Laboratories is to ensure that the methodology and the performance standards being used by each of the National Laboratories are adequate for enforcement (Decision 89/187/EEC). The National Laboratories in turn are responsible for the standards maintained in any other laboratories in their own country that are involved in the National Sampling Plan work required under the various EU Directives. The community laboratories responsible for antibiotics are the BGA (chloramphenicol) and CNEVA (antibiotics).

The EU does not recommend official methodology for routine testing, but rather prefers to have available certified reference materials (CRMs) that can be employed to check the effectiveness of the methodologies being used. These CRMs are produced on behalf of the Bureau Community Reference (BCR). The community reference laboratories have only recently received funding from the EU, and they are currently implementing programs to monitor the performance of the national laboratories.

Methods for surveillance testing may be subdivided into screening methods and confirmatory methods. Screening methods are those methods used to detect the presence of an analyte or a class of analyte at the level of interest. These methods have a high sample throughput capacity with a low incidence of false negative results and are used to sift large numbers of samples for potential positives.

Methods for confirmatory purposes are those that provide unequivocal identification of the analyte at the level of interest. They are aimed at preventing false positive results as well as providing an acceptably low probability of false negatives. These methods will usually be chemical in nature, giving qualitative and quantitative results, and are likely to be chromatographically based with either spectrophotometric or mass spectrometric (MS) identification of the analyte.

Most countries in the EU have screening and some confirmatory methodology available for antibiotics. In addition there will be reference methods, which are confirmatory methods that have been successfully collaboratively tested by the appropriate Community Reference Laboratory and have been adopted by the Community. A group of expert analysts from the EU have drawn up a Guideline Criteria for Reference Methods (Decision 89/610/EEC) that lists the standards such methodology ought to achieve to be considered acceptable for use. Although the criteria for hormonal substances have been separated from the criteria for other drug residues including antibiotics, the two sets of criteria are broadly similar and demand high standards of recovery, reproducibility, and repeatability for the ideal method (*21*).

The other topic that is currently being addressed within the EU is the elucidation of MRLs described earlier.

Methodology

The methodology discussed below refers to antibiotics (i.e., substances produced by microorganisms with bacteriocidal properties) and not to antibacterials that are synthetic substances with such activity. This is in line with the definition earlier in this chapter. The methodology will therefore encompass penicillins (including semisynthetic members), tetracyclines, aminoglycosides, chloramphenicol, macrolides, etc., but not synthetic antibiotics such as sulfonamides, nitrofurans, carbadox, or olaquindox. Current methodology can be classified as microbiological, chemical, or immunochemical, and the procedures in use will be discussed.

Microbiological methods. Antibiotics have traditionally been analyzed by microbiological methods and a good review of these methods for tissues and milk is presented in a paper by Corry et al. (*22*). For antibiotic residues in meat, there is no exception to this and this general procedure is the most widely used in Europe, at least for the National Plan work required of the Member States. The reason for this is that it allows the Member States to inexpensively cope with the large sample numbers involved. The most popular procedures are forms of Bogaerts' and Wolf's Four Plate Test (FPT) (*23*), alternatively called the Frontier Post Test. In its present form, the test is an amalgam of the Hemmstoff test previously carried out in Germany and the Sarcina lutea test previously performed in the Benelux countries (Belgium, Luxemburg, the Netherlands). As the name infers, four plates are required with three using the organism *Bacillus subtilis* and the fourth *Micrococcus luteus*. The *B. subtilis* plates are set up at pH 6.0, 8.0, and 7.2 with the last plate incorporating trimethoprim to act as a potentiator for any sulfonamide residues. The *M. luteus* plate is prepared at pH 8.0. Two discs of meat or offal (8 × 2 mm) removed from the sample (usually with a sterile cork borer) are placed on the plates in duplicate. The *B. subtilis* plates are then incubated at 30°C for 18–24 hours while the *M. luteus* plate is incubated at 37°C for 18–24 hours. Discs of standard penicillin, streptomycin, sulfadimidine, and erythromycin are used as controls to indicate that the test has been successfully performed.

A test is positive when an annular zone of greater than 2 mm appears around both the meat discs taken from the same sample. This annular zone indicates that an inhibitory substance, probably an antibiotic, is present, but it does not provide an indication of the nature of the drug or its concentration. Substances other than antibiotics can give rise to such zones, e.g. heavy metals, enzymes, or microbiological contaminants. When only one of the two discs from the same sample shows positive a repeat test is warranted, but if a similar result is obtained on the repeat test, the sample is declared negative.

The purpose of running four plates is to attempt to obtain some indication of the class of antibiotic that may be present. From the size and the pattern of the annular zone on the various plates it is suggested that some guidance can be obtained. The *B. subtilis* pH 6.0 plate is particularly sensitive to penicillins and tetracyclines, while the pH 8.0 plate responds better to aminoglycoside antibiotics such as neomycin or streptomycin. The *M. luteus* plate pH 8.0 responds best to macrolide antibiotics. The pH 7.2 plate with *B. subtilis* is specifically set up for sulfonamides whose inhibitory effect is potentiated by the presence of trimethoprim. In our experience it is not generally possible to identify the antibiotic class involved, however some practiced workers claim that good indications of antibiotic type can be obtained from the pattern of positives across the plates.

The sensitivity or minimum inhibitory concentration (MIC) for a small range of commonly used antibiotics is shown in Table 1. The FPT is not sufficiently sensitive to chloramphenicol, nitrofurans, or nitroimidazoles to be of use at the low MRL levels and an alternative test must be run.

Another disadvantage of the FPT is that it is prone to give false positive and false negative results. The false positive results are commonly found with frozen pig and horse kidney samples, which is thought to be due to the release of endogenous inhibitors from ruptured cellular tissue. A consensus view is that lysozyme could be the causative agent. In support of this, it was found that by placing a semipermeable membrane on the agar surface prior to the application of the sample disc, it was possible to greatly reduce the incidence of false positives.

Table 1 Minimum Inhibitory Concentrations of Antibiotics by Four Plate Test

Antibiotic	Lower limit of sensitivity (μg/g tissue)
	+
Penicillin	0.03
Neomycin	0.2
Chloramphenicol	10
Chlortetracycline	0.1
Oxytetracycline	0.5
Erythromycin	0.1
Bacitracin	0.2

False negative results must be suspected when the chemical tests yield a much higher incidence of sulfonamide positive tests than when the FPT is used.

The French European Community Laboratory (CNEVA) has recently modified the FPT procedure and is encouraging testing laboratories throughout the community to adopt its procedure (*24*). The modified procedure appears to be more sensitive than the original FPT as is evident when a comparison of the response to the drug standards is done. Table 2 shows a comparison of the MICs for both procedures expected to give rise to a 6 mm annular zone of inhibition.

As a backup to the FPT and to aid in identifying the antibiotic, some laboratories use electrophoretic systems. Here discs of meat are placed on a gel nutrient media and high voltage electrophoresis is carried out after a short diffusion time. Depending on the pH and the nature of the gel, antibiotics have differing mobilities. After applying the voltage for 1–2 hours the gel is overlaid with an organism-containing gel and is then incubated as before. Zones of inhibition indicate the presence of antibiotic, and the position on the plate could indicate the nature of the antibiotic.

A very thorough study of this technique was carried out by Lott and Vaughn (*25*) using agar and agarose as gel media at both pH 6.0 and 8.0 together with two organisms, *B. cereus* and *M. luteus*. They tabulated information on some 40 antibiotics in terms of their mobility (anode A or cathode C running) and the minimum levels of drug causing inhibition. Table 3 shows examples of their work.

This technique, or a less elaborate form using single gel, pH, and organism, is still used in some laboratories, although not as widely as in the past. Apart from the advantage of indicating the likely nature of an antibiotic, the electrophoretic system eliminates most of the false positives obtained from frozen pig kidneys in the FPT. The disadvantage of the electrophoretic method is its low throughput of samples, although when it is used as a backup for FPT positives this is not

Table 2 Typical Concentrations of Antibiotics Standards Giving 6 mm Zones by FPT and CNEVA Procedure

Antibiotic	FPT	CNEVA
Penicillin	0.01 I.U.	0.002 I.U.
Sulfadimidine	0.5μg	0.2 μg
Dihydrostreptomycin	0.5 μg	0.05μg

considered to be a problem. As an example from Smithers et al. (*26*), of the 5442 meat samples tested by the FPT, 34 showed positive responses but only 2 of them tested positive for recognizable antibiotics when electrophoresis was applied, while another 12 tested positive but not for recognizable antibiotics.

Variations of microbiological inhibition tests are used throughout Europe. This may simply be using fewer than the four plates of the FPT or using other variations. For example, some laboratories use two or three plates, in the Netherlands they use a single plate with samples applied on filter discs saturated with renal pelvic fluid (*27*) and in Denmark, for verification of the FPT they have a 12-plate procedure (*28*).

For milk, the situation is similar in that microbiological inhibition tests are routinely used. There are many kits commercially available that are used by the authorities. These tests normally depend on changes in pH to either colorometrically or electronically self indicate when no inhibitor is present. *Bacillus stearothermophilus* is often used because its temperature tolerance allows the test to be carried out at an elevated temperature thereby achieving a more rapid test. The indicators are usually redox indicators such as bromocresol purple or Brilliant

Table 3 Electrophoretic Mobilities and MICs of Antibiotics

	MIC (g)				Migration distance (mm)			
	B. cereus		*M. luteus*		Agar		Agarose	
Antibiotic	pH6	pH8	pH6	pH8	pH6	pH8	pH6	pH8
Penicillin	10	0.002	10	0.005	30A	149A	20A	138A
Chlortetracycline	0.004	0.02	0.04	0.2	150C	34C	87C	21A
Chloramphenicol	2	0.5	2	0.5	167C	43C	157C	40C
Dihydrostrepto-mycin	1	1	0.2	0.1	31A	234C	14C	206C
Neomycin	20	0.1	2	1	66A	190C	84A	112C
Erythromycin	0.2	0.05	0.05	0.005	250C	151C	237C	134C

MIC: minimum inhibition bioassay

Black when colorometric methods are being used. Alternatively, measurement of electronic characteristics of the media can also be performed using suitable electrodes. This involves measuring changes in resistance or conductance resulting from the metabolism of the growing organism.

As for tissue methods, the milk tests are most sensitive for penicillins and less so for other antibiotics such as tetracyclines, aminoglycosides, and chloramphenicol. The advantages are as before—low cost, high throughput of samples (the UK Milk Marketing Board test each of their 35,000 or so suppliers weekly), and the use of relatively unskilled staff. The disadvantages are similar in that the nature and therefore the concentration of the antibiotic present is not indicated.

One way to determine if the inhibitor is a penicillin is to preincubate the sample with penicillinase to inactivate the penicillin and the presence of sulfa drugs can be determined by the inclusion of *p*-aminobenzoic acid in the medium. The Charm test (*29*) for milk is also used as a backup to the primary test when the class of the inhibitory antibiotic is to be determined. This test is a receptor assay reliant on the use of receptor sites on the surfaces of microorganisms for the antibiotic classes together with competition between added radiotracer and the antibiotics in the sample. The Charm test allows for the determination of penicillins, sulfonamides, aminoglycosides, macrolides, and tetracycline classes of antibiotic as well as chloramphenicol. A modified version of this test (Charm II) can be used for tissue. It identifies the same classes of antibiotics, but it is not clear if, or how widely, this test is being used.

Last, some laboratories have experimented with several of the more recently introduced rapid microbiological testing equipment. This equipment, which depends on measuring changes in conductance of the media, usually works for standards, but when it is presented with actual samples carrying varying degrees of microbiological contamination it becomes less reliable.

Penicillin. Microbiological procedures are particularly sensitive to penicillins and by using penicillinase preparations to inactivate these antibiotics, identifying this class is made relatively straightforward if not wholly satisfactory because neither the nature or quantity of penicillin can be identified. A further test used in some European laboratories to determine penicillins is the Charm test with labeled, cold analyte competition for receptor sites on dead bacteria cells. This test is used particularly for milk analysis, but with the updated procedure suitable for tissue, wider application may be undertaken.

There are also several immunoassays available, either commercially or used within individual laboratories, however there is little data on such procedures. They are most often used as a screening technique with a simple positive or negative validation approach.

There are two published chemical approaches that appear to be receiving some use. One is the method of Meetschen and Petz (*30*) and the other the method of Boison et al. (*31*). Schemes 1 and 2 outline both procedures.

TISSUE MILK (25 g)
1. Add CH_3CN (30 mL)
2. Homogenize (1 min)
3. Add 0.5M phos. buffer pH 2.2 (5 mL) + CH_3CN (90 mL)
4. Homogenize (1 min)
5. Centrifuge (10 min, 4000 g)

SUPERNATANT
1. Add NaCl (5.5 g) + CH_2Cl_2
2. Shake 2 min, allow phases to separate.
3. Recover organic layer.

ORGANIC LAYER
1. Dry with Na_2SO_4 (5 g)
2. Filter
3. Evaporate to 1 mL approx.
4. Add CH_2Cl_2 (3 mL)
5. Evaporate
6. Dissolve pet. ether (3 mL)
7. Evaporate to 0.5 mL approx.
8. Transfer with pet. ether (3 × 3 mL)
9. Extract phos. buffer pH 7 (2 × 2.1 × 1.5 mL)

BUFFER SOLUTION
1. Wash diethyl ether (2 × 2 mL)
2. Wash CH_2Cl_2 (2 × 1 mL)
3. Add M H_3PO_4 (2 mL)
4. Extract CH_2Cl_2 (4 × 1.5 mL)
5. Recover organic phase.
6. Evaporate to dryness.

RESIDUE
1. Dissolve acetone (100 μL) + KH_2PO_4 (3 mL of 1 mg/mL)
2. Apply SAX cartridge
3. Wash H_2O (3 mL)
4. Elute 5%NaCl (3.5 mL)
5. Add methyl red soln. (0.5 mL) + phos. buffer pH 2.2 (1 mL)
6. Extract CH_2Cl_2 (1.5 mL)
7. Evaporate to dryness

RESIDUE
1. Add CH_2N_2 soln (1 mL)
2. Leave RT, 2 hrs.
3. Evaporate to dryness
4. Dissolve cyclohexane + TRI/ODZ (250 μL)

GC

Scheme 1 GC method for penicillins in tissue and milk
SOURCE: See reference (*30*).

TISSUE (5 g)
1. Add H_2O (20 mL)
2. Homogenise (2 min)
3. Add 0.17MH_2SO_4 (5 mL) + 5%$NaWO_4$ (5 mL)
4. Vortex mix (2 min)
5. Centrifuge (4500 rpm, 5 min)
6. Repeat 1–5 with H_2O (15 mL)
7. Filter

FILTRATE
1. Add 20%NaCl (10 mL)
2. Apply Bond Elut C18
3. Wash 2%NaCl (10 mL), then H_2O (10 mL)
4. Elute CH_3CN:pH6.5phos buffer (60:40) (1 mL)

ELUATE
1. Derivatise 2 M 1.2.4 triazole cont. 0.001 M $HgCl_2$
2. Heat 65°C for 30 min

LC–UV

Scheme 2 LC method for penicillins in tissue
SOURCE: See reference (*31*).

The method of Meetschen and Petz is a gas chromatographic (GC) procedure that can detect and analyze seven neutral penicillins (benzyl-penicillin, phenoxymethyl-penicillin, methicillin, oxacillin, cloxacillin, dicloxacillin, and nafcillin) in muscle, liver, kidney, and milk. The procedure is somewhat time-consuming to perform and requires a great attention to detail so it is doubtful whether it could be considered anything other than a confirmatory method. No information about the coupling of mass spectrometry (MS) to the GC for structure confirmation is available.

The method of Boison et al. is based on high-performance liquid chromatography (LC) with UV detection at 325 nm, which is the absorption maximum of a mercuric-S-complex of penicillin obtained by opening the thiazole ring of penicillin. The method is thought to be useful for the analysis of several neutral penicillins, although it has only been validated for benzyl penicillin. This method is simpler than the Meetschen & Petz method and therefore could be considered for limited surveillance work.

The structures of the members of the penicillin class, including semisynthetics, range from overall acidic through neutral to basic in nature. The cleanups of the above methods are specific for the neutral members of the class. Other than the more general microbiological or Charm test, no methods are available or apply to the polar members of this class.

Tetracyclines. For tetracyclines, chemical methods are only used for specific analysis while the more general microbiological inhibition procedures are used to screen for residues. The tetracyclines (tetracycline, chlortetracycline, and oxytetracycline) are widely used both prophylactically and therapeutically and residues are identifiable, although they are not often above the relatively high MRLs (up to 600 ppb).

The main methods used for tetracyclines are those of Ikai et al. (*32*), Farrington et al. (*33*), and Blanchflower et al. (*34*) with the latter specific for chlortetracycline and the other methods generally more applicable to tetracyclines. Schemes 3–5 outline each of these methods. The method of Ikai et al. (*32*) has been the topic of a collaborative trial that has not yet been published (MacNeil, J.D., Agriculture & Agri-Food Canada, personal communication, 1994). The procedure involves solvent extraction, solid-phase cartridge cleanup, and final LC. There are reports of poor reproducibility for this method that are often attributed to variations in batches of the solid-phase cartridges. Often laboratories suggest that using a single batch of cartridges is advisable for good results. In general this is often a good approach for trace analysis and it does seem particularly true for the tetracyclines. One reason suggested for this is that tetracyclines have the ability to chelate with metals, which is important because variations in the quantities and nature of metal ions in the silica of the cartridges could be a critical factor. The procedure of Farrington et al. (*33*) makes use of the tetracycline's chelating property during the cleanup procedure. This procedure passes the initial buffer extract through a Sepharose-chelate column containing copper ions, which retain the tetracyclines that are later stripped with the EDTA-buffer. As in the method of Ikai et al., determination is by LC with UV detection. The method has been validated for a range of species tissues and milk, and has been the topic of two collaborative trials (*35*, *36*).

The method of Blanchflower et al. (*34*), which is specific for chlortetracycline, also uses LC as the determinative step after a buffer extraction and a solid-phase cartridge cleanup. The detection is of isochlortetracycline using fluorimetric detection. This method has a throughput of 16–20 samples per analyst per day, which is generally higher than that of normal chemical analysis.

Aminoglycosides. Within the EU there is little effort expended on the routine analysis of aminoglycoside antibiotics either as individuals or as a class due to a lack of suitable methodology. Some laboratories claim to be able to identify members of this group either via FPT results or through a combination of this test and high voltage electrophoresis, but quantification must be regarded as approximate. Others have developed antibodies to the aminoglycosides, streptomycin in particular. They are using these as a backup to FPT positive identification. These

TISSUE (5 g)
1. Add 20 mL 0.1M EDTA/pH 4 Buffer
2. Homogenize
3. Centrifuge (4000 rpm, 5 min)
4. Repeat 1–3 with (20 mL), (10 mL) EDTA/buffer

SUPERNATANTS
1. Centrifuge (4000 rpm, 15 min)
2. Filter

FILTRATE
1. Apply Baker C18 cartridge
2. Wash H_2O (20 mL)
3. Air dry by aspiration
4. Elute MeOH:EtAc (5:95) (20 mL)

ELUATE
1. Evaporate to dryness
2. Dissolve 0.5 mL mobile phase

LC–UV

Scheme 3 LC method for tetracyclines in tissue
SOURCE: See reference (*32*).

TISSUE (5 g)
1. 0.1 M Glycine in 0.1 M HCl (45 mL)
2. Homogenize (1 min)
3. Centrifuge (2000 rpm, 15 min)
4. Reextract acid glycine buffer (50 mL)

SUPERNATANTS
1. Pass 10 mL of 4 above thro'Bond Elut-cyclohexyl cartridge
2. Wash H_2O (10 mL)
3. Elute MeOH (7 mL)
4. Evaporate to dryness
5. Add 2 mL mobile phase
6. Leave 3 hrs to form iso-CTC

LC–Fluorescence

Scheme 4 LC method for chlortetracycline in tissue
SOURCE: See reference (*34*).

TISSUE (10 g)
1. Add pH4 Succinate buffer (40 mL)
2. Homogenise, ultrasonicate (3 min)
3. Centrifuge (5 min, 12000 rcf)
4. Filter supernatant
5. Repeat 1–4 with 40, then 20 mL succinate buffer

FILTRATE
1. Apply to Chelate-Sepharose column
2. Wash H_2O (10 mL)
3. Wash MeOH (10 mL)
4. Wash H_2O (2 × 10 mL)
5. Elute EDTA-succinate buffer (40 mL)

ELUATE
1. Load XAD column
2. Wash H_2O (2 × 10 mL)
3. Elute MeOH (100 mL), discard first 10 mL

ELUATE
1. Evaporate to small volume
2. Add -$SHCH_2CH_2COOH$ (0.1 mL)
3. Evaporate to dryness

RESIDUE
1. Dissolve mobile phase (0.5 mL)

LC–UV

Scheme 5 LC method for tetracyclines in meat and milk
SOURCE: See reference (*33*).

approaches have not yet reached the literature, and therefore must be generally regarded as unproven for these drugs.

On the chemical front, the aminoglycosides pose problems. Their solubility properties and their lack of distinctive chromophores leave the analyst with obstacles that have been largely insuperable for tissues, and certainly for routine application. However, a very recent publication of a method for streptomycin and dihydrostreptomycin has appeared (*37*).

A method for gentamicin (*38*) has recently been published in the EU Manual (*21*) that may prove to be useful because it may cope with neomycin. However questions still exist at this stage.

Chloramphenicol. Chloramphenicol is the other major antibiotic that is specifically required to be tested for under the EC Directive 86/469/EEC. As previously mentioned, the microbiological inhibition tests, including FPT, are too insensitive to permit routine monitoring at the currently suggested MRL of 10 ppb (*21*). This requires alternative approaches, and there are a number being routinely used based on immunoassay, LC, and GC procedures.

There are three main immunoassay approaches being used: a screening Immunocard system, a radioimmunoassay (RIA), and an enzyme-linked immunosorbent assay (ELISA).

The Immunocard is a system based on the Environmental Diagnostics card or La Carte system that consists of cardboard backing, a porous antibody impregnated central layer, and a paper cover with two or more small holes a few millimeters in diameter sandwiched together. The antibody layer contains a chloramphenicol antibody that, if a sample contains chloramphenicol and is spotted on to the card, will block the available sites at the point of application. This is followed by a chloramphenicol-enzyme conjugate that will be absorbed at available sites on the antibody layer. If the initial sample is positive, no antibody sites are available in the aperture and the conjugate is absorbed within the sandwich. If no chloramphenicol is present, the conjugate is absorbed at the aperture. Spotting with the enzyme substrate follows, which reacts with the enzyme conjugate to produce a colored product. When the chloramphenicol enzyme conjugate is present at the aperture (i.e., a negative sample) a color is seen, whereas with a positive sample no obvious color development occurs.

The sensitivity of the test is of the order of 1 μg/kg (1 ppb), which is more than adequate for the current suggested EU MRL. The card can be used directly for fluids such as urine or blood fractions and for milk (after defatting). For tissue samples, some form of extraction and defatting is required as a minimum cleanup, a procedure that slows throughput considerably.

Another immunochemical approach that is well documented is the RIA procedure of Arnold et al. (*39*). This procedure has been exhaustively validated by Arnold et al. in their own laboratory, and has been successfully subjected to collaborative testing. The procedure, as shown in Scheme 6, involves the extraction of tissue and some cleanup prior to RIA. This is the basis of testing in Germany where the National Laboratory and some of the state laboratories utilize the method. As with many immunochemical methods it has a limited application because the unique reagents, the antibody, and the labelled chloramphenicol are not commercially available.

A number of EU States have indicated (*21*) that they most often use ELISA for screening samples in muscle, but also sometimes in kidney, urine, or feces. As for the aminoglycoside assays, it is suspected that all or most are "home produced for home use." Commercial sources of ELISA kits for chloramphenicol are now available, but how widely used they are is not known.

TISSUE (3 g)
1. Add CH_3CH:4% NaCl aq (1:1) (8 mL)
2. Homogenize
3. Centrifuge (4000 rpm, 10 min)
4. Decant supernatant

SUPERNATANT
1. Add Hexane (5 mL)
2. Mix and centrifuge (1700 rpm, 5 min)
3. Discard hexane layer
4. Repeat 1–3

LOWER LAYER
1. Add EtAc (3.5 mL)
2. Mix, centrifuge (1700 rpm, 5 min)
3. Recover upper layer
4. Repeat 1–3

UPPER LAYER
1. Evaporate to dryness

RESIDUE
RIA

Scheme 6 RIA method for chloramphinicol in tissue
SOURCE: See reference (*39*).

There are also chemical procedures available in the EU that are used for chloramphenicol analysis. These utilize either GC (with or without coupled MS) or LC. They would normally be used as confirmation of immunochemical screen positives or, in some cases, as the primary method. GC methods (*40, 41*) are usually applicable to a range of matrices including tissue, milk, urine, and eggs. Extraction into solvent is normally followed by varying degrees of cleanup depending on matrix and method, with the final qualitative and quantitative analysis being GC coupled with flame ionization or electron capture detection, or GC-MS of the silylated analyte. Schemes 7 and 8 are samples of typical schematic procedures.

There are a variety of methods available and used for LC (*42, 43*). Schemes 9 and 10 show two examples of these, the first being specifically for milk and the second for tissues. The former method utilizes solvent extraction and cleanup, while the latter involves buffer extraction and cleanup. This latter procedure has also been the subject of a successful collaborative trial involving international cooperation (*44*).

EGG (3 g)
1. Add CH_3CN (5 mL)
2. Homogenize
3. Centrifuge (4000 rpm, 10 min)

SUPERNATANT
1. Add Hexane (5 mL)
2. Mix and centrifuge (1700 rpm, 5 min)
3. Discard hexane layer
4. Repeat 1–3

LOWER LAYER
1. Add EtAc (3.5 mL)
2. Mix, centrifuge (1700 rpm, 5 min)
3. Recover upper layer
4. Repeat 1–3

UPPER LAYER
1. Evaporate to dryness

RESIDUE
1. Dissolve CH_3CN:H_2O (5:95) (3 mL)
2. Apply C18 Sep-Pak cartridge
3. Wash CH_3CN:H_2O (5:95) (3 mL)
4. Elute CH_3CN:H_2O (50:50) (3 mL)

ELUATE
1. Add EtAc (5 mL)
2. Mix, centrifuge (1700 rpm, 5 min)
3. Take off upper layer
4. Repeat 1–3

UPPER LAYERS
1. Evaporate to dryness
2. Derivatise with silyl reagent
3. Evaporate to dryness
4. Dissolve hexane (100 µL)

GC or GC–MS
Method can use GC with FID or ECD or GC-MS using electron impact or negative CI MS. With the latter an internal standard D5-CAP was incorporated at start of extraction.

Scheme 7 GC method for chloramphenicol in milk and eggs
SOURCE: See reference (*40*).

URINE or DEFATTED MILK (10 mL)
1. Add AcOH to pH 5.2
2. Add 0.5 mL Acetate buffer pH 5.2
3. Add 20 µL glucuronidase
4. Incubate 2 hrs, 37°C

Extrelute Column
1. Apply urine/buffer
2. Extract 2 × 25 mL EtAc
3. Reduce EtAc to dryness

Residue
1. Take up in 3 × 2 mL MeOH
2. Transfer into 5 mL H_2O
3. Reduce to 4 mL under N_2

Seppak C18 Cartridge
1. Apply from 3 above
2. Wash H_2O (5 mL)
3. Wash MeOH:H_2O (10:90) (5 mL)
4. Elute MeOH:H_2O (45:55) (5 mL)
5. Reduce to dryness

MUSCLE or EGGS (10 g)
1. Add EtAc (100 mL) + Na_2SO_4
2. Homogenise (1 min)
3. Centrifuge (30 min at 1800 g)
4. Transfer 75 mL supernatant
5. Evaporate to dryness
6. Dissolve residue CH_2Cl_2 (5 mL)

Sep-Pak Silica Cartridge
1. Apply 6 above to cartridge
2. Wash 2 × 5 mL CH_2Cl_2
3. Flush N_2
4. Elute CH_3CN:H_2O (20:80 5 mL)

Eluate
1. Extract 3 × 1 mL EtAc
2. Wash extract H_2O (1 mL)
3. Evaporate EtAc dryness
4. Dissolve residue EtOH: isooctane 3 :97 (0.3 mL)

LC (LichrosorbDiol 150 × 4.6 mm)
1. Inject 0.25 mL 4 above
2. Collect CAP fraction at appropriate time
3. Evaporate to dryness

Residue
1. Dissolve EtOH
2. Derivitise 0.1 mL BSTFA/1% TMCS 1hr/60°C
3. Evaporate to dryness
4. Dissolve isooctane 25 µL

GC or GC–MS

Scheme 8 GC method for chloramphenicol in urine, milk, muscle and eggs
SOURCE: See reference (*41*).

Other antibiotic drugs. There is little effort expended in the EU in routinely analyzing for many other drugs or drug classes that fall within the present definition of antibiotics for a variety of reasons. For example, there are few, if any, routine residue methods available for aminoglycosides, macrolides or peptide antibiotics, and the ionophores are largely used in poultry production to counter coccidiosis. Because surveillance of poultry has not yet become a requirement of

HOMOGENISED MILK (5 mL)
1. Add 5M HCl (50 mL)
2. Add EtAc (25 mL)
3. Stir for 10 min
4. Centrifuge (2400 rpm, 5 min)
5. Take 20 mL EtAc layer

MEAT (5 g)
1. Add H_2O (4.5 mL)
2. Mix 1 min
3. Add EtAc (6 mL)
4. Mix 1 min. Rest 10 min. Mix 1 min
5. Centrifuge (4000 µg, 5 min)
6. Recover upper phase (4.2 mL)

ORGANIC PHASE
1. Evaporate to dryness
2. Dissolve hexane: $CHCl_3$ (1:1) (1.6 mL)
3. Briefly stir
4. Add H_2O (0.8 mL)
5. Stir 5 min
6. Centrifuge (3300 rpm, 10 min)
7. Recover supernatant

LC

Scheme 9 LC method for chloramphenicol in milk and meat
SOURCE: See reference (*24*).

TISSUE (10 g)
1. Add H_2O (40 mL)
2. Homogenise 3 min.
3. Filter
4. Apply Extrelut cartridge
5. After 15 min elute CH_2Cl_2 (50 mL)

ELUATE
1. Evaporate N_2
2. Transfer CH_2Cl_2 (15 mL)
3. Evaporate N_2
4. Add H_2O (300 µL) + Toluene (2 mL)
5. Mix, centrifuge
6. Discard organic layer
7. Repeat steps 4–6 with toluene (1.5 mL)
8. Filter through 0.45 filter

LC

Scheme 10 LC method for chloramphenicol in meat
SOURCE: See reference (*43*).

EU legislation, activity in this area is minimal. Nevertheless, some methodology is available for the ionophores. A procedure using thin-layer chromatography (TLC) or high performance thin-layer chromatography (TLC) with either autobiographical *(45)* or on-plate fluorescence detection *(46)* after spraying with *p*-anisaldehyde has been used and, at least for monensin, has been collaboratively tested *(47)*. A method in the EU Methods Manual similar to the above, but with a modified cleanup, is being used in the Netherlands *(21)*. A method for lasalocid *(48)* involving LC with fluorescence detection is used in the author's laboratory.

Finally, the sulfonamides, nitrofurans, quinolones, as well as carbadox and olaquindox as antibacterials, fall outside the definition of antibiotics being used here. There is considerable literature on the methodology for all of these groups, particularly for the sulfonamides. The sulfonamides are specifically included in Directive 86/469/EEC as a class of compounds to be monitored separately because they are widely used and known to leave detectable residues. The methods available include TLC, LC, and GC, the latter with or without MS confirmation. Generally, TLC is used for screening and LC or GC is used for quantification.

Nitrofurans are also sometimes monitored, but because the parent drug is rapidly metabolized in vivo, and there is no clear picture of the marker metabolite's nature there is little activity in this area. There has been, however, some limited activity for quinolone and carbadox analysis.

Conclusion

The EU requires the member states to report the results of National Plans to the Commission who will possibly publish the data in the future. However, because this is supplied in a variety of formats, it will be some time before it can be published.

On the average approximately 1% of all the samples analyzed for antibiotics by FPT give a positive result. When these positives results are further investigated, the residues are most often ascribed to penicillins or tetracyclines. Residues of chloramphenicol are rarely, if ever, found. Possibly this is because the use of this drug is strongly discouraged in some states, e.g. United Kingdom, or perhaps because it is rapidly metabolized in vivo and the marker residue is not analyzed.

The actions taken for positive findings vary from country to country. In some states a positive by FPT is sufficient to have the carcass condemned. In other states a trace back to the farm is undertaken, and verbal or written advice is given to the farmer. Animals from that farm are then targeted in future surveillance.

The introduction of statutory MRLs will change the situation because it will then be possible to take legal action against the suppliers and should further reduce the number of infringements.

References

1. G. C. Brander, D. M. Pugh, R. J. Bywater, and W. L. Jenkins, *Veterinary Applied Pharmacology and Therapeutics*, 5th Ed. (London: Balliere Tindall, 1991).
2. A. Fleming, ed., *Penicillin. Its Practical Application* (London: Butterworth and Co. Ltd., 1946).
3. N. T. Crosby, *Determination of Veterinary Residues in Food* (London: Ellis Horwood, 1991).
4. R. F. Bevill, in: *Veterinary Pharmacology and Therapeutics*, ed. N.H. Booth and L. E. McDonald (Iowa State University Press, Ames: 1988), 785.
5. K. N. Woodward, *Xenobiotics and Food-Producing Animals. Metabolism and Residues*, ed. D. H. Hutson, D. R. Hawkins, G. D. Paulson, and C. B. Struble, ACS Symposium Series, no. 503 (Washington DC: American Chemical Society, 1992), 26.
6. National Office of Animal Health, *Compendium of Data Sheets for Veterinary Products 1993–94*, (London: National Office of Animal Health, Datapharm Publications, 1993).
7. W. G. Huber, *Veterinary Pharmacology and Therapeutics*, ed. N. H. Booth and L. E. McDonald (Ames, IA: Iowa State University Press 1988), 822.
8. J. E. F. Reynolds, ed., *Martindale. The Extra Pharmacopeia*, 30th ed. (London: The Pharmaceutical Press, 1993), 113.
9. E. S. Harpur, *Iatrogenic Diseases*, ed. P. F. D'Arcy and J. P. Griffin (Oxford: Oxford Medical Publications, 1986), 713.
10. C. Dollery, ed., *Therapeutic Drugs* (London: Churchill Livingstone, 1991), S100.
11. World Health Organization, *Toxicological Evaluation of Certain Veterinary Drug Residues in Food, International Programme on Chemical Safety*, WHO Food Additive Series no. 23 (Geneva: World Health Organization, 1988).
12. M. J. Shepherd, *Food Contaminants: Sources and Surveillance*, ed. C. S. Creaser and R. Purchase (Cambridge: Royal Society of Chemistry, 1991), 109.
13. World Health Organization, *"Evaluation of Certain Veterinary Drug Residues in Food," Fortieth Report of the Joint FAO/WHO Expert Committee on Food Additives*, WHO Technical Report Series no. 832 (Geneva: World Health Organization, 1993), 32.
14. D. R. Williams, *Feed Legislation* (Bakewell, UK: HGM Publications, 1993).
15. K. N. Woodward, *Xenobiotics and Food Producing Animals. Metabolism and Residues*, ed. D. H. Hutson, D. R. Hawkins, G. D. Paulson, and C. B.

Struble, ACS Symposium Series, no. 503 (Washington D.C.: American Chemical Society, 1992), 3.
16. K. N. Woodward *Food Contaminants: Sources and Surveillance*, ed. C. S. Creaser and R. Purchase, (Cambridge: Royal Society of Chemistry, 1991), 99.
17. Commission of the European Communities, *The Rules Governing Medicinal Products in the European Community*, vol. VB (Luxembourg: Commission of the European Communities, 1993).
18. Commission of the European Communities, *The Rules Governing Medicinal Products in the European Community*, vol. VI (Luxembourg: Commission of the European Communities, 1991).
19. K. N. Woodward, *General and Applied Toxicology*, ed. B. Ballantyne, T. Marrs, and P. Turner, (Basingstoke, UK: Macmillan, 1993), 1105.
20. EC Directive 86/469/EEC, *Off. J. European Comm.* L275: 36, 1986.
21. R. J. Heitzman, ed., *Residues in Food Producing Animals and their Products: Reference Materials and Methods*, "Veterinary Drug Residues" (Luxembourg: Office for Official Publications of the European Communities, L-2985, 1992).
22. J. E. L. Corry, M. R. Sharma, and M. L. Bates, *Antibiotics*, Assessment of Antimicrobial Activity and Resistance (Russell & Quesnel Acad.. Press, London: 1983), 349.
23. R. Bogaerts and F. Wolf, *Fleischwirtsch.*, **60**: 672, 1980.
24. CNEVA Labortaire des Medicaments Veterinaire, *Doc. UCM/90/04*.
25. F. Lott and D. R. Vaughn, *Antibiotics*: Assessment of Antimicrobial Activity and Resistance (Russell & Quesnel Acad.. Press, London: 1983), 349.
26. R. Smithers, A. F. Lott, R. W. Dalziel, and D. C. Ostler, *J. Hyg. Camb.*, **85**: 359, 1980.
27. J. F. M. Nouws, N. J. G. Broex, J. M. P. DenHartog, and F. Driessen, *Archiv. für Lebensmittelhyg.*, **39**: 335, 1988.
28. C. Lund, *Proceedings of the 2nd World Congress Foodborne Infections and Intoxications*, (Berlin, 1986), 819.
29. S. E. Charm and R. K. Chi, *J. Assoc. Off. Anal. Chem.*, **71**: 304, 1988.
30. U. Meetschen and M. Petz, *Z. Lebensmittel unters Forsch.*, **193**: 337, 1991.
31. J. O. Boison, C. D. C. Salisbury, W. Chan, and J. D. MacNeil, *J. Assoc. Off. Anal. Chem.*, **74**: 497, 1991.
32. Y. Ikai, H. Oka, N. Kawamura, and M. Yamada, *J. Chromatogr.*, **411**: 313, 1987.
33. W. H. H. Farrington, J. A. Tarbin, J. Bygrave, and G. Shearer, *Food Addit. Contam.*, **8**: 55, 1991.
34. J. Blanchflower, R. J. McCracken, and D. A. Rice, *Analyst*, **114**: 421, 1989.
35. M. C. Carsen, J. AOAC Int., **76**: 329, 1993.

36. P. Brereton, W. H. H. Farrington, G. Shearer, H. D. Thomas, and R. Wood, *J. Assoc. Pub. Anal.*, **30**: 55, 1994.
37. G. C. Gerhardt, C. D. C. Salisbury, and J. D. MacNeil, *J. AOAC Int.*, **77**: 334–337, 1994.
38. F. Sur, A. Nicolas, and P. Archimboult, *Anal. Chim. Acta*, **275**: 285, 1993.
39. D. Arnold, D. VomBerg, A. K. Boertz, U. Mallick, and A. Somogyi, *Archiv. für Lebensmittelhyg.*, **35**: 131, 1984.
40. D. Arnold and A. Somogyi, *J. Assoc. Off. Anal. Chem.*, **68**: 984, 1985.
41. L. A. van Ginkel, H. J. van Rossum, P. W. Zoontjes, H. van Blitterswijk, G. Ellen, E. van der Heeft, A. P. J. M. de Jong, and G. Zomer, *Anal. Chim. Acta*, **257**: 61, 1990.
42. CNEVA Laborataire des Medicaments Veterinaire, Fougers, France, Doc UCM/90/04.
43. H. J. Keukens, W. M. J. Beek, and R. M. L. Aerts, *J. Chromatogr.*, **352**: 445, 1986.
44. R. M. L. Aerts, H. J. Keukens, and G. A. Werdmuller, *J. Assoc. Off. Anal Chem.*, **72**: 570, 1989.
45. J. Okada, I. Higuchi, and S. Kondo, *J. Food Hyg. Soc. (Jpn)*, **21**: 177, 1980.
46. D. Tihova and V. Peneva, *Vet. Med. Nauki*, **19**: 52, 1982.
47. Analytical Methods Committee, *Analyst*, **111**: 1089, 1986.
48. J. A. Tarbin and G. Shearer, *J. Chromatogr.*, **579**: 177, 1992.

Chapter 4

New Test Kit Technology

Joe O. Boison and James D. MacNeil

Food Animal Chemical Residues Section, Health of Animals Laboratory, Food Inspection Branch, Agriculture and Agri-Food Canada, 116 Veterinary Road, Saskatoon, Saskatchewan Canada S7N 5E3

Contents

Introduction

Consumer demands for the production of animal-derived food products free of antibiotic residues, the desire of regulatory agencies to protect the consumer against unwanted drug residues by encouraging responsible use of veterinary drugs in food-animal production, and an expanding global trade in foods that comply with nationally established limits for chemical residues, have generated a growing need for more testing of food products. To address these concerns, public health and regulatory agencies require simpler, more cost-effective mechanisms to test more food items, thereby satisfying both consumers and trade requirements that animal-derived food products are free of unwanted drug residues.

When the demand for more testing was initially launched in the early 1970s by various interest groups, research institutions and manufacturing companies immediately responded. They developed a number of biological assay methods they claimed had the test characteristics required to meet the high-volume testing demanded by the consumer. Because these microbial assay methods failed to provide the required specificity in the detection of specific antibiotics, research efforts in the mid-1970s were directed toward the development of radioimmunoassay techniques that had previously been limited exclusively to determing large molecular weight compounds for determining antibiotic residues. The requirement to maintain a stock of high activity, often of short half-life radioisotopes, coupled with the need to assay the aminoglycoside group of antibiotics provided the impetus to develop nonisotopic immunoassays, which became the primary means of testing antibiotics in the 1980s.

Rapid tests are generally expected to provide an inexpensive yet reliable first approach to sample testing that meet the demands for high-volume testing. They are expected to provide qualitative and sometimes semiquantitative test responses that are used as a basis for determining whether a food product should be released into the food chain or detained pending the results of further laboratory confirmatory testing. Suitable rapid tests must provide same-day results using a test protocol that does not include complicated and extensive sample preparation steps. The only exception to the "simplicity to perform rule" is when a rapid test is designed exclusively for intralaboratory use by trained analysts to provide further confirmatory tests for previously screened test samples. This is because rapid tests manufactured commercially for large-scale testing are expected to be designed primarily for use where the end-user will not be expected to have the level of technical training required to handle complicated sample preparative and manipulative steps in a test protocol.

Classification of Rapid Tests

Rapid tests for detecting antibiotic residues in foods, several of which have already been mentioned in Chapters 1, 2, and 3, may be classified as either microbiological (microbial inhibition with or without a color indicator) or immunological (radioimmunoassay or nonisotopic immunoassay).

Microbiological Assays

Microbiological assays are relative rather than absolute. They are usually conducted in agar or broth media, and they are based on a definable microbiological response to a series of standard antibiotic concentrations with a strain of a test organism in determining the level of an antibiotic.

When a microbial cell is deposited onto the surface of an agar medium, soluble nutrients in the medium must first enter the microbial cell and be metabolized. The bacterium will then begin to multiply, and eventually enough progeny are produced to provide a macroscopically visible colony. As the microorganisms multiply, nutrients in the medium adjacent to the microbial growth are depleted, and growth would rapidly cease if the system was static. However, because the distribution of soluble nutrients or materials in the agar is governed by the physical laws of distribution, dissolved materials continue to diffuse from sections of the agar where they are more concentrated to the areas where they have been depleted in an attempt to restore equilibrium. Thus, microbial growth will continue as long as there is an additional supply of nutrients diffusing from adjacent agar medium to replenish what has been consumed. Simultaneously, extracellular metabolic products resulting from the growth of the organism are released into the medium and diffuse away from the area of microbial growth. A dynamic situation exists whereby metabolic products diffuse away from the area of microbial growth while soluble nutrients diffuse into the area of microbial growth.

Agar diffusion tests are conducted by inoculating a nutrient agar medium with an appropriate test organism and applying a solution containing the drug or drugs to be studied to the agar surface in one of several ways. The growth medium is then incubated for a specified period of time. Antibiotics deposited on top of the agar surface will diffuse by one-, two-, or three-dimensional mechanisms into the agar medium.

A one-dimensional microbial inhibition test uses a test tube or capillary tube in which seeded agar has been poured and allowed to harden. An antibiotic test suspension is pipetted onto the surface of the agar, and zones of inhibition in one dimension are formed. Although this technique is rarely used, it is useful in determining antibiotics under anaerobic conditions or in situations where there is very little sample fluid available for testing (*1*).

Most biological assays are the two- or three-dimensional type. In a two-dimensional assay the antibiotic diffuses directly against a wall of seeded bacteria. This may occur in one of three ways: (*a*) *well-plate method*—a well is cut in the agar that has been seeded throughout with a test organism; (*b*) *pad-plate method*—the agar is seeded throughout with a test organism and paper disks or cotton swabs impregnated with antibiotics are then placed on the agar; (*c*) *cylinder-plate method*—glass or metal cylinders filled with a fixed volume of the antibiotic solution are placed upon a uniform thickness of agar that has been seeded throughout with the test organism so that a column of sample is in direct contact with the agar medium. In a three-dimensional assay, the antibiotic added to the surface of the seeded agar diffuses from the surface of the agar through an arc 180° from its source and travels directly to the bottom of the Petri dish as well as along the surface.

For a given antimicrobial agent, the rate of diffusion (the diffusion coefficient) through the agar medium under controlled test conditions is a constant, and the distance traveled by a particular concentration of an antimicrobial agent within a specified time period is proportional to the concentration of antimicrobial agent introduced into the agar. When an antimicrobial agent is introduced into the agar, it is removed by absorption into the microbial cell, thus diminishing or effectively inhibiting the normal metabolic growth of the test organism. When this occurs, a zone of inhibition will appear around the reservoir. The size of the zone may be used to determine the concentration of drug in the reservoir. This zone of inhibition is formed when a critical concentration of drug (the amount of drug that is just required to inhibit microbial growth under the test conditions) first reaches a density of cells too large for it to inhibit. The size of the zone of inhibited growth is determined by the distance that the critical inhibitory concentration can diffuse into the agar before a particular density of cells is attained. For a more detailed treatment of the mechanisms of formation of zones of inhibited growth, however, see these excellent reviews (*2–4*). It is, nevertheless, sufficient to indicate that the formation of a zone of inhibition is dramatically influenced by several factors. The most important factors are controllable under experimental conditions and include the density of the inoculum, the nature of the sharpness of the zone's edge, the depth of the agar medium, its composition, and it's pH, the growth characteristics of the test organism, temperature and time of incubation, and the possibility of interference by naturally-occurring microbial inhibitors in the test material.

When a microbial inhibition test is designed to give qualitative test responses, one of two methods is used to measure the extent of microbial growth inhibition. Either a scaler measurement of the clear zone of inhibition, or the observation of a visible change in color of an acid-base or redox indicator incorporated into the agar medium is used.

Inhibition tests that measure clear zones of inhibition for antibiotic detection. Table 1 lists the microbial inhibition tests that measure clear zones of inhibition produced in the media in the presence of microbial inhibitors. These tests include the Live Animal Swab Test (LAST) (*5*), the Swab Test on Premises (STOP) (*6*), and the Calf Antibiotic and Sulfa Test (CAST) (*7*), used primarily for in-plant and on-farm screening in Canada and the United States, and the *Bacillus stearothermophilus* disc assay (BSDA) (*8*).

Inhibition tests that rely on color indicators. The rapid tests listed in Table 2 are those microbial inhibition tests that incorporate certain reagents such as sugars and the acid-base indicator, bromocresol purple [pH range 5.2 (yellow)–6.8 (purple)], in the test medium (*11–13*). Favoring acid production in the medium, the test organism grows rapidly during incubation in the absence of antibiotics, and the bromocresol purple indicator changes from a purple to a yellow color, which is indicative of a negative test. In the presence of an antibiotic or an

Table 1 Microbial Inhibition Tests for Detecting Antibiotics in Animal Tissues and Dairy Products that Measure Clear Zones of Microbial Inhibition

Rapid test (Ref.)	Microorganism	Test matrix
Live Animal Swab Test (LAST) (*5*)	*Bacillus subtilis*	Urine of dairy cows
Calf Antibiotic And Sulfa Test (CAST) (*6*)	*Bacillus megaterium*	Calf kidney tissue fluid and calf urine
Swab Test on Premises (STOP) (*7*)	*Bacillus subtilis*	Kidneys of cull dairy cows
B. stearothermophilus disc assay (BSDA) (*8*)	*Bacillus stearothermophilus* var. *calidolactis* strain C953	Raw milk, processed milk, and dried reconstituted milk
B. subtilus field test (AOAC field test) (*9*)	*Bacillus subtilis* (ATCC 6633)	Milk
Inhibitor test with *S. lutea (9)*	Sarcina lutea (ATCC 9341)	Raw milk, processed milk, and dried milk
Yogurt inhibitor test (*10*)	*Streptococcus thermophilus* and *Lactobacillus bulgaricus*	Liquid milk and reconstituted dried milk

inhibitory substances, however, microbial growth is inhibited, and metabolic activity to increase the acidity of the medium is reduced or completely inhibited so that the color of the indicator remains purple (a positive test).

Instead of acid–base indicators, the microbial inhibition tests listed in Table 3 incorporate redox indicators in the test medium. For the Brilliant Black Reduction Test (BBR), the absence of inhibitors in the test sample during incubation at 65°C results in the rapid growth of the test organism. The redox indicator, Brilliant Black, is reduced from its purple (oxidized) form to its yellow (reduced) form, which indicates a negative test. If antibiotics are present in the test sample during incubation, growth of the test organism will be inhibited and the purple color of the indicator will be retained, indicating a positive test. In the Arla micro test, the absence of antibiotics is indicated by a rapid growth of the test organism, which causes the redox indicator to reduce from its colorless oxidized form into its red reduced form; in the presence of antibiotics, however, growth of the test organism is inhibited and the medium remains colorless. In the Lumac method, which combines microbial inhibition with enzyme immunoassay techniques, the absence of antibiotics in the test sample during a short incubation period of 30 minutes at

Table 2 Microbial Inhibiton Tests that Rely on Acid-base Indicators for Elucidating Zones of Microbial Inhibition[a]

Rapid test	Microorganism	Test matrix
B. stearothermophilus disc assay	*Bacillus stearothermophilus* (ATCC 10149)	Raw milk, whole milk, lowfat milk, skim milk, chocolate milk, and nonfat dry milk
Charm Inhibition Assay (CIA)[a]	*Bacillus stearothermophilus*	Raw milk, market milk, and powdered milk
Accusphere (*12*)	*Streptococcus thermophilus*	Raw milk, market milk, and reconstituted dried milk
Valio T101[b]	*Streptococcus thermophilus*	Raw milk and other milk products
Fast Antibiotic Screen Test (FAST) (*13*)	*Bacillus megaterium*	Raw milk, edible animal tissues, and body fluids
Charm Farm Test[c]	*Bacillus stearothermophilus*	Raw milk, processed milk, meat, eggs, fish, serum and urine
Delvotest P and Delvotest SP[d]	*Bacillus stearothermophilus*	Raw milk, market milk, and reconstituted dried milk

[a]Indicator, bormocresol purple.
[b]Manufactured by Valio, Helsinki, Finland.
[c]Manufactured by Charm Sciences, Malden, MA, U.S.A.
[d]Manufactured by Gist-Brocades Inc., Delft, Holland.

41°C results in the bacterial test organism being metabolized leading to enzyme synthesis. If, after the incubation period, the addition of a chromogenic substrate solution containing 3-(4,5-dimethylthiazolyl-2-)2,5-diphenyltetrazolium bromide results in a color change from green (reduced form) to purple (oxidized form), then there was a significant number of enzymes formed to oxidize the substrate. In the presence of antibiotics, however, there will be little or no enzyme synthesis, and the addition of the chromogen after incubation will result in the retention of the green color. The essence of the addition of the color indicators to the test media in these microbiological assay systems is to improve the visual or colorimetric

Table 3 Microbial Inhibition Tests that Rely on Redox Indicators for Elucidating Zones of Microbial Inhibition

Rapid test	Microorganism	Test matrix	Indicator
Brilliant Black Reduction Test (BBR)[a]	*Bacillus stearothermophilus*	Raw milk, market milk, tissue fluids, honey, eggs, urine, plasma or serum, and feed	Brilliant Black
Arla Micro Test[b]	*Bacillus subtilis* (ATCC 6633)	Raw milk, frozen milk, heat treated milk	Triphenyltetrazolium chloride (TTC)
Rapid Antibiotic Test (Lumac)[c]	*Streptococcus thermophilus* OL 1010.59	Raw milk, heat treated milk	Chromogenic solution containing 3-(4,5-dimethylthiazolyl-2-) 2,5-diphenyltetrazolium bromide (MTT)

[a]Manufactured by Laboratorium Enterotox, Germany and distributed by Glengarry Biotech, Cornwall, ON, Canada.
[b]Manufactured by Arla, Department of Research and Development, Stockholm, Sweden.
[c]Available from Lumac B. V., AC Landgraff.

detection of the zones of inhibition produced in the presence of antibiotics or inhibitory substances in the test media, thereby enabling the test responses to be more accurately interpreted.

Table 4 lists microbial inhibition tests that have been officially approved to confirm (*8, 12, 14—18*) and to estimate the concentration of antibiotics detected in a test sample that has tested positive in a qualitative screening test. The *Bacillus stearothermophilus* disc assay is used worldwide for testing milk, while countries like Germany and the Netherlands prefer to use the Three-Plate Test (TPT) and the New Dutch Kidney Test (NDKT) or the New Netherlands Kidney Test (NNKT), respectively, as alternative confirmatory tests in their meat inspection programs. Countries in the European Community (EC) generally use the four-plate test (FPT) alternatively called the frontier post test (*14*). Table 5 lists the detection limits of some commercially available microbial inhibition tests for detecting antibiotic residues in raw milk.

Table 4 Confirmatory Tests for Microbial Inhibitor Tests

Confirmatory Test (Ref.)	Microorganism	Test Matrix	Test Media
Three-plate test (*12*)	*B. stearothermophilus* var. *calidolactis* C 953	Raw milk, heat treated, and reconstituted dried milk	Diffusion agar medium pH 7.9 in a 100 mm diameter Petri dish
	B. subtilis strain ATCC 6633		Diffusion agar medium pH 7.9 in a 100 mm diameter Petri dish
	B. megaterium strain ATCC 9885		Mueller-Hinton agar medium pH 7.4 in a 100 mm diameter Petri dish
Six-plate test (*14*)	*Bacillus cereus* var. *myoides* ATCC 9634	Raw milk, heat-treated milk, and reconstituted dried milk	Phosphate-peptone agar (Pederson) pH 6.5
	Bacillus subtilis BGA		Standard II nutrient agar, pH 6.0
	Bacillus subtilis BGA		Standard II nutrient agar, pH 8.0
	Sarcina lutea ATCC 9341		Standard II nutrient agar, pH 8.0
	Escherechia coli 28 PR 271		Synthetic agar medium, pH 7.2
	B. stearothermophilus var. *calidolactis* C 953		Nutrient agar (Oxoid), pH 7.0
Bacillus stearothermophilus disc assay (*8*)	*Bacillus stearothermophilus*	Raw milk, heat treated milk, reconstituted dried milk	Antibiotic medium 4, pH 6.5
Three-Plate Test (*15*)	*Bacillus subtilis* BGA	Fresh meats	Standard II nutrient agar pH 6.0, 7.2, and 8.0
Four-Plate Test (*16*)	*Micrococcus luteus* strain ATCC 9341	Fresh meats	Standard II nutrient agar, pH 8.0

Table 4 Confirmatory Tests for Microbial Inhibitor Tests–*continued*

Confirmatory Test	Microorganism	Test Matrix	Test Media
New Dutch Kidney Test (*17*)	*Bacillus subtilis* BGA	Animal kidney tissues	Standard II nutrient agar, pH 7.0
New Netherlands Kidney Test (*18*)	Animal kidneys	*Micrococcus luteus* ATCC 9341	Standard II nutrient agar

BGA, Bundesgesundheitsant (German Federal Health Centre).

Table 5 Detection Limits of Commercially Available Microbial Inhibition Tests for Antibiotic Residues Detection in Raw Milk (ppm)

Antibiotic	BSDA	CIA	Charm farm	BR	Valio T101	Lumac	Arla test
Ampicillin	0.005	0.01	0.004	0.005	0.01	0.002	—
Amoxicillin	—	—	—	—	—	0.003	—
Penicillin	0.006	0.006	0.004	0.005	0.005	0.003[a]	0.006
Cloxacillin	0.035	0.10	0.05	0.035	0.20	0.002	—
Tetracycline	0.40	0.20	0.06	0.4	0.2	0.3	0.1
Oxytetracycline	0.50	0.25	0.10	0.5	0.2	0.3	0.1
Chlortetra cycline	0.45	—	—	0.45	—	0.25	0.1
Chloram phenicol	—	—	—	—	1.0	1.0	0.6
Streptomycin	13	—	—	—	1.0	10.0	—
Dihydrostrep	22	—	0.125	—	0.5	6.0	5.0
Neomycin	2.25	—	—	—	0.5	5.0	2.5
Gentamicin	—	0.5	0.04	—	—	0.5	—
Erythromycin	—	0.25	0.25	—	0.05	1.25	0.6

BSDA, *B. Stearothermophilus* disc assay; CIA, International Units (IU)/mL.
[a]Adapted from reference (*11*).

Determining Antibiotic Residue Concentration by Microbial Inhibition Assays

Fundamental to all agar diffusion tests is the relationship between the size of the zone of inhibition and the potency of the antibiotic added (the dose-response curve). In microbial assay procedures, standard solutions containing known concentrations of a drug are tested on each assay plate and almost straight lines are observed when the $\log_{10}$[antibiotic concentration] versus the zone diameter is plotted. This is usually valid over a short range of concentrations. However, if the range of concentrations is very wide, it is more appropriate to plot the $\log_{10}$[antibiotic concentration] versus the square of the zone diameter instead. From a measurement of the zone diameter of the test sample, the potency of the drug can be estimated by interpolation from the dose-response curve.

Diffusion Test Limitations

It is well known that some of the microorganisms used in the agar diffusion tests have varying degrees of sensitivity towards different families of antibiotics. For example, *B. stearothermophilus* is very sensitive to penicillin detection but has poor sensitivity to the tetracyclines. A wide variation in method detection sensitivity is even evident within the same family of antibiotic drugs such as the β-lactams or the tetracyclines. Unlike chemical methods, which usually proceed with a preliminary extraction in order to isolate the drug(s) of interest from the biological matrix, microbial inhibition tests are generally conducted directly on the sample itself without any preliminary isolation step(s). The underlying assumption when a paper disk or cotton swab is introduced into the fluid test sample is that there will be a sufficient concentration of "free" unbound drug diffusing into the swab or paper disk to cause a detectable inhibition in the growth of the microorganism. The extent to which some of these drugs bind to protein is important in understanding why the diffusion technique works so well for some veterinary drugs but so poorly for others. The sensitivity of a diffusion test is influenced by two factors, the sensitivity of the microorganism to that particular antibiotic or family of antibiotics and the degree to which the drug binds to proteins in the test matrix. The greater the binding constant of the drug to protein, the more unlikely it will be for such a drug to be "freely" available in the inter- or extra-cellular fluid (the matrix from which the drugs diffuse into the cotton swabs and paper disks) to cause any perceptible change in the test kit result.

Penicillin is one of the drugs most frequently screened by diffusion tests, and is probably the one drug that is most readily available in the interstitial fluid. It is widely distributed throughout most body fluids and tissues, with about 65% of the penicillin in the blood being reversibly bound to plasma albumin (*19*). Unlike penicillin, which is available in the interstitial fluid, the tetracyclines are widely distributed in the body, and they concentrate in the kidneys, liver, and bones, where they exhibit protein bindings ranging from 55–80% (*20*). The aminoglycosides are distributed primarily into the extracellular fluid (ECF) because their highly polar nature prevents them from crossing cellular membranes (*21*, *22*). Less than 10% of the plasma concentration of aminoglycosides are bound to plasma (*23*). The distribution of aminoglycosides into tissue is influenced by lipid solubility and pK_a properties among other factors. Macrolides such as erythromycin, a basic drug with a high pK_a, tend to accumulate by ion-trapping in tissues, which have a lower pH than blood. Studies by Ziv and Sulman (*24*), have demonstrated that erythromycin, for example, binds to serum protein from cows and ewes at a level of 18–20% and 23–25%, respectively. This, together with the slow reversibility of tissue binding, permits erythromycin to accumulate in the liver, spleen, pancreas, kidneys, lungs, and heart. Therefore, for some of these drugs, unless chemical methods such as protein denaturization and precipitation are used to increase "free" available antibiotic in the test material, it will not be possible to improve the detection sensitivities of these diffusion tests.

When discussing the reliability of microbial inhibition tests to screen for antibiotic residues, the effect of naturally occurring substances in biological fluids and animal tissues on microbial inhibition must also be considered. It has been well documented (*25*, *26*) that lysozyme, a bactericidal compound occuring in the body fluids of man and animals, in the kidneys of all slaughtered domestic animals, and in eggs, has inhibitory effects on microorganisms and elicits false-positive test responses (*27–32*). Endogenous amino acids including lysine, arginine, ornithine, and other basic polypeptide degradation products have also demonstrated inhibitory effects on the growth of microorganisms (*33–35*). Lactoferrin, which sequesters iron and makes it unavailable for bacterial growth, has also been shown to have inhibitory effects (*36–38*). Since the 1927 report of Jones and Little (*38*) on streptococcal inhibition by bovine milk, several other reports have appeared that describe the nonspecific autoinhibition by decay products in meat (*39*–41) and certain bacteria and their metabolites including nisin. Even disinfecting agents with bacteriostatic or bactericidal properties have been implicated in inhibiting microbial growth (*42–44*). Recently coliform bacteria as well as oxytocin produced significant false-positive test responses in some test methods for the screening for antibiotic residues in milk (*45–48*). In addition, basic dyes such as methylene blue and rhodamine B, can also have an inhibitory effect on the growth of *B. subtilis* (*49*).

It is, therefore, very important to recognize that other factors, besides antibiotics, can affect the response of rapid test kits used for screening. There should always be suitable quantitative and compound-specific analytical methods with equivalent or better detection sensitivities than the test kits available that can be used to confirm the "presumptive positive" test responses obtained with any rapid tests.

Radiolabeling Techniques

Radioimmunoassay

The basic principles of immunoassays are similar regardless of whether isotopic or nonisotopic techniques are used. Nevertheless, radioimmunoassay (RIA) is significantly different from nonisotopic techniques and they will be treated under separate headings in this chapter.

Although RIA was first used for the analysis of antibiotics in the early 1970s, it had been used for several years prior to this for the analysis of hormonal drugs. RIA is based on the competition between an isotopically labeled analyte and an unlabeled analyte for a binding site on an antibody. Thus, the presence of large amounts of unlabeled analyte results in less radioactivity being bound to the antiserum. A comparison of the ratio of the bound to free labeled analyte with that obtained from a series of standards permits the quantitation of the analyte in unknown samples.

All RIA techniques (*50, 51*) involve three basic reaction steps. The first step involves three components, the radiolabeled (^{14}C, ^{3}H, or ^{125}I) antigen (Ag^{*}, antibiotic), the high titer high avidity antibody to that antigen (Ab), and the unlabeled antigen (Ag, antibiotic from the test sample). The second step involves an equilibrium reaction in which the labeled or unlabeled antigen reacts with the antibody to produce an antibody combined with a labeled antigen (Ag*Ab), or an antibody combined with an unlabeled antigen (AgAb) (Figure 1). The final step is to obtain the amount of bound antibody. Two approaches have traditionally been used to obtain the amount of bound antibody. Either separate the bound antigen or antibody complex from solution by precipitation (heterogeneous RIA), or use chemical methods to remove the unbound antigen from solution. Because antibody or antigen reactions are stoichiometric, the determination of either bound radiolabeled or unbound radiolabeled antigens will result in the direct calculation of the amount of antibiotic. Because small molecules on their own do not elicit an antibody response, and because most antibiotics have molecular weights ≤ 1000 (they are haptens), they must be attached to proteins to make them immunogenic. This has been accomplished in one of two ways. If an antibiotic has free amino functional groups, as do the aminoglycoside or nitro groups, that can easily be reduced to form free amino functional groups such as chloramphenicol, then it can be directly coupled to a protein (typically bovine serum albumin, human serum albumin, or porcine thyroglobulin) by the carbodiiamide method. It is this particular characteristic of the aminoglycosides and chloramphenicol that made these groups of antibiotics so attractive for the application of RIA techniques to their analysis. Most antibiotics, however, do not have free amino or nitro functional groups. For these antibiotics, it has been customary to use several chemical steps to introduce suitable functional groups into their chemical moieties to enable them to be coupled to proteins. For an extensive treatment of the chemical methods for

$$\underset{(A)}{Ag^*} + \underset{(B)}{Ag} + \underset{(C)}{Ab} \rightleftharpoons \underset{(D)}{Ag^*Ab} + \underset{(E)}{Ag}$$

$$\updownarrow$$

$$\underset{(F)}{Ag\,Ab} + \underset{(G)}{Ag^*}$$

Figure 1 Typical reaction sequence in radioimmunoassay
(A) Ag* labeled antigen antibiotic
(B) Ag unlabeled antigen antibiotic
(C) Ab antibody to Ag and Ag*
(D) Ag*Ab labeled antigen bound to antibody
(E) Ag free unlabeled antigen
(F) AgAb unlabeled antigen bound to antibody
(G) Ag* free labeled antigen

coupling proteins to other veteinary rugs, see an excellent treatise by Kabakoff (*52*), and the review by Blake and Gould (*53*).

Radioactivity can be detected with high sensitivity using simple methods that are usually independent of and unaffected by the sample milieu. Radiolabeling procedures are relatively simple because many radiolabeled compounds are commercially available and because radiolabels are comparatively small molecules they do not normally affect the reaction kinetics. RIA is specific and reproducible but requires an expensive beta or gamma counter. The preparation of labeled antigens involves real risks, and requires that strict safety controls be observed and adhered to as well as specially designed protective clothing be worn. The labeled antigen shows batch-to-batch variability and generally has a short shelf-life. Because separation of the reacted from unreacted labeled antigens is essential (heterogeneous RIA), other alternative approaches that that do not require any separation steps (homogeneous RIA) should be considered.

The only receptor assays that are currently in extensive use for analyzing antibiotic drug residues are those developed by Charm Sciences for the analysis of antibiotics and marketed under various names as Charm II Sequential Assay, Charm II Competitive, Charm II Transit, Charm I Sequential, and Charm II

Cowside Assays. The Charm II Receptor Assays are used for the determination of antibiotics including β-lactam, tetracycline, macrolide, aminoglycoside, and chloramphenicol in milk and dairy products (*54*, *55*), meat, eggs, and biological fluids (*56*). The assays are based on the competition for receptor or binding sites on microbial cells added to the sample. In these assays, a microorganism (*B. stearothermophilus*) containing a variety of binding sites is added and competes with antibiotics in the same drug family (with the same functional group) in the test sample. Binding is measured by a scintillation counter and is compared to counts obtained from an antibiotic-free sample. The reference to the Charm tests does not in any way suggest that there are no other RIA methods besides those developed by Charm Assays for the analysis of antibiotic residues in food. In fact, RIA methods have been described and used extensively for the quantitation of chloramphenicol in foods of animal origin including porcine muscle tissue, bovine milk, plasma, and hensl eggs (*57–60*). Similar products are commercially available for sulfonamide (antibacterials) testing but these test kits fall outside the scope of the treatment here. Table 6 lists the detection sensitivities in parts per million

Table 6 Detection Limits of the Charm Test II Receptor Assay for Detecting Antibiotics in Various Food Matrices

Antibiotic	Milk (ppm)	Tissue (ppm)	Serum (ppm)	Urine (ppm)	Fish (ppm)	Eggs (ppm)
Penicillin G	0.004IU[a]	0.02	0.02	0.02	0.02	0.02
Ampicillin	0.0025	0.04	—	—	0.04	—
Cloxacillin	0.003	—	—	—	0.15	—
Cephapirin	0.003	0.05	—	—	0.05	—
Ceftiofur	—	—	—	—	0.15	—
Tetracycline	0.001	0.05	0.2	0.2	0.05	—
Oxytetracycline	0.005	0.5	—	—	0.5	—
Chlortetracycline	0.003	0.1	—	—	0.1	—
Gentamicin	0.02	0.4	0.2	0.08	0.4	—
Streptomycin	0.02	0.15	—	0.08	0.15	0.15
Erythromycin	0.015	0.4	0.2	0.2	0.15	0.15
Chloramphenicol	—	—	0.005	0.02	—	0.005

[a]I. U./mL. Adapted from reference (*56*).

(ppm) of the Charm II Receptor Assay for common antibiotics in several food matrices.

Nonisotopic Techniques

In a nonisotopic immunoassay, a label, which can be an enzyme or fluorogenic substance, is coupled to the antigen or antibody in such a way that the activity of the parent compound is not appreciably affected. Enzyme immunoassays (EIAs) are nonisotopic immunoassays that use enzymes, whether they are fluorogenic or not, as labels. Primarily, when an enzyme substrate is used, it is the enzyme activity (i.e., the reaction velocity), which is proportional to the number of enzymes catalyzing the reaction, that is measured. Enzyme immunoassays are classified into two groups: heterogeneous EIA, in which the enzyme-labeled antigen or antibody is separated from the enzyme-labeled antigen or antibody complex before measuring the enzyme activity in either fraction, and homogeneous EIA, in which the enzyme activity of labeled antigen is measured in the presence of the labeled antigen or antibody complex.

Heterogeneous assays. Heterogeneous EIA was the first to be developed and is best exemplified by the enzyme-linked immunosorbent assay (ELISA) *(61–64)*. ELISAs differ from heterogeneous RIAs in that the antigen or antibody is labeled with an enzyme instead of a radioactive isotope. Like heterogeneous RIA, where an uncomplexed or free radiolabeled antibody or antigen is separated from the radiolabeled antigen or antibody complex, the ELISA includes a separation of an enzyme-labeled antigen or antibody complex from a free enzyme-labeled antigen or antibody. It is the enzymatic activity in the bound or free fraction, resulting from the enzyme-catalyzed conversion of a relatively nonchromatic or nonfluorogenic substrate to a highly chromatic or fluorescent product, which can be spectrophotometrically measured and quantitated as an indication of the analyte concentration. ELISAs can be further classified as either competitive or noncompetitive based on the technique involved. Competitive ELISAs involve a reaction step in which unlabeled antigen and antigen linked to an enzyme or attached to an immobilized support compete for a limited number of antibody binding sites. In noncompetitive ELISA, the antigen or antibody to be measured is allowed to react alone with an excess of the immune reactant (antibody or antigen), and the extent of the antibody or antigen reaction is measured in a second step.

Heterogeneous ELISA tests have been developed using microtiter polystyrene wells or plates, or membranes such as nylon, polypropylene, and nitrocellulose as solid phases. The membranes are immobilized onto single tubes, cups, dipsticks, or disks upon which the tests are conducted. A large number of such tests are commercially available for the analysis of veterinary drug residues, including antibiotics in milk, urine, tissues, serum, and feeds. Table 7 lists several commercially available ELISAs that have been used for screening veterinary drug residues in foods and biological fluids. Typically in competitive ELISA, monoclonal or

Table 7 Commercially Available ELISA Tests for Detecting Antibiotic Residues in Foods and Biological Fluids

Test kit	Applicable antibiotics	Test type	Manufacturer
Cite Probe	Ampi, Amox, Ceph, CTC, Genta, OTC, PG, & TTC	ELISA probe	Idexx
LacTek β-lactam	Ampi, Amox, Ceph, Clox, Genta, PG, Hetacillin, Oxacillin	ELISA tube	Idetek
LacTek Ceftiofur	Ceftiofur	ELISA tube	Idetek
LacTek Tetracycline	CTC, OTC, TTC	ELISA microwell (quantitative)	Idetek
Penzyme and Penzyme III	Amox, Ampi, Ceph, Cephalexin, Cephalosporin, Clox, Hetacillin, Nafcillin, Oxacillin, PG	EIA and quantative ELISA	SmithKline Beecham
EZ-Screen	CAP, Genta, OTC, CTC, TTC, Tylosin	ELISA card	Environmental Diagnostics
RIDASCREEN Tetracyclin	TTC, CTC, MNCY, RTCY, DMCY	Quantative ELISA	R-Biopharm GmbH
SIGNAL	Genta, Neomycin	ELISA microwell	SmithKline Beecham
SIGNAL Foresight	Gentamicin	ELISA pad	SmithKline Beecham

Ampi: Ampicillin; Amox: Amoxicillin; Ceph: Cephapirin; CTC: Chlortetracycline; OTC: Oxytetracycline; PG: Pen G; TTC: Tetracycline; Clox: Cloxacillin; CAP: Chloramphenicol; MNCY: Minocycline; RTCY: Rolitetracycline; DMCY: Demeclocycline

polyclonal antibodies (specific to the drug[s] to be analyzed) are immobilized onto a solid phase. A specified volume of the sample containing the free drug(s) of interest is added to the immobilized antibody followed by the addition of an enzyme conjugate. The mixture is allowed to incubate for a specified time, usually at room temperature, during which the free drug competes with the enzyme-labeled hapten for antibody sites. After the incubation period, any excess unbound haptens or enzyme conjugates are washed away, an enzyme substrate solution is added, and the test system is incubated for another specified period of time. The substrate is usually chosen so that it is colorless initially but yields, on degradation, a green or blue colored product. The blue or green color development indicates that there was no significant amount of antigen present in the test sample (negative result), while no color development indicates the presence of a significant enough amount of drug in the test sample to saturate the antibody sites (a positive result).

For some purposes, especially when high sensitivity is required, a fluorogenic substrate instead of a chromogen is added. Agridiagnostics' test kits which come in the EZ-Screen (card test), 20 Cup, ID BLOCK, and RIDASCREEN formats, as well as SIGNAL gentamicin, SIGNAL neomycin, and SIGNAL Fore-Site test kits (SmithKline Beecham, Philadelphia, PA, USA) are designed to be used as described above. In some variants of ELISA tests, a strong acid or alkali solution (usually referred to as a stop solution) is added after the enzyme substrate incubation step to stop the degradation reaction of the substrate, and the optical density of the resulting solution is then determined immediately. RIDASCREEN tetracycline test kit (RIDASCREEN, Danstadt, Germany), Cite Probe for antibiotics (Indexx, Burlington, NC, USA), EZ-Screen Card test for chloramphenicol (Environmental Diagnostics, Westbrook, MA, USA), LacTek tube tests for β-lactams, ceftiofur, and tetracyclines (Idetek, Sunnyvale, CA, USA) all add a solution to stop the color developing reaction just prior to reading the test results. For example, in the case of the RIDASCREEN Tetracycline test, addition of the stop reagent causes the color of the blue complex to change to yellow. For these tests, the color developed or the absorbance, which is inversely proportional to the amount of drug(s) in the sample, is measured and compared to that of a known standard.

Bioluminescent immunoassay. So far, immunoassay techniques in which a radioisotope or enzyme is used as an amplifying marker have been discussed. Another variant of immunoassay, bioluminescent immunoassay, which uses a coenzyme (the adenosine triphosphate [ATP]-dependent enzyme luciferase, from *Photobacterium fisheri* or from firefly) as an amplifying marker has been described. This technique has been used in the bioluminescence method (*65*) for the qualitative detection of inhibitory substances in milk using the *Streptococcus thermophilus* strain T.J. as a test organism. The bioluminescence method is based on the use of the luciferin-luciferase enzyme system to measure ATP that is coming from bacteria as a result of the light emission during the bioluminescent reaction. In this method, ATP is coupled to a hapten. When antihapten specific antibodies react with the ATP-hapten conjugate, the light-generating reaction catalyzed by luciferase, is diminished. The luminescence intensity produced by the reaction is proportional to the extracted ATP concentration. In practice, the milk sample to be tested is first heated, yeast extract and apyrase are added and the mixture is inoculated with a *S. thermophilus* T.J. culture in the exponential growth phase. After 20, 40, or 60 min of incubation at 45°C, a bacterial extractor, a stabilizer, and luciferin-luciferase reagent are added. The light intensity is measured with a luminometer, and the ratio of the light intensity of the samples incubated at 20 and 40 min is calculated. Inhibitory substances are present in the test sample if the measured ratio from the test sample is lower than those obtained under the same conditions with inhibitor-free milk. If confirmation is needed, the ratio of the light intensities from the samples at 20 and 60 min can also be calculated.

All the ELISA tests described above involve the formation or inhibition of a colored complex. Because most foods and biological fluids likely to be tested contain pigmented material that may interfere or even obscure the color of the complex formed, most ELISA techniques applied to the analysis of antibiotics use some preliminary sample treatment (e.g., the extraction of the drug from the matrix of interest, centrifugation and filtration, dilution with buffers, etc.) before applying the sample in a suitable form for the test. Even though the protocol for conducting ELISA tests would appear to be quite simple to a trained analyst, we have found in our experiences working with feedlot operators that they would prefer to use microbial inhibition assays instead of ELISA tests in the field, even when the ELISA test is a simple card test format.

As is common with all visual end-point tests that require an end-user to interpret the test result based on a color comparison (which may sometimes not be clearly defined) with either a negative or positive test standard, it is imperative that the analyst be properly trained to perform the test and have good visual acuity. Some manufacturers have removed the analyst-dependent interpretation of the end result for the test kits. They instead provide portable photometers or spectrophotometers calibrated to indicate whether a drug is present above or below the defined cut-off concentration is in the test sample by giving a yes or no answer.

Homogeneous Assays

Our discussion on enzyme immunoassays has been limited to heterogeneous assays. Because homogeneous assays do not require a separation step to remove free and bound ligand, they can usually be accomplished in one step. In this assay, an antibody is bound to an antigen so that the large substrate is unable to gain access to the catalytic site of the enzyme causing inhibited activity. When unlabeled antigen is added, it competes with the enzyme-labeled antigen in binding to the antibody, causing the free form of the enzyme-labeled antigen to increase and a consequent increase in enzyme activity. The enzyme activity, can therefore be correlated to the concentration of unlabeled antigen in the sample and permit the quantitation of drugs (haptens).

Enzyme-multiplied immunoassay test. The enzyme-multiplied immunoassay test (EMIT) is the most widely used homogeneous assay system. The assay, introduced by Rubenstein et al. (*66*), depends on a change in the specific enzyme activity when an antibody binds to the enzyme-labeled antigen. The activity of the unseparated assay mixture corresponds to the proportion of the enzyme-labeled antigen to which the antibody is bound. Two distinct variants of the EMIT principle have been devised and applied commercially by the Syva Company. The earliest EMIT systems (*67*), which used lysozyme as the enzyme label, were used to assay drugs of abuse in human urine (*68*). Later systems (*69*) used glucose-6-phosphate dehydrogenase or malate dehydrogenase as enzyme labels, and were used to assay therapeutic drugs (*70*) in clinical situations.

Substrate-labeled immunofluorescent assay. Another approach to homogeneous ELISA has been developed that uses an enzyme substrate as an amplifying marker. This technique, referred to as substrate-labeled immunofluorescent assay (SLIFA) and developed by Burd et al. (*71*, *72*), has been used in a variety of assay systems (*73*, *74*). In this method, neither the labeled hapten nor the antibody-bound hapten is fluorescent. However, the labeled hapten can be cleaved by enzymatic hydrolysis to release part of the labeled molecule, which is then able to fluoresce. Cleavage of the labeled hapten is sterically prevented when it is bound to the specific hapten antibody. Competition for antibody binding sites by hapten from the test sample makes more of the labeled hapten available to act as an enzyme substrate. The fluorescence intensity is, therefore, proportional to the amount of free hapten in the test sample. This technique has been used for the assay of the aminoglycoside and aminocyclitol antibiotics gentamicin (*75*), kanamycin (*76*), and dibekacin (*77*). For the assay of gentamicin, for example, galactosidase was used as the enzyme and the fluorogenic substrate was a β-galactosylumbelliferone–gentamicin conjugate (*72*). In the normal state, the labeled antibiotic does not fluoresce. However, if the antibiotic-substrate reagent is attacked by β-galactosidase, the umbelliferone moiety is released and it fluoresces. When the conjugate comes into contact with antibody against the antibiotic, a complex forms that prevents the interaction of the bound conjugate with the enzyme. As in any competitive binding assay, the free antibiotic in the test sample will compete with the drug conjugate for binding sites on the antibody. The amount of conjugate available for the reaction is, therefore, directly related to the amount of free drug in the test sample. The gentamicin assay is homogeneous, rapid, and simple to perform and is comparable to RIA in sensitivity. It requires only 1 μL of sample. The SLIFA technique using the avidin-biotin system has recently been used to screen chloramphenicol residues in swine muscle tissues (*78*).

Fluorescence polarization immunoassay. Another technique of homogeneous enzyme immunoassays is fluorescence polarization immunoassay (FPIA), which is based on the competition between free antibiotic and antibiotic labeled with fluorescein for antibody binding sites. When the system is allowed to reach equilibrium and the antibiotic binds to the antibody, the rotational relaxation time (i.e., the time interval that an oriented molecule takes to leave alignment when subjected to a beam of polarized light) of the antigen or antibody complex will be increased and thereby exceed the fluorescence lifetime (i.e., the time interval from when the molecule is excited by polarized light until it releases its emitted light

(*79*)) of the label, fluorescein. Fluorescence polarization occurs when the light of a single wavelength strikes a molecule in such a way that the light is absorbed and emitted in the same plane in the fluorescence mode (*80*). Because in FPIA the concentration of the antibody and the fluorescent-labeled antigen are kept constant, only the concentration of the unlabeled antigen from the test sample is variable. Thus, the greater the concentration of antibiotic (the unlabeled antigen) in the sample, the more this antigen will limit the binding of labeled antigen to the antibody, and the greater the extent to which the intensity of fluorescence polarization will be decreased. Therefore, as the amount of free antibiotic in the test sample decreases, the intensity of fluorescence polarization will increase. The theoretical basis for fluorescence polarization has been described by Dandliker et al. (*81*). Many of today's applications of FPIA in drug screening (*82–85*) stem, to a large extent, from the development of a specially designed microprocessor-controlled fluorometer as well as an automated fluorescence polarization analyzer by Abbott Laboratories. The fluorometer (*86*) uses collimated light from a tungsten or halogen filament that is directed through a 485 nm excitation wavelength filter. The reflected light from a transparent glass beam splitter is focused onto a reference silicon detector, the signal from which is then used to monitor the intensity of the light source (the reference channel). The transmitted light through the beam splitter (which also serves to rotate the plane of polarization) is passed through a polarizer and crystalloid transmission-type liquid crystal. The excited light is then directed onto the sample that is held in a 12×75 mm disposable glass tube, and the emitted light is collimated and passed through a 525 nm interference filter into a vertical polarizer. In this assembly, the excitation and emission optics are aligned at right-angles to each other. The liquid crystal and polarizer combination rotates in an electric field applied to the liquid crystal. When voltage is applied to the liquid crystal, no rotation occurs and horizontal light passes through the sample. When there is no voltage applied to the liquid crystal, the liquid crystal rotates through an angle of $\pi/2$ and the sample is excited by vertical light. Both the excitation and fluorescent polarization intensities (in voltages) are captured in the vertical and horizontal modes and converted into frequencies. Fluorescence intensities are presented as the ratio of the frequency-to-channel counts to the reference-to-channel counts. Because the measurement of the polarized fluorescence light intensity is critical to the success of the technique, Abbott Laboratories markets test kits that come complete with such a fluorometer.

Test Kits for Screening Antibiotic Residues in Foods

All rapid tests should have well-characterized operational parameters that the test is expected to follow, provided that the test kit is used according to the procedure described by the manufacturer. Rapid tests may be qualitative, in which case they merely provide positive or negative test reponses. When the rapid tests are designed to provide quantitative test responses, they provide, in addition to a detection of the presence or absence of the drug of interest, an estimate of the concentration of the drug present in the test sample. Unlike chemical methods, for which there are designed official procedures for their evaluation under collaborative experimental conditions in order to assign them official test status (*87*), there are no officially published guidelines for evaluating rapid tests. There is no denying the fact that during the familiarization stage of using a test kit, most scientists will usually conduct an evaluation of some sort to determine whether the test kit will work as claimed by the manufacturer. The problem is that each evaluation is done differently with no defined guidelines as to the scope of the evaluation process or to the experimental approach used to obtain statistically meaningful results.

A case in point is an evaluation exercise (a collaborative study) conducted at the Virginia Polytechnic Institute (VPI) and State University (SU) at Blacksburg, Virginia, U.S.A. (*88*). At the time of the study, the United States Food and Drug Administration (FDA) had established "safe" levels for animal drug residues in milk, and the industry was anticipating that changes were soon to be made in the Pasteurized Milk Ordinance (PMO) that would allow rapid tests other than the three rapid tests that had then been subjected to an AOAC INTERNATIONAL's collaborative study to be used for screening veterinary drug residues in raw milk (*89*). The three methods that had been collaboratively studied were the *Bacillus stearothermophilus* Disc Assay for Penicillin G (considered the official method), the Delvotest P for Penicillin G, and the Charm II Test Assay for Penicillin G and Cephapirin. The study evaluated 14 detection methods with 24 animal drug residues in milk to determine the sensitivity of each method and its ability to detect a residue at the FDA "safe" level. Tables 8A and 8B summarize the results of the VPI-SU study. Although this study served a useful purpose at the time it was commissioned, it will become apparent as we describe experiments that must be performed in conducting an exhaustive evaluation of a rapid test, that the VPI-SU study covered only a very small fraction of what would be considered a full-fledged rapid test method evaluation exercise.

Recently, AOAC (*90*) using the services of its recently established Research Institute (AOAC RI) working in collaboration with scientists from the FDA's Center for Veterinary Medicine (CVM), set up guidelines that were used to evaluate some commercially available rapid tests for detecting β-lactam veterinary drug residues in raw milk (*91*). The next section will discuss the various types of rapid tests that are available and the approaches that should be used to evaluate their test characteristics to ensure that some measure of uniformity and consistency is maintained in the way rapid test kits are evaluated. This will enable scientists and manufacturing companies who produce these test kits to become aware of the criteria used to evaluate performance characteristics of the rapid tests they develop or design. In addition, we will describe the experimental procedures that must be followed in order to properly define performance characteristics so that these

Table 8A Comparison of Claimed Detection Limits and Minimum Levels of Antibiotic Tested in the Virginia Polytechnic Institute Study

Antibiotic	Safe level[a]	Charm II Test	Charm Cowside	CITE Probe	Penzyme	LacTek
Penicillin G	10	2/2.5[b]	4.8/5	5/10	4/10	5/5
Cephapirin	20	3/5	[a]/5	8/5	[a]/10	[a]/10
Cloxacillin	10	12/20	[a]/50	80/100	[a]/50	[a]/10
Ceftiofur	1000	12/5	[a]/10	[a]/10	[a]/50	[a]/100
Amoxicillin	10	[a]/5	[a]/10	10/10	[a]/5	[a]/5
Ampicillin	10	3/5	[a]/10	8/10	[a]/10	[a]/10
Tetracycline	80	1.2/5	—	20/30	—	—
Oxytetracycline	30	30/30	—	40/30	—	—
Chlortetracycline	30	3/5	—	40/30	—	—
Erythromycin	50	20/50	—	—	—	—
Tylosin	50	10/150	—	—	—	—
Novobiocin	—	30/1000	—	—	—	—
Polymixin B	—	—	—	—	—	—
Gentamicin	30	20/30	—	30/30	30	10/30
Neomycin	150	[a]/500	—	[a]/500	10	[a]/500
Streptomycin	125	2.5/10	—	[a]/1000	—	[a]/1000
Spectinomycin	—	20/30	—	[a]/1000	—	[a]/1000

[a]There were no claimed detection levels by manufacturer. Concentrations are in ppb. FDA defined safe levels. See reference (*88*).
[b]The first entry in each segment is the claimed detection for the test. The second entry is the minimum detected level, which was the concentration at which at least 13 out of 15 analyses were found to be positive. Concentrations are in ppb.

parameters will be easily verifiable by end-users or third-party evaluators. To do this, we will examine the operational or performance characteristic parameters that must be defined by the manufacturer in order for the end-user to be able to use the test method consistently. These include specificity, selectivity, dose-response curve, ruggedness, cross-reactivity, effect of other interferences, stability, and shelf-life of the test reagents and materials.

Table 8B Comparison of Claimed Detection Limits and Minimum Levels of Antibiotic Tested in the Virginia Polytechnic Institute Study

Antibiotic	Safe level[a]	Disc assay	Charm Farm	BR	Delvo P	Delvo SP	Penzyme III
Penicillin G	10	4.8/5	3/2.5	1.5/10	2/2.5	2/5	[a]/5
Cephapirin	20	8/10	15/20	[a]/10	8/10	8/5	[a]/5
Cloxacillin	10	50/50	40/50	[a]/100	20/50	20/50	[a]/20
Ceftiofur	1000	75/100	15/50	[a]/100	50/50	50/100	[a]/50
Amoxicillin	10	8/50	[a]/5	[a]/5	3/10	3/10	[a]/5
Ampicillin	10	10/10	5/5	[a]/10	4/10	4/10	[a]/10
Tetracycline	80	980/1000	50/80	250/1000	200/1000	200/100	—
Oxytetra-cycline	30	1000/1000	80/80	[a]/1000	200/250	200/1000	—
Chlortetracy-cline	30	1000/1000	150/250	[a]/1000	500/1000	500/100	—
Erythromycin	50	1000/1000	200/1000	50/1000	400/1000	400/100	—
Tylosin	50	1000/1000	50/50	[a]/50	100/100	100/100	—
Novobiocin		1000/1000	750/750	[a]/1000	500/1000	500/100	—
Polymixin B	—	[a]/1000	[a]/1000	[a]/1000	[a]/1000	[a]/1000	—
Gentamicin	30	460/500	[a]/150	500/500	250/150	250/250	—
Neomycin	150	800/500	[a]/150	[a]/500	1000/150	1000/50	—
Streptomycin	125	1000/1000	[a]/1000	400/1000	4000/1000	4000/10	—
Spectinomycin	—	[a]/1000	[a]/1000	[a]/1000	[a]/1000	[a]/1000	—

[a]There were no claimed detection levels by manufacturer. Concentrations are in ppb. FDA defined safe levels. See reference (*88*).
[b]The first entry in each segment is the claimed detection for the test. The second entry is the minimum detected level, which was the concentration at which at least 13 out of 15 analyses were found to be positive. Concentrations are in ppb.

Rapid Tests for β-Lactam Antibiotics in Bulk Milk with the AOAC Mark of Certification

The test performances of the rapid tests listed in Table 9 were independently evaluated for their reliability and suitability to detect β-lactam drug residues in bulk milk (*92*) by the AOAC RI using its pool of expert reviewers, Arthur D. Little Incorporated (a private consulting analytical laboratory in the United States), and scientists from CVM. As a result of this evaluation, these rapid tests have, with FDA recommendation, been granted AOAC's performance tested status, and have been licensed to use AOAC's mark of certification for commercial distribution.

Table 9 Detection Limits of Commercially Available Rapid Tests Currently Certified for Detecting β-lactam Antibiotics in Milk

Screening test	Pen G	Ampi	Amox	Clox	Ceftio	Cepha
Tolerance	5	10	10	10	50	8
Snap Test	5	10	10	50	50	8
Penzyme III	5	10	8	80	80	8
Penzyme Milk	5	10	8	80	95	8
Lactek Ceftiofur	ND	ND	ND	ND	50	ND
LacTek β-Lactam	5	8	10	8	ND	16
Delvo-X-Press	5	10	10	50	10	10
DelvoTest P	3	10	8	30	50	8
Cite Probe	5	12	12	100	50	8
Charm *B. Stear.* Disk Assay	4.8	6.5	10	48	75	11
Charm I/Cowside II	4.8	10	10	50	40	8
Charm II Transit	4.8	9	10	80	13	4.5
Charm Rapid Inhibition	3	4.5	4.5	25	50	16
Charm Farm	5	10	10	40	25	20
Charm II Competitive	4.8	9	10	70	25	4.5
Charm II Sequential	4.8	8	10	50	23	4.5
Charm Quantitative	1	1.5	1.4	10	2	1

Pen G: Penicillin G; Ampi: Ampicillin; Amox: Amoxicillin; Clox: Cloxacillin; Ceftio: Ceftiofur; Cepha: Cephapirin; ND: not determined.
Adapted from reference (*92*).

Most of the test methods evaluated previously in the VPI-SU study were submitted for evaluation again. It is anticipated that additional test kits will, in the future, be evaluated in this manner and will be certified for commercial distribution. Similar evaluations will be undertaken for test kits developed for the detection of other veterinary drug residues in milk and other animal-derived foods.

Manufacturers of all the certified tests are required to indicate on their label or package inserts the individual β-lactam antibiotic(s) that the rapid test is able to detect, together with the selectivity, sensitivity, and sensitivity rates of the test for each detectable antibiotic. The label or package insert must also include data on the dose-response, the ruggedness of the test to changing test conditions, the cross-reactivity studies, and the evidence to demonstrate the stability of batch-to-batch production of rapid test reagents and material. These commercially available rapid tests were evaluated by using an independent laboratory to verify the performance characteristic data provided by each manufacturer, and they are therefore expected to operate as specified on the label, provided that the tests are conducted according to the procedures described by the manufacturer. Even under these conditions, it is incumbent on the end-user of any of these rapid tests to verify that the tests are performing according to specifications using the built-in diagnostic quality assurance parameters included with most of these rapid tests before they are used for any large-scale testing programs. Most of these test kits contain negative and positive control samples that are tested concurrently with the test material. Table 9 lists the tests that were evaluated and certified for use on raw milk matrix only, and no attempt must be made to extend this certification to other food matrices. Any manufacturer who wants to obtain certification to use any of the above tests on a matrix other than raw milk or to screen veterinary drugs other than β-lactams, will need to submit appropriate performance characteristic data to a recognized independent third party for evaluation. The next sections will describe the types of experiments that should be conducted by manufacturers or scientists when attempting to evaluate or provide performance characteristic data for the rapid tests they design or develop.

Rapid Tests Not Evaluated by an Independent Third Party

Consumer and regulatory agency demands have changed the market trend and the need for rapid tests in all sectors of the food industry. More than ever, suitable rapid tests are being sought to be incorporated into the food inspection, surveillance, and monitoring programs of various regulatory agencies in abattoir and dairy operations. There are strong indications in the livestock industry that it won't be too long before we see rapid tests made available to the primary producer or the veterinary practitioner on the farm so that they can test the animals themselves to ensure that they are free from veterinary drug residues before they are shipped to slaughter or before milk is shipped to a dairy. At the moment, only commercially available rapid tests designed for screening β-lactam drug residues in raw milk have received the performance tested status for commercial distribution. Rapid tests for screening other veterinary drugs besides β-lactam residues in milk are also being evaluated for AOAC performance tested certification. Although there are commercially available rapid tests designed for use on food matrices other than milk, these have not yet been submitted to an independent third party such as the AOAC RI, for evaluation and certification for commercial distribution. To address the apparent unavailability of suitably evaluated rapid tests for screening antibiotic residues in other food matrices than milk, scientists in the regulatory industry have been trying to modify or adapt commercially available tests. These tests are usually designed initially for screening milk, but can be modified to be used on other matrices such as hensl eggs, animal plasma or serum, tissue mass or muscle drip, urine, etc. At the Health of Animals Laboratory, we have had to adapt and evaluate test kits initially intended for milk testing for use on other food matrices including meat and eggs. For example, the BR test for milk was adapted in 1990 into a field screening test for antibiotics in eggs by Boison and MacNeil (*93*); a test that is now being used across Canada for antibiotic screening in eggs. When such test kit adaptations are made, usually in government research laboratories or in universities with the ultimate aim of using them for large-volume testing in support of food inspection programs of various regulatory agencies, the test characteristics must all be redefined and the new operational parameters must be reevaluated for the matrix under consideration. For rapid tests that fall into this category, it is the obligation of the end-user to verify most of the easily verifiable performance characteristic data such as selectivity, specificity, sensitivity, and cross-reactivity before the test is used in any extensive testing program.

Verification of Rapid Test Selectivity and Sensitivity Rates

Selectivity is the ability of a test to certify that negative samples are truly negative. The selectivity rate, an associated parameter, is defined as the percentage of negative samples that are found truly negative by the test. To meet the minimum requirements for labeling an antibiotic residue test, a test kit must demonstrate a selectivity rate of at least 90% with a confidence level of at least 95% on a zero residue (antibiotic-free) sample (*91*) of the matrix for which the test has been designed or adapted. To demonstrate that the minimum selectivity requirement is met, experiment 1 (a selectivity test) described below should be conducted. It must be emphasized that a test kit that fails the selectivity test cannot be accepted in any official certification program, and it would not be considered suitable for screening, nor would it merit any further evaluation.

Once a test kit passes the selectivity rate test, the sensitivity rate must be verified. The detection sensitivity (the detection limit) of a test kit to a given antibiotic is the lowest concentration of the antibiotic that the test can reliably detect as positive in the given matrix. The sensitivity rate is the percentage of positive samples that are found by the assay to be truly positive at a defined concentration such as the claimed detection limit of the test, the safe level, or the tolerance level. To meet the minimum requirement for detection sensitivity claim, a test kit must demonstrate a sensitivity rate of at least 90% with a minimum 95% confidence level at that concentration (*91*). This requirement is met if all the samples tested in experiment 2 (the sensitivity rate test) elicit all positive responses in the test kit.

Fundamental to the concept of using rapid tests, especially those that rely on nonspecific methods and provide only qualitative test responses for screening antibiotic residues in the field or in the abattoir, is the understanding that there are available compound-specific quantitative confirmatory methods with equivalent or better detection limits, which can be used to verify the presumptive positive or negative test reponses obtained with the qualitative rapid tests. It is very important to have such sensitive analytical methods available before evaluations of any rapid tests are contemplated. The reasons for this will become apparent when we discuss the experimental procedure for conducting sensitivity rate tests using incurred samples and when we perform the cross reactivity experiments.

Experiment 1—Selectivity Rate

To demonstrate the selectivity rate requirement, perform 30 replicate analyses of an antibiotic-free sample. The requirement is satisfied if all 30 samples elicit all negative responses in the test. To demonstrate the specificity of the test (i.e., the ability to detect truly negative samples regardless of the source of the sample), the selectivity rate test must be repeated for antibiotic-free samples taken from no less than six different sources. Test responses for all the samples should be negative. This test is of utmost importance if the rapid test is to be used for testing samples from individual animals (cowside testing). For biological fluids or animal tissues, it is advisable to pool together antibiotic-free samples taken from no less than six different sources. Perform the selectivity test using sub-samples from this pool to demonstrate that the specificity requirement is also satisfied on pooled (bulk) drug-free samples.

Experiment 2A—Sensitivity Rate on Fortified Samples

To demonstrate the minimum sensitivity rate requirements for a test kit, perform 30 replicate analyses of drug-free samples of the specified matrix fortified with the antibiotic of interest at its claimed detection limit. The 30 samples must all elicit positive results in the test kit. This experiment, conducted on drug-free samples intentionally fortified with the drug of interest at the specified concentration, must also be repeated for samples containing physiological concentrations of the drug (incurred residue) at the claimed detection limit.

Experiment 2B—Sensitivity Rate on Incurred Residue Samples

This can be done for the incurred residue sample using an illustrative test kit that has a claimed detection limit of 5 ppb for penicillin G in raw milk. The raw milk sample is obtained from a healthy dairy cow with no previous antibiotic treatment history that has been administered, for example, the label dose of a commercial formulation of penicillin G in order to generate physiological concentrations of this antibiotic in the milk. Collect milk from this animal just prior to the administration of the drug to obtain antibiotic-free milk, and then shortly after drug administration to obtain penicillin-incurred milk. Determine the concentration of penicillin in the incurred milk sample by a quantitative, penicillin-specific analytical method. Once the level of penicillin in the incurred sample has been determined, the drug-free milk samples obtained from the same animal can be used to serially dilute the penicillin-incurred milk sample to the desired concentration of 5 ppb. The same analytical method must be used to determine and to confirm that the concentration of penicillin in the diluted milk sample is indeed 5 ppb. Upon verification of the drug concentration by this technique, samples of this milk can then be used to perform the selectivity rate experiment on incurred residues. Metabolic processes take place under physiological conditions and may result in the production of active metabolites of the drug after the drug has been absorbed into the blood stream and into the tissues of the animal. These metabolic processes may not be apparent when a drug is intentionally added by fortification to the sample in the laboratory. Therefore, the determination of the actual residual concentration of the active drug in the incurred sample, especially in the presence of active metabolites, requires the availability of sensitive compound-specific quantitative analytical methods. This is one of the reasons why it is imperative to have compound-specific, highly sensitive, and accurate quantitative analytical methods available for rapid test kit evaluations. Without such a sensitive analytical laboratory method, it would be impossible to determine whether the test response being observed is due to the presence of the drug alone or to the cumulative effect of both the drug and its metabolites. It would also be difficult to confirm the actual concentration of penicillin in the incurred sample after serial dilution. It is strongly advised that experiments 1 and 2 be conducted simultaneously in order to eliminate any bias in the outcome of the resulting test responses. The samples (both the control and the fortified) to be tested in experiments 1 and 2 should be independently prepared by one analyst, coded, then randomly mixed, and blind-analyzed by another analyst. Where test results are based on subjective interpretation such as a color change, it is advisable to have the test results read and test results recorded independently by two different analysts. The tests should be spread over more than one day.

Additional Experiments

To meet the minimum sensitivity or selectivity rate of 90% with a minimum 95% confidence level criterion, a test kit must demonstrate no false-negative or false-positive test responses in the 30 coded and blind-analyzed samples. Test kits fail if three or more failures occur in experiments 1 and 2. Such rapid tests will not be considered worthy of further evaluation because any test kit that fails the selectivity rate test will not be considered suitable for use in any screening program.

Test kits that produce one or two failures in experiments 1 and 2 on residue-free samples or incurred residue samples require further testing. This additional testing involves the analyses of a new independent set of 60 residue-free and fortified samples, coded, randomized, and analyzed blind by an analyst. The test kit meets the stated minimum requirements if all 60 samples elicit the anticipated test response or if only one failure out of the 60 tests occurs. Test kits fail if they elicit two or more false-positive or false-negative responses in this second independent set of 60 analyses. It must be emphasized that the testing scheme suggested in experiments 1 and 2 represents the minimum number of samples that can be used to establish 90% sensitivity with 95% confidence. Investigators are, however, encouraged to use much larger sample sizes (e.g., a single-phase testing with 120 samples or the two-phase testing procedure described above).

Special Cases

Test kits with lower than established tolerance or MRL limits. One way to protect the consumer from unwanted drug residues is by defining toxicologically acceptable or allowable levels of any drug(s) in foods intended for human consumption. By using the concept of tolerance levels (i.e., the level of antibiotic in a defined food matrix judged to be safe for human consumption after extensive testing for toxicity, teratogenicity, and carcinogenicity), or maximum residue limits (MRL, i.e., the maximum concentration of antibiotic resulting from the use of a veterinary drug that has been recommended by the Codex Alimentarius Commission, or that has been established by a national jurisdiction, to be legally permitted as acceptable in food), regulatory agencies can enforce legislation regulating the use of veterinary drugs in food production. As explained in detail in Chapter 1, test kits for screening antibiotic residues incorporated into food inspection programs in the United States and Canada usually provide first-hand indications of whether the food sample is drug-free and fit to be released into the food chain or whether it is suspected to be contaminated by microbial inhibitors, which have elicited a positive test response in the screening test, and should therefore be detained pending laboratory analyses of the implicated sample.

If a test kit is designed so that its detection limit is slightly lower numerically than the tolerance level or MRL established for the drug of interest in that specified matrix, then verifying the detection limit claim with experiment 2 is enough to demonstrate that the test kit will perform reliably at the MRL level. If the detection limit of the test kit, however, is numerically lower than the MRL by more than 20%, it is necessary to verify the detection limit claim but, even more importantly, a precautionary note must be included in the package insert to alert end-users that this test kit has the potential to detect as positive nonviolative samples. If a test kit has a detection limit that is numerically greater than established tolerance limits,

MRL's, or action level for a specified antibiotic by more than 20%, it will not be considered suitable for screening the specific drug at the regulatory level and its detection limit claim would not warrant any verification. It is unlikely that such a rapid test will be adopted by any regulatory body for screening this specific drug.

Test kits with no defined tolerances or MRLs. There are some antibiotics approved for use in food-animal production where no tolerance limits or MRLs have been established in specified food matrices. The reasons are many and varied, and fall outside of the scope of this discussion. However, in Canada when an MRL or tolerance level for the specific antibiotic has not been established, the drug is usually considered to have a zero tolerance level in that matrix and the detection of such an antibiotic in the matrix at any concentration with the most sensitive available method would constitute a violation. It is, therefore, to be expected that under these circumstances the best rule to follow when selecting a test kit to screen such an antibiotic is the more sensitive the test kit, the better. In selecting the most sensitive test to be used for screening antibiotics that fall into this category, however, caution must be exercised in selecting only those rapid tests which have detection limits that can be matched by available compound-specific, quantitative analytical methods.

So far, we have focused our discussions on evaluating commercially available rapid tests which have their operational characteristics already defined by the manufacturer. We have also described in detail how to conduct relevant experiments to verify performance characteristic claims in any evaluation exercise. Very often, however, scientists will design and develop rapid tests whose performance characteristics need to be determined. Assuming that the test has met the minimum selectivity rate requirement described earlier, this next section describes in detail experiments that need to be conducted in order to define the dose-response curve when estimating the detection limit of the rapid test for the antibiotic(s) of interest.

Dose-Response Curve

Assume, for example, that an adapted or modified test kit can detect ampicillin in bovine muscle tissue. Assume also, that a tolerance or MRL of 50 ppb has been established for ampicillin in this matrix.

Experiment 3—Constructing an Approximate Dose-Response Curve

Conduct 30 replicate tests at each concentration using drug-free muscle tissue fortified with the antibiotic under study at concentrations ranging from 10 times below to 10 times above the tolerance level. For our ampicillin example with a defined tolerance of 50 ppb, the following concentrations may be considered: 5, 10, 15, 25, 50, 100, 200, and 500 ppb. Calculate the percent positives found at each concentration. Assume that after the test, it is found that all test results at a concentration of 5 ppb were negative, but all test results at 15 ppb and above were positive. The results of this experiment suggest that the test kit has an approximate detection limit of 15 ppb, which satisfies the minimum detection sensitivity rate criterion.

Experiment 4—Constructing an Accurate Dose-Response Curve

To construct an accurate dose-response curve using the preliminary data obtained from experiment 3, conduct 30 replicate tests on antibiotic-free tissue samples fortified with the specified antibiotic at each of four evenly-spaced increments of concentration between the concentration that elicited all negative responses and the concentration that elicited all positive responses in experiment 3. Also conduct 30 replicate analyses of antibiotic-free tissue sample fortified with the antibiotic at 20% above the test concentration that elicited all positive responses. For the ampicillin example, these concentrations would be 6, 8, 10, 12, 15, and 18 ppb. For a qualitative test, use PROBIT analysis (*94*) to compute the percent positives versus the concentration curve and its 95% confidence limits. The lowest concentration in the dose-response curve that satisfies the minimum selectivity rate criterion is the detection sensitivity for the rapid test for the drug under investigation. For a quantitative test, calculate the mean test response at each concentration and plot the mean response versus the concentration, and use the usual statistical definitions for the detection limit to determine this parameter from the dose-response data. The samples being tested in experiments 3 and 4 should be coded by one analyst, randomized, and blind-analyzed by another analyst. In addition, if the test response requires a subjective interpretation, it is advisable that the test result be independently read and recorded by two analysts, and that the tests should be conducted over a period of more than 1 day.

Experiment 5—Test Kit Ruggedness

Some tests are designed to perform satisfactorily only under very stringent and carefully controlled experimental conditions. Such tests would not be considered robust enough to be recommended for field use because experimental conditions in the field can be uncontrolled and vary from day to day. These tests, however, may be suitable in the laboratory where experimental conditions can be adequately controlled by trained analysts. Other tests may have operational characteristics such that small changes or deviations from specified operating conditions will not adversely affect test responses. Such tests are more likely to be recommended for use in the field for screening antibiotic drug residues. For example, if a rapid test developed in a country in the temperate zone is designed to be conducted at room temperature, and the room temperature is not specified, this test kit might not work somewhere in the tropics where room temperatures could vary from 25–40°C. Each test is, therefore, to be ruggedness tested in order to reveal any critical control steps in its experimental procedure or operational characteristics. Youden's procedure for conducting a ruggedness evaluation has been adequately described by Wernimont in AOAC's statistical manual (*95*) and should be consulted when performing this test.

Determining Interference on Rapid Test Kit Responses

Interference of Test Responses by Other Drug Residues

Table 10 shows the most commonly used antibiotics in agriculture for food production, which all belong to one of the following drug families: β-lactams, tetracyclines, aminoglycosides, macrolides, and glycopeptides. There are some antibiotics such as chloramphenicol and bacitracin that do not belong to any of the common groups and we have, for convenience, grouped them together under miscellaneous drugs. There are also other veterinary drugs such as the sulfa drugs that are not antibiotics, but they are considered to be antibacterial and are used extensively in food-animal production. These have been placed together under sulfonamides. The groups of veterinary drugs listed in Table 10 will constitute drug mixtures I, II, III,...VII containing each individual drug at a final concentration of 100 ppb.

If a rapid test is designed to detect a specific antibiotic such as chloramphenicol or a group of antibiotics belonging to the same family such as the penicillins, cephalosporins, or tetracyclines (this usually happens with RIA or ELISA test kits), it is important to determine whether other antibiotics either belonging to the same drug family or to other drug families will cross react. Three basic experiments need to be conducted to determine the extent of cross reactivity. These are direct interference (producing positive results on an otherwise residue-free sample) by other potentially cross-reactive drugs on antibiotic-free sample, masking effect (masking the otherwise positive test result expected to be obtained with the target drug in the presence of the other drugs), and the effect of other drugs (approved or illegal) within the same family on the test response in the presence of the target drug.

Experiment 6A—Direct Interference by Other Potentially Cross-Reactive Drugs

In Table 10, representative veterinary drugs from each of the families of antimicrobials commonly used in food animal production have been grouped together into a set of drug mixtures. If a test kit is claimed to be able to detect only β-lactam antibiotics, for example, this implies that all the other drugs belonging to the other families are potentially capable of cross-reacting, and the extent to

Table 10 Representative Veterinary Drugs Commonly Used in Food Animal Production[a]

Family group	Drug mixture	Commonly used veterinary drugs
β-Lactams	I	Amoxicillin, ampicillin, cloxacillin, penicillin-g, cephapirin, ceftiofur
Tetracyclines	II	Tetracycline, oxytetracycline, chlortetracycline, doxycycline
Aminoglycosides	III	Gentamicin, streptomycin, neomycin, dihydrostreptomycin
Macrolides	IV	Erythromycin, tylosin, tilmicosin
Glycopeptides	V	Lincomycin, novobiocin
Sulfonamides	VI	Sulfadiazine, sulfamethazine, sulfathiazole, sulfadimethoxine, sulfapyridine, sulfanilamide
Miscellaneous	VII	Oxytocin, phenylbutazone, dexamethasone, dipyrone, chloramphenicol, bacitracin, dapsone

[a]This list is by no means exhaustive, and other drugs may be added or deleted depending on the particular evaluation that is being conducted.

which they do this must be investigated and defined so that precautionary methods are used in the interpretation of the test results.

Experiment 6B—Direct Interference

Fortify a zero residue sample with each set of mixtures containing the listed drugs from the families other than the one the drug of interest belongs to, and assay each fortified sample in triplicate with the test kit. If any samples elicit a positive response in the test kit, the drug(s) in that drug mixture responsible for the positive effect shoud be identified and retested at lower concentrations of the mixture to determine the extent of direct interference.

Experiment 6C—Masking Effect

This experiment is to determine whether the presence of other potentially cross-reacting drugs in Table 10 will mask the expected positive test response when a test sample containing the target drug at its detection sensitivity, or some toxicologically defined concentration, is analyzed. To do this, fortify drug-free samples with the target drug at its detection limit if the sample is numerically lower than a toxicologically defined concentration such as the tolerance level or MRL.

Split the fortified sample into seven sub-samples, and fortify each sample with one of the mixtures of potentially cross-reactive drugs so that the final concentration for each drug in the fortified mixture is 100 ppb. Assay in triplicate with the test kit. All the test responses should be positive. If any negative test responses result (a positive masking effect), the drug(s) responsible for the masking effect must be identified. The test is repeated at lower concentrations of the implicating drug mixture to establish the critical concentration where no interference occurs.

Experiment 6D—Cross-Reactivity by Drugs Within the Same Family

This test is to determine the extent of cross-reactivity for other drugs within the same drug family including their active metabolites. This test is also intended to determine the extent of cross reactivity even by drugs in that family that have not been approved for food-animal production but that are suspected of being used. To do this, fortify zero residue sample with each potentially cross-reactive drug in the family at 100 ppb, and assay in triplicate. A negative test response is expected for a zero residue sample and so, if a positive test is found with any of the drug(s), the extent of cross-reactivity should be determined by retesting at lower interfering drug concentrations until a critical concentration is found where no interference occurs.

Experiments 6A–6D attempt to determine which other veterinary drugs, and to what extent, these chemicals may have on the test responses obtained with the test kit. Food samples obtained from healthy animals may contain naturally-occurring substances that are known to interfere with some of the test responses of several of the test kits for screening antibiotic drug residues available on the market today. The range of such naturally-occurring compounds likely to interfere with some test kit responses increases significantly when samples are obtained from sick animals. For example, it is well known that naturally occurring free fatty acids, lysozymes, and lactoferrins in raw milk behave as antibacterials, and they have been shown to elicit positive test (false-positive) responses on "apparently antibiotic-free milk samples". It is the presence of such naturally occurring substances in the matrices of some of the foods to be tested and their possible interference on test kit responses that makes it imperative to determine the selectivity rate criterion for all rapid tests.

Interference from Bacteria and Somatic Cells

Because it is not possible to cover the entire spectrum of nonchemical interferents, we shall, for simplicity sake, consider the special case of raw milk in the dairy industry. We will examine how rapid test kit responses could be influenced by the presence of large amounts of bacteria and somatic cells in the product. Bacteria and somatic cell counts, are two indicators frequently used to measure the quality of milk. One must have a good knowledge of microbiological agents (pathogens, etc.) that might be responsible for lowering food quality, causing spoilage, or be of a public health concern in each particular tested food matrix in order to be able to select the proper agents whose interference must be investigated. This is expected to vary from one food matrix to another. If there are regulatory limits established for these agents in those food matrices, then it is incumbent on the manufacturer or test kit evaluator to determine whether an excess of these agents will interfere with the normal responses of the rapid test.

Dairy cow productivity improves with a lowering of the somatic cell count (SCC) because of the negative correlation between the SCC and milk, fat, casein, and lactose production (*96, 97*). A low level of bulk milk SCC is also associated with a low prevalence of infection with major mastitic pathogens. Thus, the SCC is very frequently used as an indicator of the health status of cows udders (*98*). Additionally, a low SCC increases milk quality and dairy product yields (*99*). In an effort to improve product quality and farm productivity, regulatory limits on milk quality, as measured by bulk milk SCC, have become quite stringent world-wide, and efforts are still being made towards reducing these regulatory levels even further. In the EU, for example, the regulatory limit was set in January 1992 at 400,000 cells/mL. In the United States, the federal regulatory limit of 1,000,000 cells/mL was reduced to 750,000 cells/mL in July 1993. In Ontario, Canada, a six-year step-wise program was implemented in 1989 to lower the regulatory limits from 800,000 to 500,000 cells/mL.

When test kits are used to screen dairy products for antibiotic residues, it is imperative that the effect, if any, of bacteria and the SCC at counts in excess of regulatory limits on test kit responses be determined. This will be especially true for sick animals that have undergone antibiotic therapy and are being reintroduced into food production. Even though SCC regulatory limits are currently based on regional practices and different countries are using different SCC values as a measure of milk quality, the AOAC RI used the previous U.S. regulatory limit of 1,000,000 cells/mL in their evaluatory experiment to determine the effect of SCC on test kit responses.

Experiment 7—Interference from Somatic Cells

In the United States, for example, the SCC of up to 10^6 cells/mL were allowed in milk according to their Pasteurized Milk Ordinance (PMO) until 1993. To demonstrate the effect of a high SCC on the test response, fortify antibiotic-free milk with somatic cells to a concentration of 10^6 cells/mL, and perform the selectivity rate test (Experiment 1). If the rapid test passes this test (i.e., all 30 drug-free samples elicit ALL negative test responses), the sensitivity rate test (Experiment 2) must also be conducted using an antibiotic-free sample fortified with the target drug at its detection sensitivity, and fortified with somatic cells to a concentration of 10^6 cells/mL. All 30 samples should also elicit the expected positive test responses in the test kit.

Experiment 8—Interference from Bacteria

The response of the test kit in the presence of Gram-negative, Gram-positive, and mixed cultures of Gram-positive and Gram-negative bacteria at a concentration of 250,000 cfu/mL must be tested. This can be done by conducting the selectivity and sensitivity rate experiments with a bacterial load of Gram-positive, Gram-negative, or mixed Gram-positive/Gram-negative organisms at a concentration of 250,000 cfu/mL. Once again, all the selectivity samples should elicit all negative test responses, while the sensitivity test samples should all elicit positive test responses in the presence of the bacterial load.

Experiment 9—Shelf-Life and Batch-To-Batch Stability of Test Kit Reagents and Materials

Most commercially available rapid tests will have some indication as to the useful shelf-life of the test kit when stored at room temperature or refrigerated or frozen. If test kits are to be used over an extended period of time, it is imperative that this claim to stability of test kit materials and reagents be verified. For example, if a test kit package, once opened, claims to have a useful shelf-life of 14 days when stored at 2-4°C, it is advisable to conduct the selectivity and sensitivity rate experiments using both incurred and fortified test samples on the day the test kit package was opened, on the 7th, and on the 14th day thereafter to determine the inter-assay variability of the test kit over the recommended period. Because the shelf-life stability hinges on the storage temperature, it is important to ensure during the course of this exercise that the temperature of the storage facility is held within the specified range. The test for all the samples in the selectivity experiment should be negative, while all the samples tested in the sensitivity experiment should be positive.

No analyst will tell you that it is a comforting revelation when they find out that a commercially available product (for example, a test kit) they had used successfully for several years suddenly stopped performing as before because the company changed the original formulation or composition of the product without informing users. It is the primary responsibility of the manufacturing company to have quality assurance and quality control methods in place to ensure that their products manufactured within the same batch, as well as those from batch-to-batch production, meet the necessary statistical constraints imposed on the resulting product. However, it is incumbent on the analyst who intends to use the product for long-term, large-volume testing to request enough test kits produced in the same batch and in different batches that will be used in performing the selectivity and sensitivity rate experiments. All the test kits used for the selectivity test should show negative responses on residue-free samples, and all positive responses on intentionally fortified samples and samples containing physiologically induced concentration(s) of the drug(s) of interest.

Conclusion

An evaluation exercise requires the involvement of several analysts and requires the verification of several parameters that extend beyond the two primary ones, selectivity and sensitivity rates, that have usually been verified in previous evaluation exercises. Because end-users cannot, and should not, be expected to continuously monitor the batch-to-batch variability in test kit materials and reagents, it was made clear to the manufacturers of the β-lactam test kits, which have been granted performance tested status, to notify the end-users of any changes in their protocol, materials, or reagents that may affect test kit performance or face the loss of the AOAC RI certification and the FDA recommendation. Additionally, if they fail to take corrective action relative to poor performance of the test kit as reported by test kit users, the certification status may be revoked. The role played by the FDA and the AOAC RI in the certification program of the sort described above makes it obligatory for manufacturers to continue to maintain their product quality and to continue to demonstrate that their product, which has been certified to support a specific program, will continue to perform as specified. Hopefully, this will limit the use of such evaluation exercises to work out the "bugs" or limitations in rapid tests as a means to optimize their performance before they become available for commercial use.

Acknowledgments

We wish to acknowledge the extensive time and effort contributed towards the formulation of the "Specific data submission requirements package for test kits for β-lactam residues in milk:

AOAC Research Institute's Test Kit Performance Testing Program by Mr. Scott Coates, Managing Director, AOAC Research Institute

Dr. Joe Settepani, Director of Scientific Research, Food and Drug Administration, BARC-EAST, Beltsville, Maryland

Professor Walther Heeschen, Institute for Hygiene, Germany

Professor Lennart Björck, Swedish University of Agricultural Sciences, Sweden

Professor Michael Murphy, University of Minnesota

Susan Bright, USDA-APHIS, Hyattsville, Maryland

Mr. Daniel Denman, Statistics Laboratory, University of Maryland, Maryland.

We would also like to acknowledge the work done by the staff at Arthur D. Little, Inc., and the staff at the Dairy Quality Control Institute Minnesota, for conducting the laboratory verification of the test parameters of the submitted test kits during the evaluation exercise.

We hereby indicate that the list of test kits named in this chapter is by no means exhaustive and that any reference or omission to cite any commercially available rapid test does not in any way constitute a judgement, endorsement, or recognition by the authors or Agriculture and Agri-Food Canada.

References

1. L. P. Garrod, H. P. Lambert, and F. O'Grady, *Antibiotic and Chemotherapy,* (Edinburgh: Churchill Livingstone, 1973).
2. K. E. Cooper *Analytical Microbiology,* ed. F. Kavanagh (New York: Academic Press, 1964), 1.
3. K. E. Cooper and A. H. Linton, *J. Gen. Microbiol.*, **7**: 8, 1952.
4. K. E. Cooper, A. H. Linton, and S. N. Sehgal, *J. Gen. Microbiol.*, **18**: 670, 1958.
5. Environmental Diagnostics, Inc., *Performing the Live Animal Swab Test* (Burlington, NC: Environmental Diagnostics, Inc., 1979).
6. United States Department of Agriculture, *Performing the Calf Antibiotic and Sulfa Test* (Washington, D.C.: United States Department of Agriculture, Food Safety and Inspection Service, Meat and Poultry Program, 1984).
7. R. W. Johnston, R. H. Reamer, E. W. Harris, H. G. Fugate, and B. Schwab, *J. Food Prot.*, **44**: 828, 1981.
8. Association of Official Analytical Chemists, *Official Methods of Analysis*, 15th ed. (Arlington, VA: AOAC, 1990), secs **982.16A–982.16D**.
9. G. H. Richardson, *Standard Methods of Analysis*, 15th ed. (Washington, D.C.: American Public Health Association, 1985).
10. M. C. Goel, D. C. Kulshrestha, E. H. Marth, D. W. Francis, J. G. Bradshaw, and R. B. Read, *J. Milk Food Technol.*, **34**: 54, 1971.
11. International Dairy Federation, *Bulletin of the International Dairy Federation*, no. 258, 2nd ed. (Brussels, Belgium: International Dairy Federation, 1991), 26.
12. International Dairy Federation, *Bulletin of the International Dairy Federation* no. 258, 2nd ed. (Brussels, Belgium: International Dairy Federation, 1991), 48.
13. S. A. Bright, S. L. Nickerson, and N. H. Thaker, "Fast Antibiotic Sulfonamide Test," (poster presentation at the AOAC Annual Meeting, St. Louis, MO, 1989).
14. International Dairy Federation, *Bulletin of the International Dairy Federation* no. 258, 2nd ed. (Brussels, Belgium: International Dairy Federation, 1991), 59.
15. F. Untermann, *Arch. Tierheik.*, **128**: 307, 1986.
16. R. Bogaerts and F. Wolf, *Fleischwirtsch.*, **60**: 667, 1980.

17. J. F. M. Nouws, N. J. G. Broex, J. M. P. Hartog, and F. Driessens, *Arch. Lebensmittelhyg.*, **39**: 133, 1988.
18. J. F. M. Nouws, "Tissue Distribution and Residues of Some Antimicrobial Drugs in Normal and Emergency-Slaughtered Ruminants," master's thesis, University of Utrecht, 1978.
19. T. R. B. Barr, *Antibiotics, Sulfonamides, and Public Health*, vol. 1, ed. J. H. Steele and G. W. Beran (Boca Raton, FL: CRC Press, 1979), 235.
20. J. F. Timoney, *Antibiotics, Sulfonamides, and Public Health*, vol. 1, ed. J. H. Steele and G. W. Beran (Boca Raton, FL: CRC Press, 1979), 267.
21. J. M. Bowen, *Antibiotics, Sulfonamides, and Public Health*, vol. 1, ed. J. H. Steele and G. W. Beran (Boca Raton, FL: CRC Press, 1979), 289.
22. M. A. Sande and G. L. Mandell, *The Pharmacological Basis of Therapeutics*, ed. A. G. Gilman, L. S. Goodman, and A. Gilman, (New York: Macmillan, 1980), 1162.
23. M. Barza and R. T. Sheife, *J. Maine Med. Assoc.*, **68**: 194, 1977.
24. G. Ziv and F. G. Sulman, *Antimicrob. Agents Chemother.*, **2**: 206, 1972.
25. J. F. M. Nouws, *Tijschr. Diergeneesk.*, **100**: 662, 1975.
26. W. Wiesser, *Tieräerztl. Umsch.*, **6**: 276, 1976.
27. E. Forschner, *Arch. Lebensmittelhyg.*, **24**: 159, 1973.
28. E. Forschner and W. Glende, *Fleischwirtsch.*, **56**: 226, 1976.
29. H. H. Heinert, G. Van Der Wall, and H. Brehmer, *Arch. Lebensmittelhyg.*, **27**: 55, 1976.
30. F. V. Kosikowski, *J. Dairy Sci.*, **46**: 95, 1963.
31. A. Carlsson and L. Bjorck, *Milchwissenschaft*, **42**: 282, 1987.
32. A. G. Wolin and F. V. Kosikowski, *J. Dairy Sci.*, **41**: 34, 1958.
33. W. Watson and W. L. Bloom, *Proc. Soc. Exp. Biol. Med.*, **81**: 29, 1952.
34. E. Katachalski, L. Bichovski, and B. E. Volcani, *Nature (London)*, **196**: 1095, 1952.
35. W. C. Burger and M. A. Stahlmann, *Arch. Biochem. Biophys.*, **39**: 27, 1952.
36. W. E. Van Heyningen, *J. Gen. Microbiol.*, **5**: 758, 1951.
37. H. G. Helmbold, *"Studies on the Antibacterial Effect of Hematin, Acid Metheglobulin, and Hemolyzed Thickened Blood, Tested on Seven Organisms of Significance in Food Hygiene,"* master's thesis, Hannover, Tierärztl., Hochschule, 1978.
38. F. S. Jones and R. B. Little, *J. Exp. Med.*, **45**: 319, 1927.
39. E. Schaale, *Fleischwirtsch.*, **7**: 561, 1961.
40. B. Meyer, W. Thiel, and W. Weis, *Fleischwirtsch.*, **49**: 203, 1969.
41. F. S. Jones and H. S. Simms, *J. Exp. Med.*, **51**: 327, 1930.
42. R. Levetzow and E. Wiese, *Schlacht-Viehhof-Ztg.*, **74**: 329, 1974.
43. G. Shellhaas, *Schlacht-Viehhof-Ztg.*, **75**: 223, 1975.
44. G. Poli, A. Balsari, W. Ponti, A. Ceccarelli, and E. Fesce, *Arch. Lebensmittelhyg.*, **30**: 120, 1979.

45. A. Carlsson and L. Bjorck, *J. Dairy Sci.*, **72**: 3166, 1989.
46. J. Egan and W. J. Meaney, *Ir. J. Food Sci. Technol.*, **8**: 115, 1984.
47. Y. Okada, *J. Jpn. Vet. Med. Assoc.*, **39**: 97, 1986.
48. J. S. Cullor, A. Van Eenennaam, J. Dellinger, L. Perani, W. Smith, and L. Jensen, *Vet. Med.*, May: 477, 1992.
49. T. Ogawa, M. Shibata, C. Yatome, and E. Idaka, *Bull. Environ. Contam. Toxicol.*, **40**: 545, 1988.
50. S. A. Berson and R. S. Yallow, *Radioimmunoassay*, ed. R. A. Good and D. W. Fisher (Stanford, CA: Sinauer Publishers, 1971).
51. R. Luft and R. S. Yallow, *Radioimmunoassay Methodology and Applications in Physiology and in Clinical Studies* Stuttgard: George Thieme Verlag, 1974).
52. D. S. Kabakoff, *Enzyme-Immunoassays*, ed. E. T. Maggio (Boca Raton, FL: CRC Press, 1980), 71.
53. C. Blake and B. J. Gould, *Analyst*, **109**: 533, 1984.
54. S. E. Charm and C. K. Ruey, *J. Assoc. Off. Anal. Chem.*, **65**: 1186, 1982.
55. S. E. Charm and C. K. Ruey, *J. Assoc. Off. Anal. Chem.*, **71**: 304, 1988.
56. S. E. Charm, *Analysis of Antibiotic Residues in Food Products of Animal Origin,* ed. V. K. Agarwal (New York: Plenum Press, 1992), 31–56.
57. D. Arnold, D. vom Berg, A. K. Doertz, U. Mallick, and A. Somogyi, *Arch. Lebensmittelhyg.*, **35**: 131, 1984.
58. D. Arnold and A. Somogyi, *J. Assoc. Off. Anal. Chem.*, **68**: 984, 1985.
59. K. W. Freebairn and N. T. Crosby, *Immunoassays for Veterinary and Food Analysis*, ed. B. A. Morris, M. N. Clifford and R. Jackman (New York, NY: Elsevier Applied Science, 1988).
60. G. Balizs and D. Arnold, *Chromatographia*, **27**: 489, 1989.
61. B. K. Van Weeman and A. H. W. M. Schuurs, *FEBS Lett.*, **15**: 232, 1971.
62. E. Engvall and P. Perlmann, *Immunochemistry*, **8**: 871, 1971.
63. E. Engvall, K. Johsson, and P. Perlmann, *Biochim. Biophys. Acta*, **251**: 427, 1971.
64. L. Belanger, D. Hamel, D. Dufour, and M. Pouliot, *Clin. Chem.*, **22**: 198, 1976.
65. H. Fricke, C. J. Strasburger, and W. G. Wood, *J. Clin. Chem. Clin. Biochem.*, **20**: 91, 1982.
66. K. E. Rubenstein, R. S. Schneider, and E. F. Ullman, *Biochem. Biophys. Res. Commun.*, **47**: 846, 1972.
67. R. S. Scheider, P. Lidquist, E. T. Wong, K. E. Rubenstein, and E. F. Ullman, *Clin. Chem.*, **19**: 821, 1973.
68. G. L. Rowley, K. E. Rubenstein, J. Huisjen, and E. F. Ullman, J. Biol. Chem., **250**: 3759, 1975.
69. E. F. Ullman, R. A. Yoshida, J. I. Blakemore, E. Maggio, and R. Leute, *Biochim. Biophys. Acta.*, **567**: 66, 1979.

70. C. E. Pippenger, R. J. Bastiani, and R. S. Schneider, *Clinical Pharmacology of Antiepileptic Drugs,* ed. H. Schneider, D. Janz, C. Gardener-Thorpe, H. Meinard, and A. L. Sherwin (New York: Springer-Verlag, 1975), 331.
71. J. F. Burd, R. J. Carrico, M. C. Fetter, R. T. Buckler, R. D. Johnson, R. C. Boguslaki, and J. E. Christner, *Anal. Biochem.*, **77**: 56, 1977.
72. J. F. Burd, R. C. Wong, J. E. Feeney, R. J. Carrico, and R. C. Boguslaski, *Clin. Chem.*, **23**: 1402, 1977.
73. R. C. Wong, J. F. Burd, R. J. Carrico, R. T. Buckler, J. Thoma, and R. C. Boguslaski, *Clin. Chem.*, **25**: 686, 1979.
74. T. T. Ngo, R. J. Carrico, R. C. Boguslaski, and J. F. Burd, *J. Immunol. Methods*, **42**: 93, 1981.
75. J. De Louvois, *J. Med. Microbiol.*, **7**: 11, 1974.
76. A. F. DeCastro, J. D. Place, C. T. Lam, and C. Patel, *Antimicrob. Agents and Chemother.*, **29**: 961, 1986.
77. J. D. Place and S. G. Thompson, *Antimicrob. Agents and Chemother.*, **24**: 240, 1983.
78. C. Van De Water and N. Haagsma, *J. Assoc. Off. Anal. Chem.*, **73**: 534, 1990.
79. R. A. Watson, J. Landon, E. J. Show, and D. S. Smith, *Clin. Chim. Acta*, **73**: 51, 1976.
80 S. Bakerman, *Lab. Manage.*, July: 16–18, 1983.
81. W. B. Dandliker, R. J. Kelly, and J. Dandliker, *Immunochemistry*, **10**: 219, 1973.
82. L. J. Blecka, *Ther. Drug Monit.*, March: 1–6, 1983.
83. W. B. Dandliker, H. C. Schapiro, J. W. Meduski, R. Alonso, G. A. Feigen, and J. R. Hamrick, Jr., *Immunochemistry*, **1**: 165, 1964.
84. M. A. Pesce and S. H. Bodourian, *Clin. Chem.*, **29**: 1199, 1983.
85. S. A. Brown, D. R. Newkirk, R. P. Hunter, G. G. Smith, and K. Sugimoto, *J. Assoc. Off. Anal. Chem.*, **73**: 479, 1990.
86. S. R. Popelko, D. M. Miller, J. T. Holen, and D. M. Kelso, *Clin. Chem.*, **27**: 1198, 1981.
87. Association of Official Analytical Chemists, "Guidelines for Collaborative Study," *J. Assoc. Off. Anal. Chem.*, **74**: 694, 1989.
88. J. R. Bishop, S. E. Duncan, G. M. Jones, and W. D. Whittier, "Evaluation of Animal Drug Residue Detection Methods" a research report presented at the NCIMS Meeting, Louisville, KY, 1991.
89. Association of Official Analytical Chemists, *Official Methods of Analysis*, 15th ed., (Arlington, VA: AOAC, 1984), secs **982.17**.
90. Association of Official Analytical Chemists, "AOAC announces New Test Kit Rapid Verification Program," *J. Assoc. Off. Anal. Chem.*, **74**: 143A, 1991.

91 AOAC Research Institute Test Kit Performance Testing Program: Specific Data Submission Requirements for Test Kits to Test β-Lactam Residues in Milk (1993) (unpublished procedure).

92. β-Lactam Test Kits. *FDA Document M-I-93-3.*

93. J. O. Boison and J. D. MacNeil, *Performing the Brilliant Black Reduction Test For Antibiotic Screening In Eggs* (Ottawa, Canada: Agriculture and Agri-Food Canada Technical Document, 1990).

94. D. J. Finney, *Statistical Method In Biological Assay*, 3rd ed., (New York: MacMillan Publishing Co., 1978).

95. G. T. Wernimont, *Use Of Statistics To Develop and Evaluate Analytical Methods: AOAC Statistical Manual*, ed. W. Spendley, (Arlington, VA: AOAC, 1985).

96. P. C. Bartlett, G. Y. Miller, C. R. Anderson, and J. H. Kirk, *Dairy*, **73**: 2794, 1990.

97. H. DeLuyker, *Flemish Vet. J.*, **62**: 207, 1991.

98. U. Emanuelson and H. Funke, *Dairy*, **74**: 2479, 1991.

99. G. E. Mitchell, S. A. Rodgers, D. B. Houlihan, and V. C. Tucker, *Aust. J. Dairy Technol.*, **41**: 9, 1986.

Chapter 5

Chemical Analysis of Polyether Antibiotics

Hirohiko Asukabe and Ken-Ichi Harada
Faculty of Pharmacy, Meijo University, Tempaku-ku,
Nagoya 468 (Japan)

Contents

Introduction

Ionophore antibiotics form a coordinate bond with metal ions and can transport cations across membranes against a concentration gradient. Polyether antibiotics (PEs) (see Figure 1), are chemically characterized by a single carboxylic acid group, a number of cyclic ethers, several hydroxyls, and a large number of C-alkyl groups. Biologically, they are characterized by their in vitro activity against Gram-positive bacteria and mycobacteria. These compounds are effective in treating coccidial infections in poultry, and they act as a growth promoter for cattle to increase feed conversion efficiency (*1*, *2*). Salinomycin (SL), monensin (MN), and lasalocid (LA) have been administrated to poultry at concentrations of 80, 50, and 75 parts per million (ppm), respectively, since 1978 in Japan. Narasin (NA) and maduramicin (MA) have been also used in the United States, the European Union (EU), and other countries.

The structures of PEs have been elucidated by X-ray analysis of their heavy metal salts. In the crystalline states, PE molecules exist in a cyclic conformation with the carboxyl group on one end held by a hydrogen bond to a tertiary hydroxyl group on the terminal cyclic ether. The alkyl groups extend outside the molecule, imparting their hydrophobic property with the oxygen functions are concentrated in the center. The cation is held by ligands within the formed pocket, rendering the salt virtually insoluble in water but very soluble in almost all organic solvents. The free acids and salts have similar solubility characteristics (*3*).

In order to establish chemical methods of analysis for PEs, we must take the following chemical properties of PEs into consideration:

(*a*) most PEs have no chromophore and are produced as sodium salts,

(*b*) PEs are unstable under acidic conditions because cyclic ketals are labile and hemiketal functions form ketals very rapidly, and

(*c*) the polarity of PEs is comparatively low because the alkyl groups in these compounds extend outside.

Properties (*a*) and (*b*) are closely related to detection systems; that is, the carboxylate of PEs can be utilized as a target functional group to introduce a chromophore and increase the detection sensitivity under nonacidic reaction conditions. In addition, property (*c*) is also related to separation and cleanup systems. Normal-phase chromatography would be more effective than reversed-phase chromatography for the separation and the cleanup of these low polarity compounds. In this chapter, a strategy for the chemical analysis of the PEs (MN, SL, LA, and NA) and a simultaneous method for the determination of these antibiotics are described.

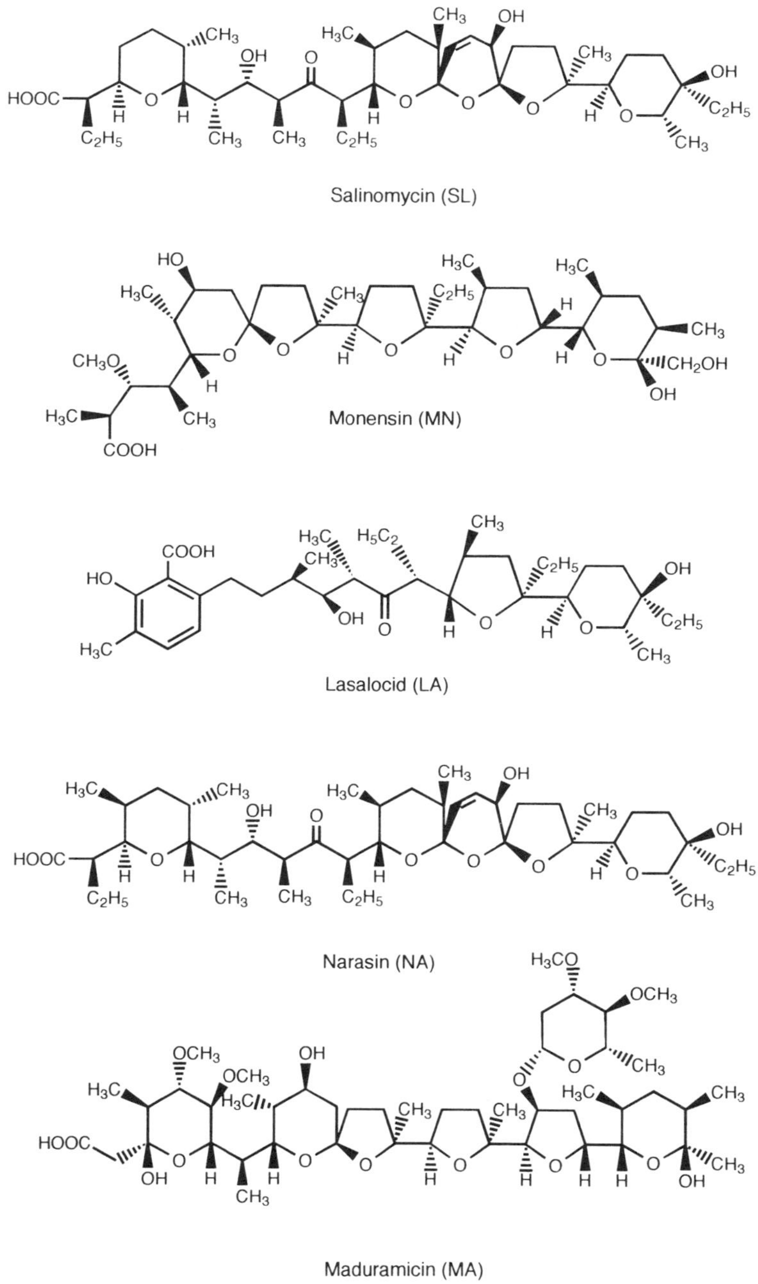

Figure 1 Polyether antibiotics

Overview of Polyether Antibiotics Analysis

Detection

Historically, the progress of analytical methods in this field has been regarded as that of detection methods, because PEs have no appropriate chromophore to be used for detection. Detection is classified into four categories (*a*) the use of an intact molecule, (*b*) the degradation method, (*c*) the derivatization method, and (*d*) other methods. Category (*b*) involves strategies to put color on PEs, and category (*c*) utilizes the carboxylate of PEs to introduce a chromophore. Table 1 summarizes the characteristic features of each method.

Using the intact molecule. Table 2 gives the detection methods using the base molecule. To detect intact PE molecules, refractive index (RI), ultraviolet absorption spectroscopy (UV AS) at wavelengths near 210 nm, and mass spectrometry (MS) can be used, although they lack sensitivity. Several groups performed high-performance liquid chromatographic (LC) determinations of MN (*4*, *5*), SL (*6*, *7*), NA (*8*), and laidlomycin (see Figure 2) (*9*) using RI (*4*, *6–9*), or UV AS detection at 214 nm (*7*) and 215 nm (*5*). These methods were applied to bulk and premix analyses of feeds with a linear range of 40–160 µg (*4*) and 12.5–200 µg (*8*), respectively.

MS is a universal detector for organic compounds, and electron impact (EI), chemical ionization (CI), and fast atom bombardment (FAB) MS techniques have been used for determining PEs (*10–15*). Characteristic features of the MS of PEs are that it is difficult to obtain a molecular ion with EI because of the high molecular weight of PEs, negative CI displays very strong $[M\text{-}H]^-$ ions, and FAB gives metal ion-adduct molecular ion species by adding metal salt to the matrix instead of the protonated molecule. MS has also been combined with chromatographic methods for the analysis of PEs. Stout et al. (*16*) and Horii et al. (*17*) established liquid chromatography-mass spectrometric (LC–MS) methods for MA in chicken fat and LA in chicken liver, respectively, using thermospray LC–MS–

Table 1 Summary of the Detection Method of Polyether Antibiotics

Detection method	Simplicity	Sensitivity	Applicability
Intact molecule	O	X	Δ
Degradation	Δ	Δ	X
Derivatization	X	O	O

O: excellent; Δ: fair; X: poor

Table 2 Detection of Polyether Antibiotics Using an Intact Molecule

PEs	Sample	Method, detection	Stationary phase	Mobile phase	Applicable	Ref.
MN	Premix	LC, RI	Zorbax C8	Methanol–water (9:1)	40–160 μg	*(4)*
SL	Premix	LC, RI	Ultrasphere C8	Water–phosphoric acid–acetonitrile (150:0.1:850)	—	*(6)*
NA	Premix	LC, RI	DuPont C8	Acetonitrile–pH 2.9 phosphate buffer (65:35)	12.5–200 μg	*(8)*
LI	Broth	LC, RI	Separon SGX C18	Acetonitrile–water–phosphoric acid (850:150:0.1, pH 3)	>50 μg/mL	*(9)*
SL	Bulk	LC, RI, UV (214 nm)	Lichrosorb RP-8	0.5% acetic acid (pH 2.8)	—	*(7)*
LY	Broth	LC, UV (259 nm)	Zorbax ODS	Methanol–water (80:20) containing 80 mg/L of cetylpyridinium chloride and 14 mg/L NaH_2PO_4	>500 ng	*(85)*
MN	Bulk	LC, UV (215 nm)	Separon C18	Methanol–water (88:12)	—	*(5)*
LA	Feed	LC, UV (254, 313 nm)	Wakosil 5C18-200	Gradient elution system from ammonium buffer to methanol	>1.5 ppm	*(21)*
LA	Feed, food	LC, UV (320 nm)	Hypersil ODS	Methanol–pH 4.0 phosphate buffer (900:100)	>1 μg/Kg	*(22)*
LA	Premix	LC, UV (254 nm)	μ-Porasil	Heptane–methanol–acetic acid (97.6:2:0.4)	15 μg–3 ng	*(61)*

Table 2 Detection of Polyether Antibiotics Using an Intact Molecule—*continued*

PEs	Sample	Method, detection	Stationary phase	Mobile phase	Applicable	Ref.
MA	Feed	LC, atomic absorption spectrometry (589 nm)	Ultrastyragel	Xylene–tetrahydrofuran (85:15)	>100 ng	*(86)*
MN, SL, NA, LA	Bulk	TLC, vapor phase fluorometry (Ex 366 nm, Em >400 nm)	Silica gel TLC	Ethyl acetate–water (100:3)	>350 ng/spot	*(60)*
			RP-18 TLC	Methanol–5% acetic acid (9:1) twice developed		
Nigericin, abierixin, grisorixin	Bulk	TLC, FID	Silica gel TLC	Chloroform–methanol–formic acid (97:4:0.6)	1–10 μg/spot	*(87)*
MN	Residue	TLC, visualization	Silica gel TLC	Chloroform–methanol (95:5)	>0.003 ppm	*(88)*
MN	Premix, broth	EI-MS	—	—	>100 ng	*(10)*
Ionophore A (MW 939) Ionophore B (MW 997)	Bulk	CI-MS, SFC	Hypersil ODS	Carbon dioxide–methanol (2:98) carbon dioxide–2-propanol (2:98)	—	*(11)*
Ionophore	Bulk	SFC	—	—	—	*(89)*
MA	Chicken fat	thermospray LCMS–MS	Partisil C8	Methanol–0.1*N* ammonium acetate (90:10)	>0.33 ppm	*(16)*

Table 2 Detection of Polyether Antibiotics Using an Intact Molecule—*continued*

PEs	Sample	Method, detection	Stationary phase	Mobile phase	Applicable	Ref
LA	Chicken liver	thermospray LC–MS–MS	μ-Bondasphere C18	Acetonitrile–methanol–0.1M ammonium acetate (2:2:1)	>0.06 ppm	*(17)*
semduramicin	Chicken liver	electrospray LC–MS	Hypersil ODS	Aqueous–organic (10:90), aqueous portion; 5 mM sodium acetate organic portion; acetonitrile–tetrahydrofuran (85:15)	>0.03 ppm	*(18)*
MA	Bulk	FABMS, FABMS–MS	—	—	—	*(12)*
nigericin	Bulk	FABMS	—	—	—	*(13)*
MN	Chicken fat	TLC–FABMS	Silica gel TLC	Carbon tetrachloride–benzene–methyl cellosolve (80:10:10)	>10 ng/g	*(14)*
LA, MN						
Septamycin	Bulk	TLC–FABMS	Silica gel TLC	—	>0.1 μg	*(15)*

MN: monensin; SL: salinomycin; NA: narasin; LI: laidlomycin; LY: lysocellin; LA: lasalocid; MA: maduramicin; FID: flame ionization detection; SFC: supercrical fluid chromatography; FABMS: fast atom bombardment mass spectrometry

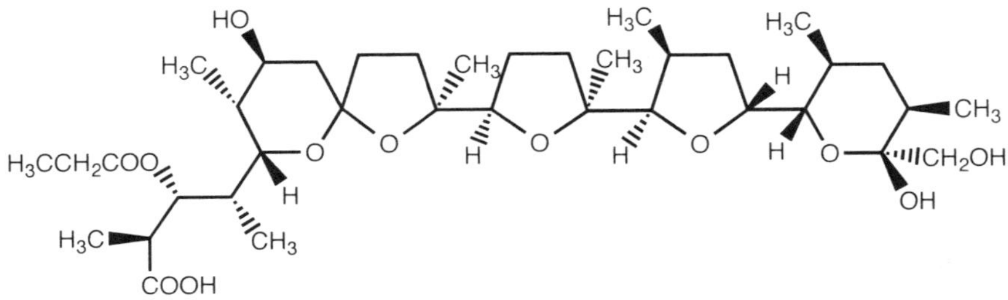

Laidlomycin

Figure 2 Structure of laidlomycin

MS. Schneider et al., meanwhile, reported electrospray LC–MS of semduramicin (see Figure 3) (*18*). However, these methods were only suitable for confirming PEs.

Thin-layer chromatography/fast atom bombardment is a useful technique for the direct measurement of a FAB mass spectrum of a desired spot on a thin-layer chromatographic (TLC) plate. Two methods were reported, one introduced the adsorbent directly into the ion source (*15*), and the other eluted the absorbed material from the adsorbent with a suitable solvent and evaporated it prior to introduction into the MS (*14*). However, in the case of the first method, when a

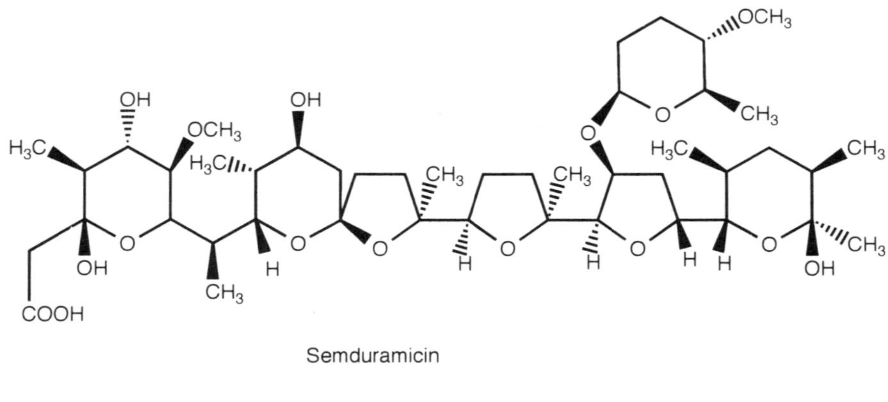

Semduramicin

Figure 3 Structure of semduramicin

matrix is deposited on a sample spot on a TLC plate, diffusion of the sample occurs, usually along with the matrix used, resulting in a lower concentration than on the initial spot. As a result, no satisfactory spectrum is obtained with good sensitivity, unless a large amount of the sample is applied to the TLC plate. In the case of the second method, a stepwise operation, including individual separation by TLC and a measurement of the mass spectrum, is laborious and time-consuming because it needs extra steps such as extracting the desired compound from the TLC plate and eliminating adsorbents.

Degradation method. Table 3 summarizes the detection of PEs by the degradation method colorimetry, Table 4 summarizes densitometry, and Table 5 summarizes postcolumn derivatization. In 1973, Golab et al. (*19*) reported using colorimetry to analyze MN in premix. The chemical analysis had been extensively used for MN and SL, with vanillin or *p*-dimethylaminobenzaldehyde employed as a condensation reagent to form colored compound(s) with these antibiotics. Under anhydrous acidic conditions, MN and SL decompose and their decomposition product(s) react with vanillin or *p*-dimethylaminobenzaldehyde to form products with strong absorption maxima in the visible region. This is known as the Komarowsky reaction and a mechanism whereby colored materials are formed by aldol condensation was proposed for the first time in 1947 by Duke (*20*). The formation of the colored materials consists of two stages, decomposition of the PEs with sulfuric acid and condensation of the degradation product(s) with a benzaldehyde. Because the reaction conditions are so drastic, the decomposition is not always reproducible and the exact reaction mechanism is not clearly understood. The reaction products of MN and SL have not been identified, so the analytical values obtained are not reliable. However, this method has been still used as a basis for methods based on colorimetry, spectrodensitometry, and postcolumn derivatization techniques for analyzing MN and SL, because there is no available functional group for the detection of the intact antibiotics.

The general reaction and determination conditions for this approach are: reagent, 3% vanillin in methanol containing 0.5% sulfuric acid; reaction temperature, 60–70°C; reaction time, 20–25 min; and detection, 518 nm for MN and 600 nm for SL. This colorimetry method can be applied to analyzing feed for MN and SL, but it is difficult to determine LA by this method because it is less sensitive to vanillin and *p*-dimethylaminobenzaldehyde reagents than MN and SL (*3*). However, colorimetry is not a suitable method for feed (*21*) and residue analysis because in addition to the PEs, other compounds containing alcohol and ketone groups in the sample matrix also show a positive response.

The densitometric method (Table 4) is based on the same principle as the colorimetry method. After developing the TLC plate, MN and SL are detected by densitometric determination or visualization after being sprayed with vanillin reagent. The developed TLC plate is heated for about 5 min in a drying oven at 60°C before densitometric determination or visualization. MN yields a pink spot,

Table 3 Colorimetric Detection of Polyether Antibiotics

PEs	Sample	Method	Detection reagent	Reaction condition and detection	Applicable	Ref.
MN	Premix, feed	Colorimetry	Vanillin	60°C, 25 min, 518 nm	>1 μg/mL	*(19)*
Lonomycin	Powder, broth	Colorimetry	Vanillin	60°C, 25 min, 518 nm	>10 μg/g (for powder) >50 μg/g (for broth)	*(90)*
MN, SL	Feed	Colorimetry	*p*-Dimethyl aminobenzaldehyde	70°C, 20 min, 578 nm (for MN)	80 ppm (for MN) 50 ppm (for SL)	*(91)*
				70°C, 20 min, 600 nm (for SL)	30–40 ppm (for MN) 10–30 ppm (for SL)	*(94)* *(95)*
MN	Feed	Colorimetry	50 mg vanillin and 50 mg 5-sulfosalicylic acid	60°C	—	*(92)*
MN	Feed	Colorimetry	Vanillin	518 nm	>20 mg/kg	*(93)*
MN	Premix, feed	Colorimetry	Vanillin	518 nm	>10 mg/kg	*(96)*
MN, SL	Premix, feed	Flow injection	*p*-Dimethyl aminobenzaldehyde	78°C, 578 nm (for MN) 600 nm (for SL)	20–80 mg/kg	*(97)*

MN: monensin; SL: salinomycin

Table 4 Detection of Polyether Antibiotics Using Densitometry

PEs	Sample	Method	TLC solvent system	Detection reagent	Applicable	Ref.
SL	Bulk	Densitometry	ethyl acetate–benzene (9:1) ethyl acetate–chloroform (8:2)	—	—	*(7)*
MN	Premix, feed	Visualization	carbon tetrachloride–methanol (95:5), ethyl acetate, benzene–methanol (95:5), carbon tetrachloride–diethylamine (95:5)	Vanillin	>1 μg	*(19)*
MN	Feed	Visualization	carbon tetrachloride–methanol (97:3)	Vanillin	>20 mg/kg	*(93)*
MN	Feed	Densitometry	carbon tetrachloride–methanol (80:20)	Vanillin	30–130 ppm	*(98)*
MN	Meat	Visualization	1) chloroform–ethanol–benzene (36:1:4) 2) ethyl acetate	Vanillin	>0.005 mg/kg	*(99)*
MN, NA, SL	Premix, feed	Visualization	ethyl acetate–water (100:3)	Vanillin	3–100 mg/kg	*(23)*
MN, nigericin	Premix, feed	Densitometry	chloroform–methanol (9:1)	Vanillin	0.05–10%	*(100)*
MN	Milk	Visualization	1) chloroform–benzene–ethanol (36:4:1) 2) ethyl acetate	Vanillin	>0.008 mg/L	*(101)*
MN	Chicken	Visualization	ethyl acetate–water (97:3)	*p*-Anisaldehyde	>0.1 mg/kg	*(59)*

MN: monensin; NA: narasin; SL: salinomycin

Table 5 Detection of Polyether Antibiotics Using Post-Column Derivatization

PEs	Sample	Column	Mobile phase	Reagent, reaction temp., detection	Applicable	Ref.
MA, NA, SL	Feed, food	Hypersil ODS	Methanol–pH 4.0 phosphate buffer (900:100)	*p*-Dimethylamino-benzaldehyde, 90°C, 600 nm	>1–2 µg/kg	*(22)*
SL	Feed	Zorbax Sil	Ethyl acetate–isooctane–acetic acid–triethylamine (750:250:4:2)	Vanillin, 95°C, 527 nm	>1 ppm	*(24)*
MN, NA, SL	Feed	Partisil ODS	Methanol–water–glacial acetic acid (940:59:1)	Vanillin, 70°C, 520 nm	>0.25 mg/kg (for MN), >0.5 mg/kg (for NA, SL)	*(29)*
MN, NA, SL	Feed	Hewlett-Packard C18	Methanol–water (9:1), water contains 5% acetic acid	Vanillin, 95°C, 520 nm	3–100 ppm	*(30)*
NA	Broth	Little Champ ODS	Methanol–pH 4 buffer (93:7)	Vanillin, 120°C, 520 nm	1–300 ppm	*(25)*
MN	Broth	Little Champ ODS	Methanol–pH 3 buffer (85:15)	Vanillin, 120°C, 520 nm	0.25–2.0 mg/mL	*(26)*
MN	Bulk, premix feed	Partisil 5 ODS-3	Methanol–water–acetic acid (940:60:1)	Vanillin, 80°C, 520 nm	10–200 ppm	*(27)*
SL	Feed	Chromspher C18	Methanol–water (94:60) contained 1.5% glacial acetic acid	Vanillin, 80°C, 520 nm	>1 mg/kg	*(28)*

MA: maduramicin; NA: narasin; SL: salinomycin; MN: monensin

and SL and NA yield purple spots (*23*). However, because this method is insensitive and interference from other materials cannot always be excluded, it is not suitable for feed and residue analysis.

Since Goras et al. (*24*) reported an LC determination of SL using a postcolumn derivatization with vanillin, many workers have used the technique for the determination of SL, MN, and NA (*25–28*) and for their simultaneous LC determination in feed (*22, 29, 30*) (Table 5). Neely achieved LC determination of MN and NA in fermentation broth using postcolumn derivatization and optimized the following reaction conditions: (*a*) vanillin reacts rapidly to produce a red color among a variety of substituted benzaldehydes (i.e., vanillin, benzaldehyde, 4-benzyloxybenzaldehyde, 3,4-dimethoxy-benzaldehyde, 4-methoxybenzaldehyde, 4-methylbenzaldehyde); (*b*) derivatization conditions are that the reagent solution contains 3% vanillin and 3% sulfuric acid, and the reaction temperature is 120°C (*25, 26*). These postcolumn derivatization systems require column heating and two pumps, one of which is used for pumping concentrated sulfuric acid (which may cause corrosion of some pump components).

Derivatization methods. Detection methods based on derivatization are shown in Table 6. Since Martinez and Shimoda (*31*) attempted to establish an LC method for MN with fluorescence detection derivatizing with 9-anthryldiazomethane (ADAM) (Figure 4, left), several groups have performed an LC determination of SL, MN, LA, NA, and MA using fluorescence detection coupled with precolumn derivatization techniques (*31–37*). ADAM is a known fluorescence derivatization reagent for carboxylic acids, and has been applied to the analysis of prostaglandins and fatty acids (FAs) because of its high reactivity (*38, 39*). Although ADAM reacts rapidly with free carboxylic acids, it does not react with the carboxylate of PEs, which are produced as sodium salts. PEs, therefore, must first be converted to the corresponding free carboxylic acid by treating them with a phosphate buffer (pH 3) (*33*). However, PEs are unstable under lower pH conditions. Martinez et al. (*31*) applied this reagent to the determination of MN in which MN was first acetylated followed by esterification with ADAM. The reason for using acetylation in this situation is not clear to us.

Other methods. Other detection methods are presented in Table 7. Unlike other PEs, spectrofluorometry (*40–43*) and LC with fluorescence detection methods (*44–52*) have been established for LA because it has a unique cresotic acid moiety that produces significant fluorescence in organic solvent. LA exhibits fluorescence with excitation λ_{max} 308–315 nm, and emission λ_{max} 400–430 nm, depending upon the solvents used. The spectrofluorometric and the LC fluorescence detection methods were applied to premix (*40, 44*), feed (*40, 41, 44*), residual (*45, 47–50*), and blood (*42, 46*) analyses. Other characteristic analytical methods of LA, unlike other PEs, are gas chromatography (GC) and gas chromatography–mass spectrometry (GC–MS) (*53–56*). As indicated in Figure 5, a cleavage of the β-hydroxy

Table 6 Detection of Polyether Antibiotic Using the Derivatization Method

PEs	Sample	Column	Mobile phase	Reagent	Reaction condition	Applicable	Ref.
MN	Beef liver	RP–C8, RP–C18	Acetonitrile, water–acetonitrile (9:1) gradient program	ADAM	30 min, dark place	>50 ppb	*(31)*
MN	Chicken	Nucleosil C18	Methanol–water (95:1)	ADAM	60 min, dark room	>0.01 ppm	*(32)*
MN	Chicken	Nova-Pak C18 μ-Porasil	Acetonitrile–water (1:1), Dichloromethane–methanol (19:1)	ADAM	room temperature, over night	>1 ppb	*(33)*
MN, SL, NA, LA	Beef liver	RP–C8, RP–C18	Water–acetonitrile gradient program	ADAM	30 min, dark place	>0.15 ppm	*(34)*
MN	Feed	Shim-pack CLC–ODS	10 mM magnesium acetate in 95% aqueous methanol	ADAM	120 min, room temperature	>0.1 μg/mL	*(35)*
MN	Feed, meat	—	—	ADAM	—	—	*(36)*
polyether antibiotic (e.g., MN)	Feed, meat	—	—	ADAM	—	—	*(37)*
SL, MN, LA, NA	Bulk	Develosil ODS	Methanol–water (97:3)	1-BAP	50°C, 90 min	0.2–100 μg/mL	*(83)*
SL, MN, LA	Feed	Develosil ODS	Methanol–water (97:3)	1-BAP	50°C, 90 min	25–100 ppm	*(84)*

MN: monensin; ADAM: 9-anthryldiazomethane; SL: salinomycin; NA: narasin; LA: lasalocid; 1-BAP: 1-bromoacethylpyrene

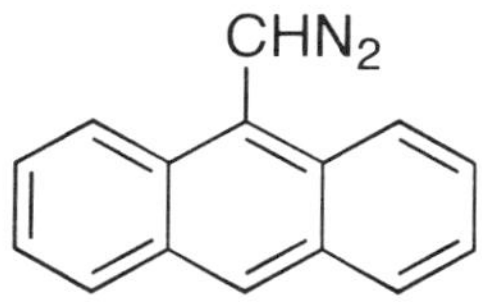

9-anthryldiazomethan (ADAM)

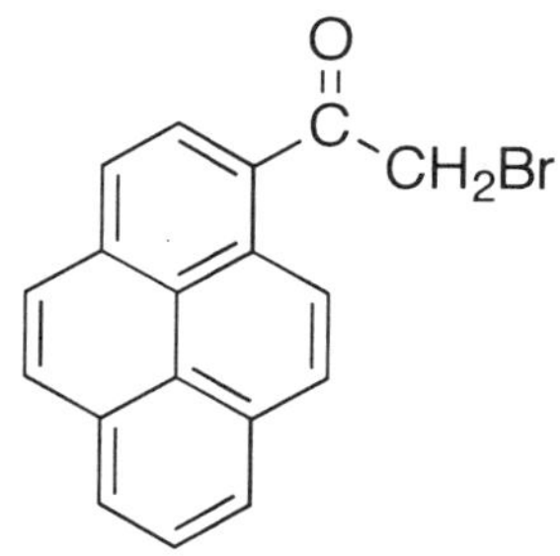

1-bromoacetylpyrene (1-BAP)

Figure 4 Fluorescence derivatization reagents for carboxylic acid

ketone of LA occurs completely in the injection port of a gas chromatograph and LA is degraded to retroaldol ketone and retroaldol aldehyde. This was applied in a pyrolysis GC–MS confirmatory method for LA in bovine liver (*56*). However, it is difficult to apply these methods to SL and MN due to their involatility and instability, and they have not been used for simultaneous determinations of SL, MN, and LA.

Because MA possesses a cyclic hemiketal, it reacts with dansylhydrazine to give a fluorescent hydrazone. This reaction has been proposed for LC determination of MA in feed (*57, 58*) (Figure 6). However, this method is not suitable for feed analysis because there are two peaks in the chromatogram derived from a formation of another ketal derivative.

Separation

Thin-layer chromatography. To separate PEs, most methods use silica gel thin-layer chromatographic (TLC) plates with chloroform–methanol (95:5, 90:10), benzene–methanol (95:5), or carbon tetrachloride–methanol (97:3, 95:5, 80:20) as solvent systems (Table 4). These solvent systems are composed of a mixture of low polarity solvents, such as chloroform, benzene, carbon tetrahloride and a high polarity solvent, methanol, in a ratio between 5 and 20%. These solvent systems show that PEs have a low polarity compared to other antibiotics, i.e., penicillins, tetracyclines, and peptides. The Analytical Method Committee (*59*),

Table 7 Detection of Polyether Antibiotics Using Other Methods

PEs	Sample	Method	Column	Mobile phase	Detection	Applicable	Ref.
LA	Premix, feed	Spectrofluorometry	—	—	Ex 310 nm, Em 419 nm	>0.0037%	*(40)*
LA	Feed	Spectrofluorometry	—	—	Ex 310 nm, Em 419 nm	>0.0064%	*(41)*
LA	Blood	Spectrofluorometry	—	—	Ex 310 nm, Em 415 nm	>0.5 μg/mL	*(42)*
LA	Bulk	Spectrofluorometry	—	—	Ex 310 nm, —	—	*(43)*
LA	Premix, feed	LC fluorescence detection	Partisil PXS-10/25	A) tetrahydrofuran–methanol–ammonium hydroxide–hexane (150:30:10:180), (150:15:10:825), B) tetrahydrofuran–methanol–hexane (150:30:820), (150:15:835), Combine mixture A with mixture B (1:1)	Ex 310 nm, KV 418 emission fliter	3–120 ppm	*(44)*
LA	Cattle liver	LC fluorescence detection	Partisil PXS-10/25	A) tetrahydrofuran–methanol–ammonium hydroxide–hexane (150:30:10:810), top layer, B) tetrahydrofuran–methanol–hexane (150:30:820), Combine mixture A with mixture B (3:1)	Ex 310 nm, Em 430 nm	>25 ppb	*(45)*

Table 7 Detection of Polyether Antibiotics Using Other Methods—*continued*

PEs	Sample	Method	Column	Mobile phase	Detection	Applicable	Ref.
LA	Animal blood	LC fluorescence detection	Partisil PXS-10/25	hexane–tetrahydrofuran–methanol–triethylamine–ammonium hydroxide (810:140:20:20:10)	Ex 310 nm, Em 430 nm	5 ppb–5 ppm	*(46)*
LA	Bovine liver	LC fluorescence detection, GC–MS	Partisil PXS-10/25	A) tetrahydrofuran–methanol–ammonium hydroxide–hexane (150:30:10:810), top layer, B) tetrahydrofuran–methanol–hexane (150:30:820), Combine mixture A with mixture B (3:1)	Ex 310 nm, Em 430 nm	>0.7 ppm	*(47)*
LA	Chicken skin	LC fluorescence detection	Partisil PXS-5/25	A) tetrahydrofuran–methanol–ammonium hydroxide–hexane (150:30:10:810), top layer, B) tetrahydrofuran–methanol–hexane (150:30:820), Combine mixture A with mixture B (3:1)	Ex 310 nm, Em 430 nm	0.1–0.6 ppm	*(48)*
LA	Chicken tissue egg	LC fluorescence detection	Styrene-divinyl-benzene copolymer column	0.01M disodium tetraborate pH 10–acetonitrile (40:60)	Ex 310 nm, Em 430 nm	0.002–0.05 mg/kg	*(49)*

Table 7 Detection of Polyether Antibiotics Using Other Methods—*continued*

PEs	Sample	Method	Column	Mobile phase	Detection	Applicable	Ref.
LA	Chicken tissue egg	LC fluorescence detection	Porous graphitic carbon column	1,1,3,3-tetramethylguanidine–acetonitrile (5:100)	Ex 310 nm, Em 430 nm	0.002–0.05 mg/kg	*(49)*
LA	Chicken tissue	LC fluorescence detection	Unisil pack C18	Methanol–20 mM phosphate buffer, pH 3.0 (9:1)	Ex 310 nm, Em 420 nm	>0.005 ppm	*(50)*
LA	Animal tissue	spectrofluorometry, LC fluorescence detection, pyrolysis GC	—	—	—	—	*(51, 52)*
LA-Br	Plasma	LC fluorescence detection	μ-Bonda-pak C18	Methanol–water–2-methoxyethanol–1M potassium phosphate buffer, pH 3 (90:10:2.5:0.2)	Ex 215 nm, Em 370 nm	>0.1 μg/mL	*(102)*
LA	Bulk	Pyrolysis GC	—	—	FID	—	*(53)*
LA	Bulk, broth	GC	10% OV-17 Gas Chrom Q	—	FID	—	*(54)*
LA, LA-Br	Bulk, premix, mycelial cake	GC	3% methyl silicone	—	FID	—	*(55)*

Table 7 Detection of Polyether Antibiotics Using Other Methods—*continued*

PEs	Sample	Method	Column	Mobile phase	Detection	Applicable	Ref.
LA	Bovine liver	Pyrolysis GC–MS	3% SE-30 Gas-Chrom Q	—	FID	>51 ppb	*(56)*
MA	Feed	LC fluorescence derivatization	Hypersil ODS	water–acetonitrile–tetrabutyl-ammonium hydrogen sulphate (180 mL:820 mL:0.25 g)	Ex 220 nm, Em 470 nm	>5 ppm	*(57)*
MA	Feed	LC fluorescence derivatization	IBM C18	water–acetonitrile–tetrabutyl- ammonium hydrogen sulphate (250 mL:750 mL:0.25 g)	Ex 210 nm, Em 320 nm	>5 ppm	*(58)*
SL	Chicken skin, fat	LC UV detection precolumn oxidation	Ultrasphere ODS	water–tetrahydrofuran–acetonitrile–phosphoric acid (60:40900:0.1)	225 nm	>100 ppb	*(103)*

LA: lasalocid, LA–Br: bromo-lasalocid, MA: maduramicin, SL: salinomycin

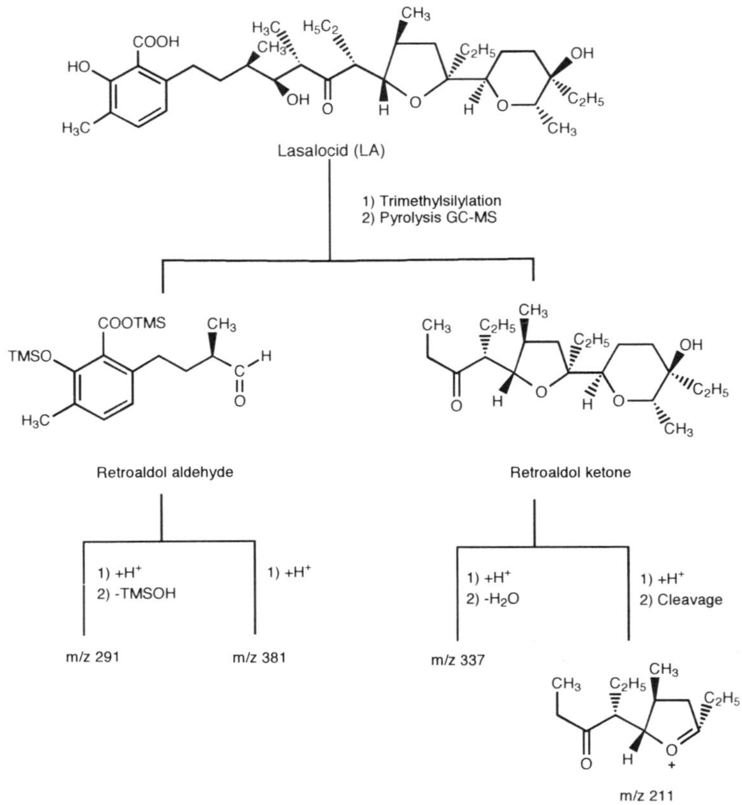

Figure 5 Proposed scheme for the gas chromatographic pyrolysis and mass spectral fragmentation of trimethylsilylated lasalocid.

SOURCE: **Reproduced by permission of the American Chemical Society. See reference (*56*).**

Owles (*23*), and Maxwell et al. (*60*) used water-saturated ethyl acetate or ethyl acetate–water (100:3) as a solvent system. Owles reported that 3% water in ethyl acetate was quite important for the separation of SL and NA (*23*). Although an example was found using a reversed-phase RP-18 high-performance thin-layer chromatographic (TLC) plate coupled with methanol–5% acetic acid (9:1) as a solvent system for the separation of MN, SL, NA, and LA, no chromatogram was shown for their simultaneous separation (*60*).

High-performance liquid chromatography. The first paper on the LC determination of PEs appeared in 1978, and described the determination of LA (*61*) (Tables 2, 7). At that time, a normal-phase silica gel column was used with a mobile phase composed of heptane–methanol–acetic acid (97.6:2.0:0.4) and a detection

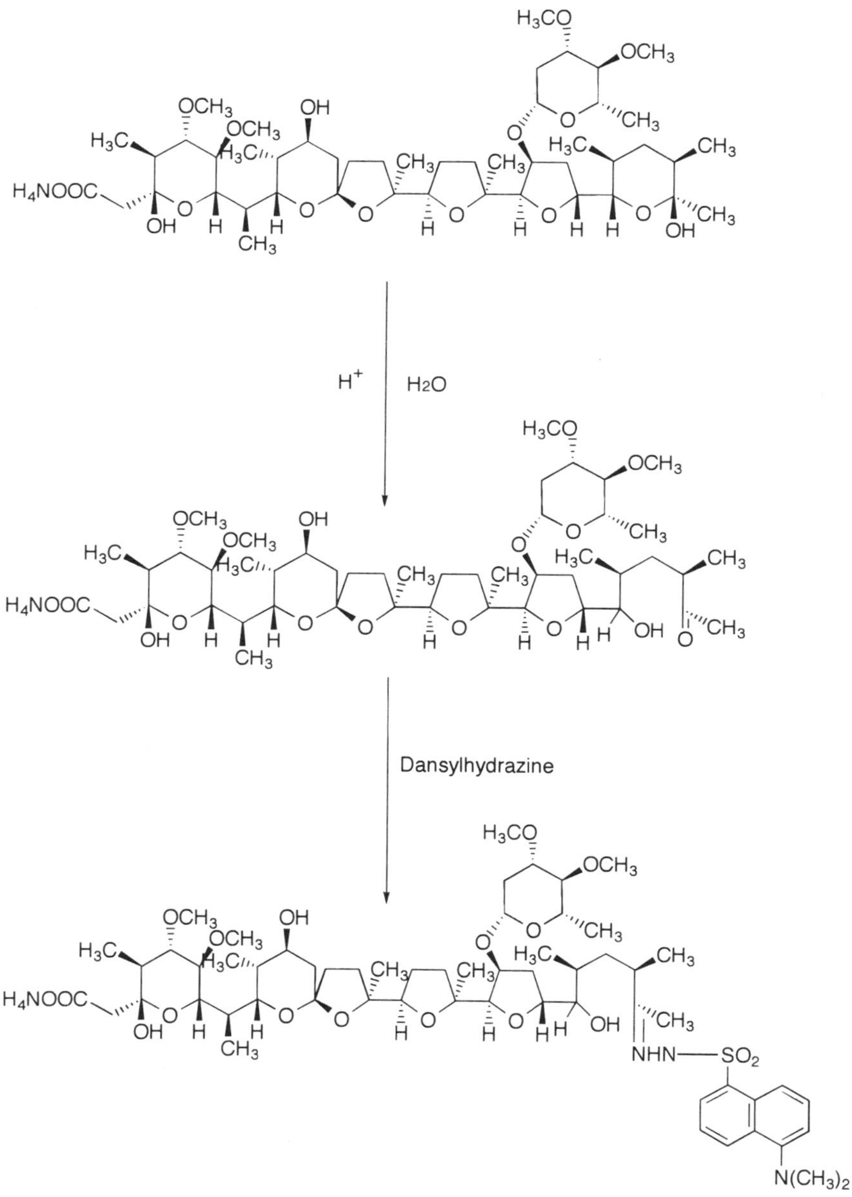

Figure 6 Derivatization of maduramycin with dansylhydrazine.

SOURCE: **Reproduced by permission of The Royal Society of Chemistry. See reference (57).**

at 254 nm (*61*). In Frank et al.'s 1989 paper (*48*) describing the LC determination of LA, a normal-phase column coupled with a complicated mobile phase was also used, but LA was detected with fluorescence instead of UV AS. The reason for the use of these mobile phases is that the fluorescence of LA is most intense when the carboxylic acid moiety is in the ionized form, which makes it necessary to use a basic mobile phase to obtain an optimum sensitivity. However, the basic nature of such mobile phases has a corrosive effect on silica-based columns.

A number of separation systems for MN, SL, LA, NA, and MA have been reported using RP-18 or RP-8 columns coupled with RI, UV AS detection, and LC–MS (Table 2). The ratios between aqueous and organic phases of these solvent systems are approximately 10–20:90–80, due to the comparatively low polarity of PEs. The solvent systems with higher volatility may influence the reproducibility of LC-MS.

Extraction and Cleanup

Methanol, methanol–water (90:10), and acetonitrile are the usual choices for the extraction of PEs in various matrices, including bulk, feeds, premix, and residue samples. Selecting an extraction solvent is critical in establishing an analytical method because it is closely related to the cleanup systems. PEs are hydrophobic compounds that have a high solubility in most organic solvents, and even *n*-hexane has been used as an extraction solvent for the feed analysis of SL (*24*). When methanol and methanol–water (90:10) are used for extraction, it is necessary to consider how to eliminate coextractives, because methanol has the ability to extract both hydrophobic and hydrophilic compounds.

There have been several reports of methanol–water (90:10 or 80:20) as an extraction solvent, coupled with alumina cleanup, being applied to feed and residual analysis of MN, SL, and NA (*23, 31, 34*). Although alumina is effective in removing fatty acids in the microbiological analysis of SL and MN (*62–67*), LA is completely retained on it because of its stronger adsorbing power (*23*). Cleanup systems using alumina, therefore, are not suitable for the simultaneous determination of SL, MN, NA, and LA.

Application

There are two papers on the simultaneous determination of SL, MN, LA, and NA. Johannsen (*22*) established a sequential detection system composed of UV detection (320 nm) for LA and a postcolumn derivatization system with UV detection at 600 nm for SL, MA, and NA. However, the system requires three pumps and two detectors and no actual analyses have been performed using this method. Martinez et al. (*34*) tried a multiresidue determination of SL, MN, LA, and NA. LA was directly esterified with ADAM, but SL, MN, and NA were first acetylated then esterified with ADAM. Although the reason why acetylation is required for SL, MN, and NA is not apparent, the method would probably not be directly applied to an actual mixture composed of these PEs. As stated above, no practical analytical method for a simultaneous LC determination of PEs including SL, MN, LA, and NA has been established to date.

Establishing a Simultaneous Analytical Method

Design for the Method

As previously mentioned, PEs are predominatly used for treating coccidial infection in poultry and as growth promoters for beef cattle (*1*). The increasing use of PEs may cause induction of resistant bacteria, residue problems in agricultural products, and environmental pollution. Therefore, a simple, rapid, and reliable analytical method is required for their precise determination. However, there are no available analytical methods applicable to any PEs that are contained in various matrices such as bulk, broth, premix, feed, and residue samples. The approximate concentrations of PEs in these matrices are as follows: bulk and broth, >1%; premix and feed, >1 ppm; and residue samples, <1 ppm. We started the present work to develop an analytical method applicable to any PEs in any matrix, and so decided to use fluorescence detection because a higher sensitivity is required for the analysis of PEs in residue samples. In the near future the method may be used in conjunction with chemiluminescence detection offering a higher sensitivity than fluorescence detection.

Usually, when establishing an analytical method for trace amounts of a target analyte in a complicated matrix, it is necessary to optimize the operating conditions of the extraction and cleanup, separation, and detection steps.

Detection. An appropriate fluorescent chromophore must be chemically introduced into PEs because they have no chromophore. One effective approach is to utilize the PEs carboxylate moiety and select 1-bromoacetylpyrene (1-BAP) as a fluorescence derivatization reagent because it provides high sensitivity and no chemical treatment of the PE is needed prior to derivatization. Although there have been several papers on the derivatization, there has been no detailed discussion about the reactivity of the reagent. Usually, there are differences in the reactivity of a derivatization reagent between a standard target compound and the compound in a matrix. In such cases, it is desirable to calculate a reaction yield using a derivatized authentic sample that has been synthesized and purified.

However, it is almost impossible to establish derivatization conditions to always obtain a quantitative reaction yield for the analysis of the target compound in a complicated matrix. We introduced as internal standards structural analogs, which have a similar reactivity toward the derivatization reagent and allow us to correct for the reactivity. In the present study, plural internal standards were prepared, an approach that is intended to be applicable in any case.

Separation. It is necessary to have a versatile separation process because it is difficult to establish a complete cleanup system. An established multiple separation system consists of normal- and reversed-phase TLC and LC, which would be used in a complementary fashion, depending on the samples. It is also necessary to separate PE derivatives from derivatization reagents that are usually used in excess for low concentrations of the analytes.

Extraction and cleanup. Alumina is not always suitable for a simultaneous cleanup system including LA because of its stronger adsorbing power. We use a silica gel cleanup system, in consideration of the low polarity of PEs. However, several difficult points are left to be resolved in the establishment of the cleanup system. There are many coextracting materials that have a similar polarity to PEs in broth, premix, feed, and residue samples. It is, therefore, necessary to obtain as much information as possible concerning the chemical properties of these compounds. A more promising approach for establishing the desired cleanup system is to understand the structural characteristics of the coextracting materials, and to carefully investigate separation conditions between them and the PEs. The information obtained would lead to a reliable cleanup system.

Detection

Because SL and MN are produced as sodium salts, we undertook the fluorescence derivatization of the carboxylate group in these compounds. Pederson (*68–70*) first reported that crown ethers complex many cations and hold the cation in the center of the main ring. The complexing ability depends on the relative sizes of the cation and the diameter of the hole in the macrocycle. As a result, the anion becomes very reactive, especially carboxylate anions. The cation complex aids the extraction of the naked anion into an aprotic solvent, so that a reaction proceeds rapidly even under mild conditions (Figure 7) (*71–75*). Kawahara et al. (*76*) reported the synthesis and reactivity of 1-BAP (Figure 4, right) as a fluorescence derivatization reagent for carboxylate groups, and Tsuji et al. (*77*) and Kamada and co-workers (*78–80*) investigated the LC of bile acids and their conjugates derivatized with the reagent in acetonitrile using dicyclohexyl-18-crown-6.

Although the reactivity of SL with 1-BAP was first investigated using dicyclohexyl-18-crown-6 at room temperature, the reaction yield was very low and was less than 10% for other crown ethers tested. Several Kryptofixes (Kryptofix 5, 21, 22, and 222) were also examined. Of these, Kryptofix-222 (K-222) effectively converted SL into its pyrenacyl ester in a high reaction yield. We therefore selected K-222 as a catalyst instead of dicyclohexyl-18-crown-6 for our study (Figure 7).

1-BAP has been used in about 50 to 200-fold molar excess with dicyclohexyl-18-crown-6 as a catalyst with a reaction temperature of 40–80°C in the derivatization of bile acids and carboxylic acids (*77–80*). The relationship between the reaction yield of the corresponding derivatives of SL and the molar ratio, the reaction temperature, and the reaction time were examined using derivatized pyrenacyl ester as a standard to calculate a reaction yield. The following optimum conditions were obtained for producing salinomycin pyrenacyl ester (SL-PY): a molar ratio of SL to 1-BAP to K-222 = 1:3:2; reaction temperature, 25°C (room temperature); reaction time, 90 min. These results demonstrated that the derivatization proceeds quantitatively at room temperature using K-222 as a catalyst and that there was a linear relationship between the fluorescence intensity and the amount of SL between 2 and 14 ng (*81*).

Separation

We developed this method to simultaneously determine SL, MN, and LA because LA was approved as feed additive in Japan. For this purpose the following two approaches were devised: (*a*) to introduce an internal standard for simultaneous determinations, and (*b*) to introduce TLC in both normal- and reversed-phase modes. It is difficult to obtain a constant derivatization yield in feed and residual analyses. An appropriate means to eliminate such a problem is to introduce an internal standard that has a similar reactivity for pyrenacylation, displays a similar R_f value on a TLC plate to those of SL, MN, and LA, and is not contained in a

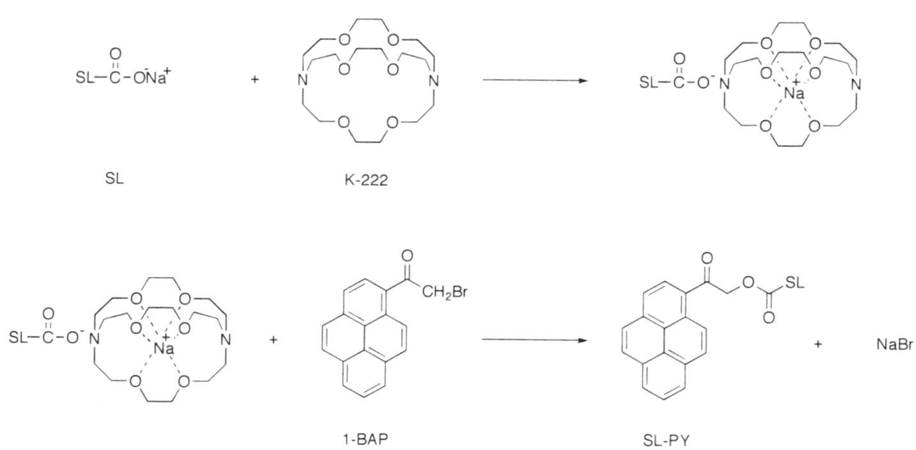

Figure 7 Effect of Kryptofix 222 on the pyrenacylation

sample as an impurity. A compound that fulfills these conditions is a structural analog of the three antibiotics. Thus, the SL derivatives, 18,19-dihydrosalinomycin (DSL) and 18,19-dihydro-20-ketosalinomycin (DKSL), were expected to be effective as internal standards (Figure 8). We investigated separation conditions for five polyether antibiotics–pyrenacyl esters (PEs–PYs). A satisfactory result was obtained on a silica gel TLC plate using carbon tetrachloride–ethyl acetate–acetonitrile (50:5:10) as a solvent system (Figure 9). The addition of acetonitrile into the system brought about a satisfactory and reproducible separation between SL-PY and DSL-PY. In the case of RP-18 TLC, dichloromethane–ethyl acetate–acetone–acetonitrile (15:2:1:55) (Figure 10) yielded good resolution of the five PYs, with the small amount of acetone used to obtain clear spots.

We successfully attempted to simultaneously determine SL, MN, and LA with a constant concentration of DSL and DKSL, and we found that DSL was suitable for the simultaneous determination of SL and MN, and DKSL for LA (Table 8). A linear relationship was obtained between the fluorescence intensity and the amount of each antibiotic in a range of 2–14 ng. In this way, an TLC method was established for a simultaneous fluorodensitometric determination of SL, MN, and LA based on fluorescence labeling with 1-BAP using DSK and DKSL as internal standards (*82*).

Although the TLC method has proven simple and reliable, it is not always sensitive enough to detect small amounts of these compounds with precision.

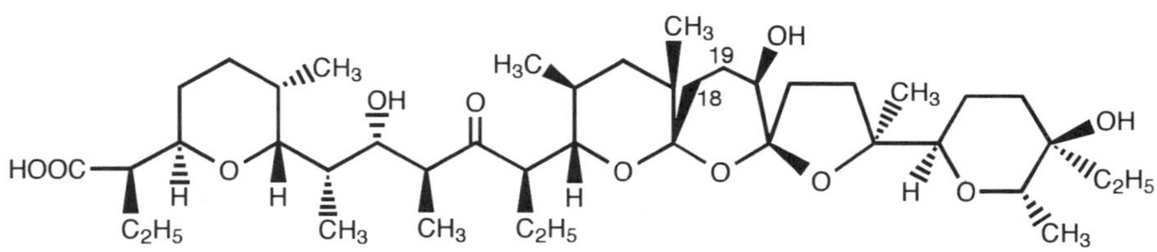

18, 19-Dihydrosalinomycin (DSL)

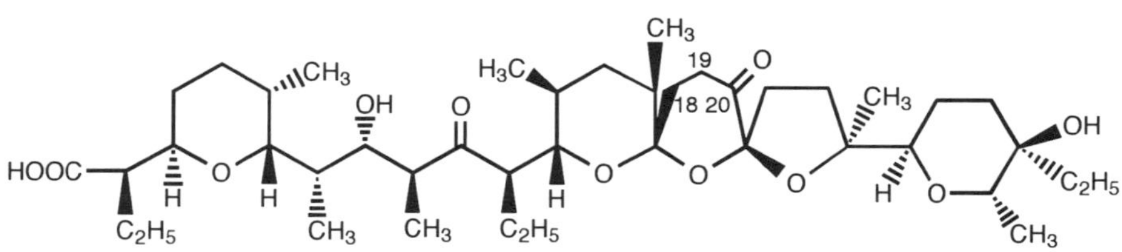

18, 19-Dihydro-20-ketosalinomycin (DKSL)

Figure 8 Internal standards for simultaneous determination
SOURCE: Reproduced by permission of Elsevier Science Publishers B.V., see reference (*82, 83*), and of the American Chemical Society. See reference (*84*).

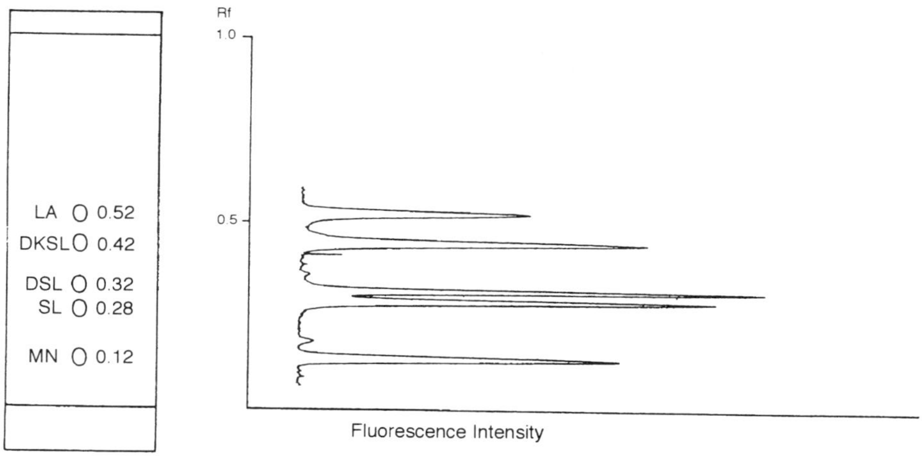

Figure 9 Separation of pyrenacyl esters of polyether antibiotics on silica gel (E. Merck, 5641) high-performance thin-layer chromatography and the densitometric profile. Solvent system: carbon tetrachloride–ethyl acetate–acetonitrile (50:5:10).
SOURCE: Reproduced by permission of Elsevier Science Publishers B.V. See reference (*82*).

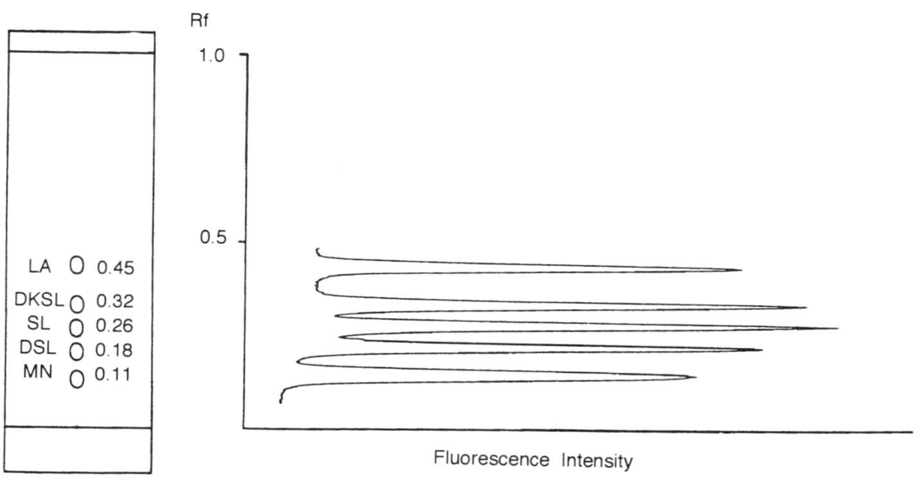

Figure 10 Separation of pyrenacyl esters of polyether antibiotics on RP-18 (E. Merck, 5914) high-performance thin-layer chromatography and the densitometric profile. Solvent system: dichloromethane-ethyl acetate-acetone-acetonitrile (15:2:1:55).
SOURCE: Reproduced by permission of Elsevier Science Publishers B.V. See reference (*82*).

Table 8 Characteristics of Standard Curves on Silica Gel and RP-18 High-Performance Thin-Layer Chromatography Plates Using Internal Standards

Sample	Internal standard	Silica gel TLC: Regression equation	Silica gel TLC: Correlation coefficient	RP-18 TLC: Regression equation	RP-18 TLC: Correlation coefficient
Salinomycin	DSL	$y = 0.060x + 0.019$	0.9999	$y = 0.074x - 0.039$	0.9999
	DKSL	$y = 0.087x + 0.031$	0.9993	$y = 0.064x - 0.032$	0.9994
Monensin	DSL	$y = 0.058x - 0.022$	0.9999	$y = 0.077x + 0.038$	0.9999
	DKSL	$y = 0.078x + 0.222$	0.9993	$y = 0.060x + 0.107$	0.9985
Lasalocid	DSL	$y = 0.032x - 0.056$	0.9995	$y = 0.065x - 0.048$	0.9985
	DKSL	$y = 0.044x - 0.062$	0.9997	$y = 0.056x - 0.027$	0.9996

DSL: 18,19-dihydrosalinomycin; DKSL: 18,19-dihydro-20-ketosalinomycin; TLC: high-performance thin-layer chromatography.
SOURCE: Reproduced by permission of Elsevier Science Publishers B. V. See reference (*82*).

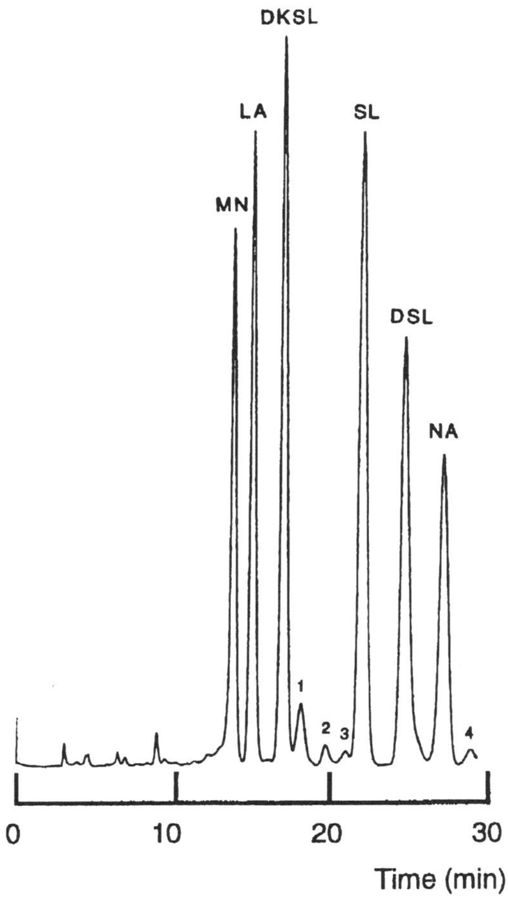

Figure 11 Liquid chromatographic separation of pyrenacyl derivatives of polyether antibiotics. Column, Develosil 5C18 (250 × 4.6 mm i.d.); mobile phase, methanol-water (97:3); flow-rate, 1 mL/min; detection, λ_{ex} = 360 nm and λ_{em} = 420 nm.

SOURCE: Reproduced by permission of Elsevier Science Publishers B.V. See reference (*83*).

Furthermore, we have not applied the method to NA, which has not been approved as a poultry feed additive in Japan. On the basis of the above considerations, an LC method was attempted so a precise analysis of the four prederivatized PEs (*83*) could be established.

In order to optimize the separation conditions, authentic PEs–PYs of six PEs were first prepared according to our method (*81*, *82*) and their structures were confirmed by MS. Separation conditions for six standards of PEs–PYs, including those of the internal standards, DSL and DKSL, were then investigated and the following conditions were optimized: column, Develosil 5C-18, 250 × 4.6 mm i.d.; mobile phase, methanol–water, 97:3; flow rate, 1.0 mL/min; and detection, λ_{ex} 360 nm and λ_{em} 420 nm (Figure 11). After derivatization of the six PEs at a concentration of 100 μg/mL under the conditions described in Asukabe et al.'s

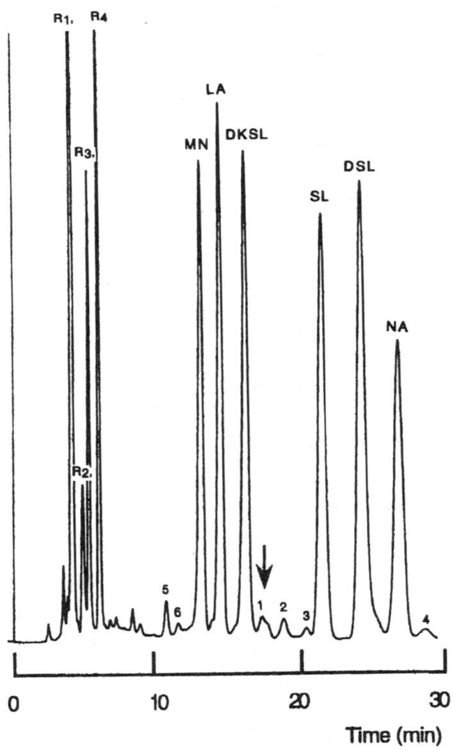

Figure 12 Typical Liquid chromatogram of simultaneous determination (100 μg/mL).
SOURCE: **Reproduced by permission of Elsevier Science Publishers B.V. See reference (*83*).**

paper (*82*), the reaction mixture was directly injected into the liquid chromatograph. As shown in Figure 12, good separation of the PEs–PYs was obtained except for the peaks around 5 min, which are derived from the reagent. The minor peaks, including peaks 1–6, did not interfere with the determination of PEs. These results demonstrate that a combination of a C-18 column and methanol:water (97:3) was suitable for this study.

In the paper by Asukabe et al. (*82*) derivatization conditions used include: a molar excess of reagent (1-BAP), 50-fold; a molar excess of catalyst (K-222), 10-fold; reaction temperature, 50°C; and reaction time, 90 min. We adopted these conditions to the simultaneous derivatization of the six PEs at concentrations of 100 µg/mL, 10 µg/mL, and 1.0 µg/mL and estimated each reaction yield (Table 9). If the derivatization is applied to feed and residual analyses, it is necessary to obtain a high reaction yield even at low concentrations of PEs. Although 10- to 100-fold excess amounts of reagent and catalyst as the amounts used as 100 µg/mL were used to obtain a higher reaction yield at the concentrations of 10 µg/mL and 1 µg/mL, no peaks of the six kinds of PEs–PYs were found due to peaks of unused reagents present on the chromatogram.

To eliminate excess reagent, the differing polarities of PEs–PYs and 1-BAP were efficiently employed for a cleanup using a silica gel cartridge. The separation procedure was optimized as follows. After derivatization, the reaction mixture was evaporated to dryness. The residue was dissolved in 5 mL of chloroform–benzene (1:1) solution. The solution containing PEs–PYs and 1-BAP was applied to the silica gel cartridge and rinsed twice with 5 mL of the same solvent. The PEs–PYs were then eluted using two washes with 10 mL of benzene–acetone (7:3). We

Table 9 Derivatization Conditions and Reaction Yield of Polyether Antibiotics with 1-BAP

Amount of PEs, µg/mL	Molar excess		Yield (%)[a]					
	Reagent	Catalyst	SL	MN	LA	NA	DSL	DKSL
100	50-fold	10-fold	96	100	94	102	101	90
10	50-fold	10-fold	96	85	50	105	100	91
1	50-fold	10-fold	80	6	2	76	72	59
10	500-fold	100-fold	93	98	98	104	99	94
1	5000-fold	1000-fold	84	85	68	85	82	80

SL: salinomycin; MN: monensin; LA: lasalocid; NA: narasin; DSL: 18,19-dihydrosalinomycin; DKSL: 18,19-dihydro-20-ketosalinomycin

[a]Reaction yield was calculated from standard curves of authentic pyrenacyl esters.

SOURCE : Reproduced by permission of Elsevier Science Publishers B. V. See reference (*83*).

achieved a simultaneous determination of SL, MN, LA, and NA at concentrations ranging from 20 to 100 μg/mL, 2 to 10 μg/mL, and 0.2 to 1.0 μg/mL under the conditions mentioned above using DSL and DKSL as internal standards (Figure 13) (*83*). The data shown in Table 10 indicate a good correlation between the four PEs and the two internal standards showing that they can be used in a complementary fashion, depending on the nature of the sample and its interfering substances. These results indicate that this derivatization system is more reliable than that with ADAM.

Extraction and Cleanup

Five grams of each of the four kinds of feed was extracted with 50 mL each of methanol, ethanol, ethyl acetate, acetone, acetonitrile, chloroform, benzene, and *n*-hexane. After evaporating the supernant to dryness, the previous reported derivatization system (*83*) was applied without any cleanup, and the reaction mixtures were injected into a liquid chromatogram with fluorescence detection. Approximately ten peaks of interfering substances appeared on the chromatogram, as indicated on the left of Figure 14. These peaks were observed to be present in chromatograms of the extracts from all of the feeds.

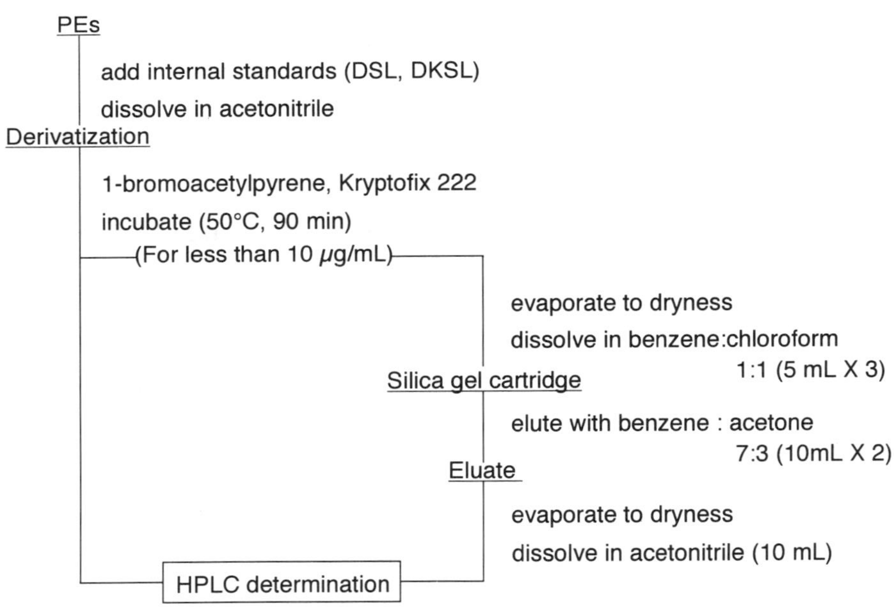

Figure 13 Analytical procedure for polyether antibiotics prederivatized with 1-bromoacetylpyrene.
SOURCE: Reproduced by permission of Elsevier Science Publishers B.V. See reference (*83*)

Table 10 Characteristics of Standard Curves of Polyether Antibiotics Using Internal Standards

		20 μg/mL ~ 100 μg/mL		2μg/mL ~ 10 μg/mL		0.2 μg/mL ~ 1.0 μg/mL	
Sample	Internal standard	Regression equation	Correlation coefficient	Regression equation	Correlation coefficient	Regression equation	Correlation coefficient
Salino-mycin	DSL	$y = 0.015x + 0.810$	0.9995	$y = 0.171x + 0.006$	1.0000	$y = 1.682x + 0.063$	0.9996
	DKSL	$y = 0.016x + 0.081$	0.9998	$y = 0.147x - 0.000$	0.9999	$y = 1.476x + 0.028$	0.9995
Monensin	DSL	$y = 0.013x + 0.015$	0.9999	$y = 0.152x - 0.091$	0.9996	$y = 1.370x + 0.030$	0.9993
	DKSL	$y = 0.015x + 0.005$	0.9998	$y = 0.131x - 0.018$	0.9997	$y = 1.202x + 0.004$	0.9993
Lasalocid	DSL	$y = 0.017x + 0.028$	0.9996	$y = 0.174x + 0.002$	0.9996	$y = 1.698x - 0.011$	0.9998
	DKSL	$y = 0.018x + 0.024$	0.9998	$y = 0.015x - 0.003$	0.9995	$y = 1.489x - 0.035$	0.9989
Narasin	DSL	$y = 0.011x + 0.032$	0.9998	$y = 0.125x + 0.018$	1.0000	$y = 1.253x + 0.044$	0.9998
	DKSL	$y = 0.012x + 0.029$	1.0000	$y = 0.108x + 0.012$	1.0000	$y = 1.099x + 0.018$	0.9994

DSL: 18,19-dihydrosalinomycin; DSKL: 18,19-dihydro-20-ketosalinomycin

SOURCE: Reproduced by permission of Elsevier Science Publishers B.V. See reference (*83*).

A more promising approach to establishing a cleanup system was to understand the structural characteristics of the coexisting materials in feeds. A GC–MS and LC technique was used to identify the interfering substances on the chromatogram (Figure 14, left). The results indicate that the interfering substances are PYs of fatty acids (FAs). We studied the separation conditions between PEs and FAs using Mega Bond-Elut silica gel cartridges and found that PEs could be separated from FAs by applying the sample solution (chloroform) to a Mega Bond-Elut silica gel cartridge, washing the cartridge with chloroform (10 mL × 2) to eliminate FAs, and eluting the PEs with ethyl acetate (10 mL × 2). The peaks of interfering substances are almost eliminated by the first step of the cleanup system when compared to the chromatogram (Figure 14, middle), but a few peaks derived from FAs still appear on the chromatogram. Because the first step of the cleanup system was not sufficient for separating PEs and FAs, the different chromatographic behavior of PEs–PYs and fatty acid–pyrenacyl esters (FAs–PYs) was efficiently applied to the second step of the cleanup.

The separation of PEs–PYs from FAs–PYs was also studied using Mega Bond-Elut silica gel cartridges where PEs–PYs could be separated from FAs–PYs by applying the sample solution (benzene) to a Mega Bond Elut silica gel cartridge, washing the cartridge with benzene (10 mL × 2) to eliminate FAs–PYs, and eluting PEs–PYs with benzene–acetone (7:3, 10 mL × 2). The peaks of interfering substances in feeds were completely eliminated by the second cleanup system

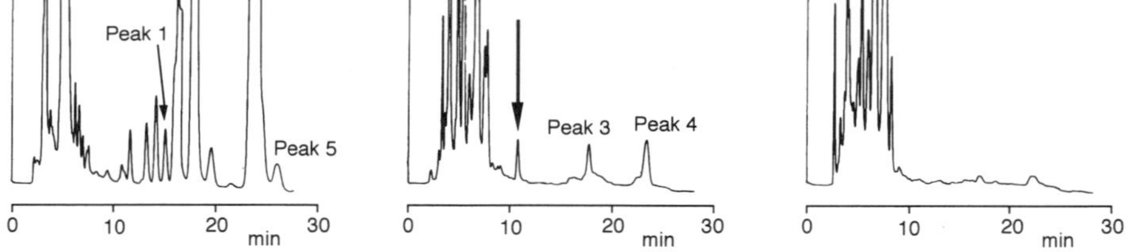

Figure 14 Effect of cleanup with silica gel on liquid chromatograms of backgrounds. Left: without cleanup. Middle: after the first step cleanup. Right: after the second cleanup.
SOURCE: Reproduced by permission of American Chemical Society. See reference *(84)*.

(Figure 14, right). The cleanup system including the two steps with Mega Bond-Elut silica gel cartridges was quite effective for the determination of PEs in feed.

Five grams of sample free from PEs were extracted with 50 mL of acetonitrile and the first step of the cleanup system was performed. After adding each PE to the ethyl acetate fraction, the reactivity of the PEs was examined at 50°C for 90 min, followed by the second step of the cleanup system. The low reactivity of MN and DKSL was unchanged, even if the molar excess of reagent increased to 100-fold. This was attributed to the partial conversion of the carboxylate groups in the molecules to free carboxylic acid through the first step cleanup with the silica gel. Adding triethylamine improved the reactivity of MN (Table 11).

Finally, the following derivatization conditions applicable to feeds were chosen: internal standard, DSL; molar excess of 1-BAP 100-fold; molar excess of K-222, 20-fold; amount of triethylamine, 2 μL; reaction temperature, 50°C; and reaction time, 90 min. To confirm the applicability of the cleanup and derivatization conditions, the experiment shown in Figure 15 was performed. Good recoveries were obtained without interfering peaks. The reactivity of DSL was almost the same as that for SL, MN, and LA, showing that DSL gives satisfactory results as an internal standard for the simultaneous determination of SL, MN, and LA in feeds.

Table 11 Derivatization of Polyether Antibiotics in Feed Extract with 1-Bromoacetylpyrene

Molar excess			Yield (%)				
Reagent	Catalyst	Triethylamine, μL	SL	MN	LA	DSL	DKSL
50-fold	10-fold	—	100	59	100	100	88
		1	100	89	99	100	87
		2	98	92	97	98	88
100-fold	10-fold	—	100	53	100	100	79
		1	99	87	99	100	91
		2	98	84	97	98	86
100-fold	20-fold	—	100	90	99	100	86
		1	98	99	98	99	89
		2	98	100	98	98	87
		3	98	100	98	98	87

SL: salinomycin; MN: monensin; LA: lasalocid; DSL: 18,19-dihydrosalomycin; DSKL: 18,19-dihydro-20-ketosalinomycin

SOURCE : Reproduced by permission of the American Chemical Society. See reference (*84*).

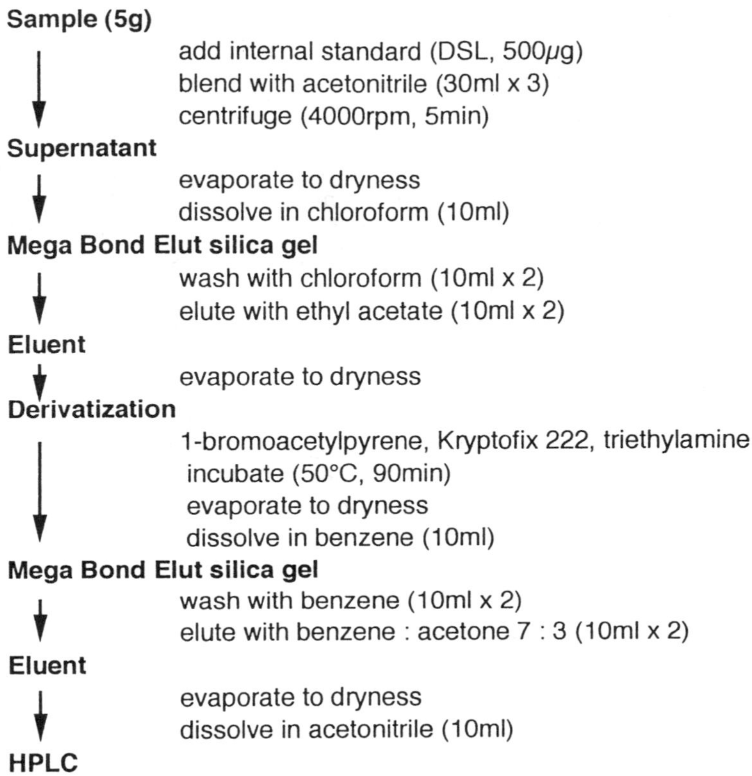

Figure 15 Established analytical procedure for salinomycin, monensin, and lasalocid in feed samples.
SOURCE: **Reproduced by permission of American Chemical Society. See reference (*84*).**

Five samples of 5 g each (each spiked at 100 ppm) were extracted with each of the solvents: methanol, acetone, ethyl acetate, acetonitrile, benzene, and *n*-hexane. Of these, acetonitrile was found to be the most suitable for the extraction. PEs (each of 125–500 μg) were spiked into 5 g of each feed, and the recoveries of PEs were investigated at 25–100 ppm. As shown in Table 12, satisfactory recoveries were obtained even at a low concentration levels of PEs. The typical high-performance liquid chromatograms of these feeds are shown in Figure 16. No interfering peaks appeared on the chromatograms.

A simple and rapid technique has been established for the simultaneous analysis of PEs in feeds (*84*) where: acetonitrile is suitable to extract PEs from feeds; the internal standard, DSL, is effective for the simultaneous determination of PEs in feeds; two-step silica gel cartridge cleanup system provides a complete

Table 12 Recovery of Polyether Antibiotics From Feed Sample

	Recovery (%)		
Fortified (ppm)	SL	MN	LA
25	99	98	96
50	99	100	95
75	98	100	95
100	100	104	94

SL: salinomycin; MN: monensin; LA: lasalocid
SOURCE : Reproduced by permission of the American Chemical Society. See reference (*84*).

separation of PEs from FAs; the addition of triethylamine to the reaction solution is essential to derivatize efficiently; and the recoveries of SL, MN, and LA from feeds spiked at 25–100 ppm are 99%, 100%, and 100%, respectively.

Conclusion

Since 1972, more than 100 papers have been published on the chemical analysis of PEs, which, except for LA, have no chromophores. As of now, the analytical methods for these compounds have been presented in the following order: colorimetric determination, densitometric determination, and postcolumn derivatization for MN and SL; and for LA, spectrofluorometic determination and LC fluorescence detection. However, no practical analytical method for a simultaneous determination of PEs has been established so far. In our previous work, we established techniques for the simultaneous determination of PEs using silica gel and RP-18 TLC, followed by fluorodensitometry and LC with fluorescence detection using 1-BAP as a fluorescence derivatization reagent and K-222 as a catalyst (*81–83*). Furthermore, we established a simultaneous LC determination of three PEs (SL, MN, and LA) in feed using these derivatization conditions and prepacked silica gel cartridges for cleanup systems (*84*). Our technique is summarized as follows:

Detection: 1-BAP was a suitable fluorescence derivatization reagent of PEs, which are produced as sodium salts. K-222 was very effective in the transformation of a carboxylate group into its pyrenacyl ester.

Separation: A silica gel TLC plate with carbon tetrachloride–ethyl acetate–acetonitrile (50:5:10) as a solvent system and a RP-18 TLC plate with dichlo-

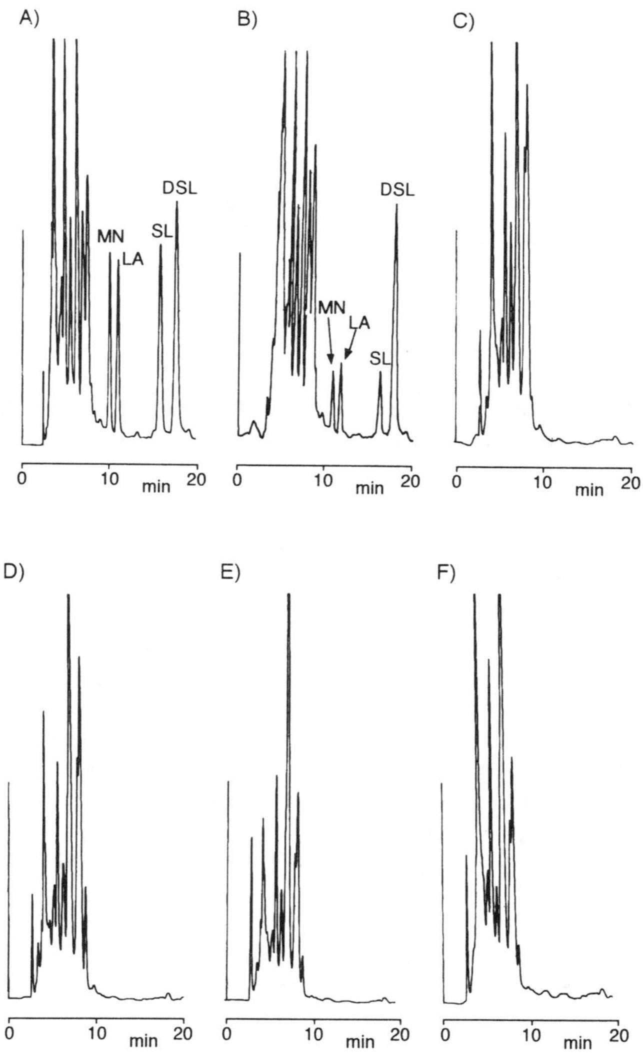

Figure 16 Typical chromatograms of broiler finisher feed extracts. (A) Fortified broiler finisher (100 ppm); (B) fortified broiler finisher (25 ppm); (C) broiler finisher; (D) broiler starter; (E) chicken finisher; (F) chicken starter SOURCE: Reproduced by permission of the American Chemical Society. See reference (*84*).

romethane–ethyl acetate–acetone–acetonitrile (15:2:1:55) gave satisfactory separations of SL, MN, LA, DSL, and DKSL. Under these conditions DSL was suitable as an internal standard for simultaneously determining SL and MN, and DKSL for LA.

A reversed-phase LC system, coupled with Develosil 5C-18 and methanol–water (97:3) as the mobile phase, was effective for the separation of SL, MN, LA, NA, DSL, and DKSL. The simultaneous determination of SL, MN, LA, and NA was achieved at concentrations 0.2–100 μg/mL using DSL and DKSL as internal standards.

Extraction and cleanup: Acetonitrile was most suitable to extract PEs from feed, and a two-step silica gel cartridge cleanup system permits the separation of PEs from FAs.

Acknowledgment

We are grateful to Makoto Suzuki, Meijo University for his encouragement.

References

1. J. W. Westley, *Polyether Antibiotics,* vol. 1, *Biology* (New York: Marcel Dekker, 1982), chap. 6.
2. J. W. Westley, *Polyether Antibiotics,* vol. 2, *Chemistry*, (New York: Marcel Dekker, 1982), chap. 2.
3. M. J. Weinstein and G. H. Wagman, *J. Chromatogr. Lib.*, **15**: 479, 1978.
4. T. D. Macy and A. Loh, *J. Assoc. Off. Anal. Chem.*, **66**: 284, 1983.
5. M. Beran, J. Tax, V. Schon, Z. Vanek, and M. Podojil, *J. Chromatogr.*, **268**: 315, 1983.
6. G. P. Dimenda, J. A. Creegan, L. B. Turnbull, and G. J. Wright, *J. Assoc. Off. Anal. Chem.*, **70**: 504, 1987.
7. M. Vuzharova, E. Tomov, I. Dorosiev, and P. Papazova, *Farmatsiya (Sofia)*, **37**: 7, 1987.
8. R. L. Hussey, T. D. Macy, J. Moran, and A. Loh, *J. Assoc. Off. Anal. Chem.*, **68**: 417, 1985.
9. M. Beran, R. Zielke, K. Dornberger, C. Stengel, J. Zima, and U. Graefe, *Chromatographia*, **31**: 603, 1991.
10. M. Beran, J. Zima, and V. Schon, *Biomed. Environ. Mass Spectrom.*, **15**: 153, 1988.
11. H. T. Kalinoski, B. W. Wright, and R. D. Smith, *Biomed. Environ. Mass Spectrom.*, **15**: 239, 1988.
12. M. M. Siegel, W. J. McGahren, K. B. Tomer, and T. T. Chang, *Biomed. Environ. Mass Spectrom.*, **14**: 29, 1987.

13. B. N. Pramanik, P. R. Das, and A. K. Bose, *J. Nat. Prod.*, **52**: 534, 1989.
14. G. B. Blomkvist, K. M. Jansson, E. R. Ryhage, and B. Oesterdahl, *J. Agric. Food Chem.*, **34**: 274, 1986.
15. T. T. Chang and F. Andrawes, *Proc. Int. Symp. Instrum. High-Perform. Thin-Layer Chromatogr.*, **3**: 427, 1985.
16. S. J. Stout, L. A. Wilson, A. I. Kleiner, A. R. DaCunha, and T. J. Francl, *Biomed. Environ. Mass Spectrom.*, **18**: 57, 1989.
17. S. Horii, K. Miyahara, and T. Maruyama, *Shokuhin Eiseigaku Zasshi*, **32**: 30, 1991.
18. R. P. Schneider, M. J. Lynch, J. F. Ericson, and H. G. Fouda, *Anal. Chem.*, **63**: 1789, 1991.
19. T. Golab, S. J. Barton, and R. E. Scroggs, *J. Assoc. Off. Anal. Chem.*, **56**: 171, 1973.
20. F. R. Duke, *Anal. Chem.*, **19**: 661, 1947.
21. E. Ishikuro, *Shiryo Kenkyu Hokoku (Tokyo Hishiryo Kensasho)*, **16**: 128, 1991.
22. F. H. Johannsen, *Agric. Biol. Res.*, **44**: 79, 1991.
23. P. J. Owles, *Analyst*, **109**: 1331, 1984.
24. J. T. Goras and W. R. Lacourse, *J. Assoc. Off. Anal. Chem.*, **67**: 701, 1984.
25. F. L. Neely, *Chromatographia*, **31**: 277, 1991.
26. F. L. Neely, *J. Liq. Chromatogr.*, **15**: 1513, 1992.
27. J. M. Rodewald, J. W. Moran, A. L. Donoho, and M. R. Coleman, *J. AOAC Int.*, **75**: 272, 1992.
28. M. Sokolic and M. Pokorny, *J. Pharm. Biomed. Anal.*, **9**: 1047, 1991.
29. W. J. Blanchflower, D. A. Rice, and J. T. G. Hamilton, *Analyst*, **110**: 1283, 1985.
30. M. R. Lapointe and H. Cohen, *J. Assoc. Off. Anal. Chem.*, **71**: 480, 1988.
31. E. E. Martinez and W. Shimoda, *J. Assoc. Off. Anal. Chem.*, **68**: 1149, 1985.
32. Y. Hoshino, M. Horie, N. Nose, and H. Iwasaki, *Shokuhin Eiseigaku Zasshi.*, **26**: 585, 1985.
33. K. Takatsuki, S. Suzuki, and I. Ushizawa, *J. Assoc. Off. Anal. Chem.*, **69**: 443, 1986.
34. E. E. Martinez and W. Shimoda, *J. Assoc. Off. Anal. Chem.*, **69**: 637, 1986.
35. K. Gamoh and E. Okada, *Bunseki Kagaku.*, **37**: 324, 1988.
36. E. Okada and K. Gamoh, *Jpn. Kokai Tokkyo Koho*, JP 01,152,362 [89,152,362], 1989.
37. K. Gamoh, *Jpn. Kokai Tokkyo Koho*, JP 01,152,363 [89,152,363], 1989.
38. N. Nimura and T. Kinoshita, *Anal. Lett.*, **13**(A3): 191, 1980.
39. S. A. Barker, J. A. Monti, S. T. Christian, F. Benington, and R. D. Morin, *Anal. Biochem.*, **107**: 116, 1980.
40. M. Osadca and M. Araujo, *J. Assoc. Off. Anal. Chem.*, **57**: 636, 1974.
41. M. Osadca and M. Araujo, *J. Assoc. Off. Anal. Chem.*, **58**: 507, 1975.

42. M. A. Brooks, L. D'Arconte, J. A. F. De Silva, G. Chen, and C. Crowley, *J. Pharm. Sci.*, **64**: 1874, 1975.
43. A. K. Mitra and M. M. Narurkar, *J. Org. Chem.*, **49**: 1293, 1984.
44. M. Osadca and M. Araujo, *J. Assoc. Off. Anal. Chem.*, **61**: 1074, 1978.
45. G. Weiss, N. R. Felicito, M. Kaykaty, G. Chen, A. Caruso, E. Hargroves, C. Crowley, and A. MacDonald, *J. Agric. Food Chem.*, **31**: 75, 1983.
46. M. Kaykaty and G. Weiss, *J. Agric. Food Chem.*, **31**: 81, 1983.
47. D. R. Newkirk and C. J. Barnes, *J. Assoc. Off. Anal. Chem.*, **72**: 581, 1989.
48. L. R. Frank and C. J. Barnes, *J. Assoc. Off. Anal. Chem.*, **72**: 584, 1989.
49. J. A. Tarbin and G. Shearer, *J. Chromatogr.*, **579**: 177, 1992.
50. E. Ishikuro, *Shokuhin Eiseigaku Zasshi.*, **30**: 160, 1989.
51. A. MacDonald, *J. Assoc. Off. Anal. Chem.*, **61**: 1214, 1978.
52. G. Weiss and A. MacDonald, *J. Assoc. Off. Anal. Chem.*, **68**: 971, 1985.
53. J. W. Westley Jr., R. H. Evans, T. Williams, and A. Stempel, *J. Org. Chem.*, **38**: 3431, 1973.
54. J. W. Westley Jr., R. H. Evans, and A. Stempel, *Anal. Biochem.*, **59**: 574, 1974.
55. G. Manius and V. Viswanathan, *J. Pharm. Sci.*, **66**: 769, 1977.
56. G. Weiss, M. Kaykaty, and B. Miwa, *J. Agric. Food Chem.*, **31**: 78, 1983.
57. M. J. Gliddon, D. Wright, A. Markantonatos, and W. Groth, *Analyst*, **113**: 813, 1988.
58. A. Markantonatos, *J. Liq. Chromatogr.*, **11**: 877, 1988.
59. Analytical Methods Committee, *Analyst*, **111**: 1089, 1986.
60. R. J. Maxwell and J. Unruh, *J. Planar Chromatogr.–Mod. TLC.*, **5**: 35, 1992.
61. R. B. Hagel, *J. Assoc. Off. Anal. Chem.*, **61**: 1070, 1978.
62. Analytical Methods Committee, *Analyst*, **102**: 206, 1977.
63. E. E. Martinez and W. Shimoda, *J. Assoc. Off. Anal. Chem.*, **66**: 1505, 1983.
64. E. E. Martinez and W. Shimoda, *J. Assoc. Off. Anal. Chem.*, **67**: 845, 1984.
65. T. Kusama, *Shiryo Kenkyu Hokoku (Tokyo Hishiryo Kensasho)*, **11**: 107, 1986.
66. T. Kusama and H. Shimada, *Shiryo Kenkyu Hokoku (Tokyo Hishiryo Kensasho)*, **11**: 124, 1986.
67. K. Takeba, F. Murakami, K. Zinbo, R. Kiritani, and M. Matsumoto, *Kenkyu Nenpo–Tokyo-toritsu Eisei Kenkyusho.*, **37**: 141, 1986.
68. C. J. Pederson, *J. Am. Chem. Soc.*, **89**: 7017, 1967.
69. C. J. Pederson, *J. Am. Chem. Soc.*, **92**: 386, 1970.
70. C. J. Pederson, *J. Am. Chem. Soc.*, **92**: 391, 1970.
71. H. D. Durst, *Tetrahedron Lett.*, **28**: 2421, 1974.
72. C. L. Liotta, H. P. Harris, M. McDermott, T. Gonzalez, and K. Smith, *Tetrahedron Lett.*, **28**: 2417, 1974.

73. C. L. Liotta and H. P. Harris, *J. Am. Chem. Soc.*, **96**: 2250, 1974.
74. H. D. Durst, M. Milano Jr., E. J. Kikta, S. A. Connelly, and E. Grushka, *Anal. Chem.*, **47**: 1797, 1975.
75. K. Blau and G. S. King, *Handbook of Derivatives for Chromatography* (London: Heyden and Son Ltd., 1978).
76. Y. Kawahara, A. Inage, T. Morioka, Y. Shibano, and N. Kasuya, *The 100th Annual Meeting of Pharmaceutical Society of Japan, Tokyo* Abstract 1980, 568.
77. A. Tsuji and Y. Kawahara, *Jpn. Kokai Tokkyo Koho,* JP 81,126,744, 1981.
78. S. Kamada, M. Maeda, A. Tsuji and Y. Kawahara, *The 100th Annual Meeting of Pharmaceutical Society of Japan, Tokyo* Abstract 1980, 568.
79. S. Kamada, M. Maeda and A. Tsuji, *The 101st Annual Meeting of Pharmaceutical Society of Japan, Kumamoto* Abstract 1981, 132.
80. S. Kamada, M. Maeda, and A. Tsuji, *J. Chromatogr.*, **272**: 29, 1983.
81. H. Asukabe, T. Sasaki, K. I. Harada, M. Suzuki, and H. Oka, *J. Chromatogr.*, **295**: 435, 1984.
82. H. Asukabe, H. Yoneyama, Y. Mori, K. -I. Harada, M. Suzuki, and H. Oka, *J. Chromatogr.*, **396**: 261, 1987.
83. H. Asukabe, H. Murata, K. -I. Harada, M. Suzuki, H. Oka, and Y. Ikai, *J. Chromatogr.*, **657**: 349, 1993.
84. H. Asukabe, H. Murata, K. -I. Harada, M. Suzuki, H. Oka, and Y. Ikai, *J. Agric. Food Chem.*, **42**: 112, 1994.
85. W. H. Williams, G. D. Ash, and M. A. Heady, *Analyst*, **114**: 887, 1989.
86. N. A. Johnson, *J. Assoc. Off. Anal. Chem.*, **72**: 235, 1989.
87. S. Auboiron, D. Bauchart, and L. David, *J. Chromatogr.*, **547**: 411, 1991.
88. D. Tikhova and V. Peneva, *Vet.–Med. Nauki.*, **19**: 52, 1982.
89. C. M. White, D. R. Gere, D. Boyer, F. Pacholec, and L. K. Wong, *J. High Resolut. Chromatogr. Chromatogr. Commun.*, **11**: 94, 1988.
90. R. L. Brown, *J. Assoc. Off. Anal. Chem.*, **62**: 266, 1979.
91. T. Kono and S. Yamamoto, *Shiryo Kenkyu Hokoku (Tokyo Hishiryo Kensasho)*, **5**: 174, 1979.
92. G. Marten and W. Hans, *Ger. Offen.*, DE 3,318,597, 1983.
93. A. Kozak, *Bromatol. Chem. Toksykol.*, **17**: 321, 1984.
94. T. Oyama, *Shiryo Kenkyu Hokoku (Tokyo Hishiryo Kensasho)*, **11**: 209, 1986.
95. T. Suhara, *Shiryo Kenkyu Hokoku (Tokyo Hishiryo Kensasho)*, **11**: 214, 1986.
96. M. Illing and B. Mueller, *Monatsh. Veterinaermed.*, **41**: 854, 1986.
97. Y. Suzuki, N. Arai, and M. Okamoto, *Anal. Sci.*, **3**: 441, 1987.
98. D. C. Koufidis, *Chem. Chron.*, **5**: 239, 1976.
99. I. Karkocha, *Rocz. Panstw. Zakl. Hig.*, **34**: 281, 1983.
100. K. Kovacs-Hadady and K. Kupas, *Acta Vet. Hung.*, **32**: 97, 1984.

101. I. Karkocha, *Rocz. Panstw. Zakl. Hig.*, **36**: 310, 1985.
102. M. A. Brooks, N. Strojny, M. R. Hackman, and J. A. De Silva, *J. Chromatogr.*, **229**: 167, 1982.
103. G. P. Dimenna, J. A. Creegan, L. B. Turnbull, and G. J. Wright, *J. Agric. Food Chem.*, **34**: 805, 1986.

Chapter 6

Chemical Analysis of Macrolide Antibiotics

Masakazu Horie
Saitama Prefectural Institute of Public Health 639-1, Kamiokubo, Urawa, Saitama 338 (Japan)

Contents

Introduction

Macrolide antibiotics are a very important class of antibacterial compounds widely used in medical and veterinary practices. Erythromycin, isolated in 1952 from *Streptomyces erythraeus*, was the first macrolide antibiotic utilized in medicine (*1*). The chemical characteristics of erythromycin and of other related antibiotics such as carbomycin, kitasamycin (leucomycin), spiramycin, and oleandomycin all show a common feature, a macrocyclic lactone ring attached to amino sugar(s) as a basic structure. Therefore, in 1957, Woodward (*2*) proposed the term "macrolide" to describe this class of compounds.

Macrolide antibiotics are considered to be medium-spectrum antibiotics. They are highly active against a wide range of Gram-positive bacteria such as *Mycoplasma* and *Chlamydia*, but have only limited activity against most Gram-negative bacteria. The macrolides are the most effective medicine against diseases produced by *Mycoplasmas* (*3*). These drugs are well absorbed after oral administration and are distributed extensively in tissues, especially in the lungs, the liver, and kidneys, with high tissue/plasma ratios (*4*). In addition, although the macrolides generally exhibit low toxicity (*4*), carbomycin is no longer available because of its strong toxicity (*5*). These characteristics make the macrolide antibiotics suitable for the treatment of a wide range of infections. Therefore, macrolide antibiotics such as tylosin, tilmicosin, spiramycin, erythromycin, kitasamycin, oleandomycin, josamycin, mirosamicin, and sedecamycin have been used in medical and veterinary fields. Of these compounds, mirosamicin and sedecamycin are mainly used in Japan, and with tylosin, are used exclusively in veterinary medicine.

Concerns about food hygiene have arisen regarding the presence of drug residues in livestock products (*6*, *7*). Antibiotic residues may have direct toxic effects on consumers (e.g., allergic reactions) or they may indirectly cause problems through the induction of resistant strains of bacteria. In Japan, according to the Food Sanitation Law, no food should contain antibiotics. On the other hand, in the United States, as in many other countries, tolerance limits for macrolide antibiotics (*8*) in livestock products have been set as shown in Table 1. Therefore, simple and reliable analytical methods are required to monitor these drug residues in edible tissues of livestock animals.

The determination of antibiotics, including the macrolides, is carried out predominantly by microbiological assays (*9, 10*). These assays excel as a qualitative means by which samples may be screened for residual amounts of antibacterial substances. However, they tend to lack specificity, and it is difficult to confirm what kinds of drugs remain in the animal tissues. In addition, microbiological assays cannot distinguish among the various components in the macrolide antibiotics, such as tylosin and spiramycin.

In order to overcome these problems, chemical methods such as gas chromatography (GC) and high-performance liquid chromatography (LC) have been used to determine the macrolide antibiotics. Many chemical methods have been developed to determine macrolides in drug formulations, biological fluids and livestock products. In this chapter, chemical analysis methods for the above-mentioned nine macrolide antibiotics are described.

Chemical and Biological Properties

Macrolide antibiotics are produced by strains of *Streptomyces* or *Micromonospora*. Antibiotics obtained from these macrolide-producing organisms commonly consist of several homologous components.

A characteristic structural feature of macrolide antibiotics is a large lactone ring that may contain isolated or conjugated double bonds. The molecules are built from a 12-, 14-, 16- or 17-macrocyclic lactone ring attached to amino sugars as

Table 1 Tolerances for Residues of Macrolide Antibiotics in Food

	Tolerance levels (ppm)				
Drug	Cattle	Swine	Poultry	Milk	Eggs
Carbomycin	—	—	0 Et	—	—
Erythromycin	0 Et	0.1 Et	0.125 Et	0	0.025
Oleando-mycin	—	0.15 Et	0.15 Et	—	—
Tylosin	0.2	0.2	0.2	0.05	0.2

Et: edible tissue

the basic structure. At one time, macrolide antibiotics were classified by ultraviolet absorption properties (*11*), but now they are classified according to the size of the lactone ring of the aglycone as 12-, 14-, 16- or 17-membered ring macrolides. However, 12-membered ring macrolides are not used in veterinary practices.

Macrolide antibiotics are unstable in acidic solutions, with almost all macrolide antibiotics being converted by acid to degradation products at pH <4.5. Also, the antibacterial activities of macrolides are extremely affected by the pH value of the assay medium.

There are a number of reviews concerning the physico-chemical and biological properties of macrolide antibiotics (*12*, *13*).

Fourteen-Membered Macrolides

Erythromycin (Figure 1) is currently one of the most important macrolide antibiotics in medical and veterinary practices. It is a mixture of three closely related compounds, namely erythromycin A, B and C; A is the major and most important component. On the other hand, oleandomycin (Figure 1) consists of a single component. Erythromycin is the most effective against various Gram-positive bacteria. Oleandomycin shows an antibacterial spectrum similar to that of erythromycin, but its antibacterial activities are weaker than erythromycin. Erythromycin and oleandomycin do not have a specific ultraviolet (UV) chromophore.

Sixteen-Membered Macrolides

Spiramycin, kitasamycin, josamycin, tylosin, tilmicosin, and mirosamicin belong to the class of sixteen-membered macrolides. Sixteen-membered ring macrolide antibiotics are produced mainly as complex mixtures of related components. Spiramycin consists of three components, spiramycin I, II, and III (Figure 2); spiramycin I is the main component. The antimicrobial spectrum of spiramycin is similar to that of other macrolide antibiotics but its activity against *Treponema* is superior to erythromycin and oleandomycin.

Kitasamycin consists of several components, leucomycin A_1, $A_{3\text{-}9}$ and A_{13}, with leucomycin A_5 as the major component (Figure 2). The structural differences in the leucomycins are as follows: leucomycin molecules of the Fr-group, which consists of the components A_1, A_5, A_7, and A_9, have an OH in the 3-position of the macrocyclic lactone ring, whereas those of the Ac-group (A_3, A_4, A_6, and A_8) have an O-acetyl group instead of the OH group. In general, the antibacterial activity of the Fr-group is higher than that of the Ac-group, although the antimycoplasma activity of the Fr-group is lower than that of the Ac-group (*14*).

The antibacterial activity of known macrolides differs from their antimycoplasma activity because of their macrolide structure. This is assumed to be due to the difference in membrane structure between the two kinds of microorganisms; mycoplasma have no cell wall and are enveloped with only a cell membrane.

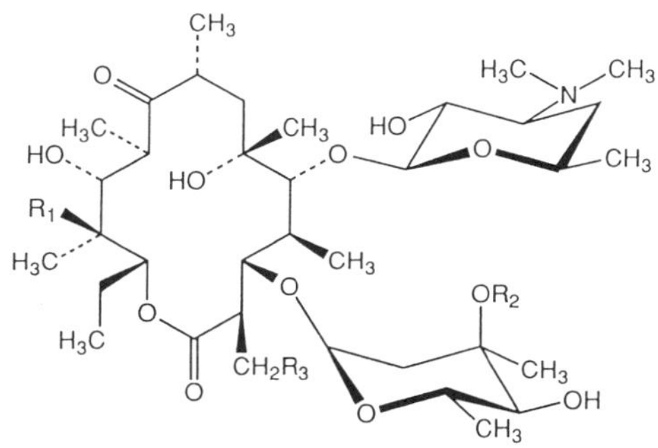

	R_1	R_2	R_3
Erythromycin A	-OH	$-CH_3$	-H
Erythromycin B	-H	$-CH_3$	-H
Erythromycin C	-OH	-H	-H

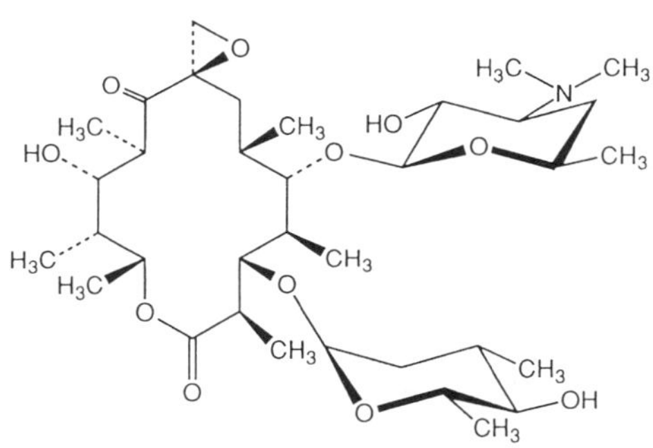

Oleandomycin

Figure 1 Chemical structures of erythromycin and oleandomycin

Josamycin was discovered by Umezawa et al. (*15*) in 1967, and is identical to leucomycin A3. Spiramycin, kitasamycin, and josamycin show strong UV absorption peaks around 232 nm in 30% acetonitrile.

Tylosin consists of four components, tylosin A, B, C, and D (Figure 3); tylosin A is the major and most important component. This drug is widely used both as a feed additive for growth promotion and a therapy in many countries. Tylosin is the most effective antibacterial agent against *Mycoplasma* among the macrolide

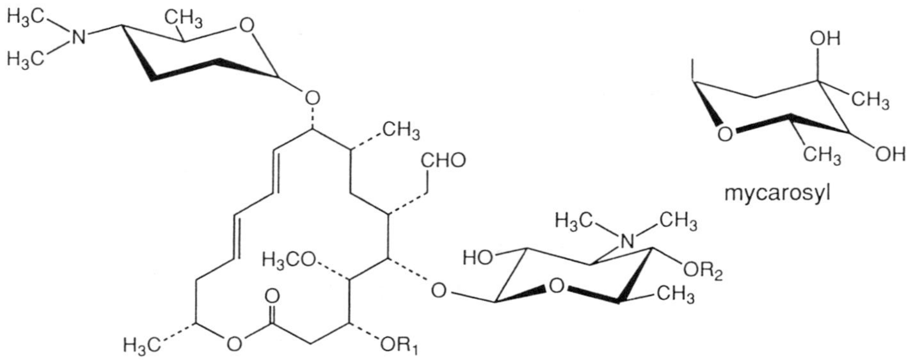

	R_1	R_2
Spiramycin I	-H	-mycarosyl
Spiramycin II	$-COCH_3$	-mycarosyl
Spiramycin III	$-COCH_2CH_3$	-mycarosyl
Neospiramycin I	-H	-H
Neospiramycin II	$-COCH_3$	-H
Neospiramycin III	$-COCH_2CH_3$	-H

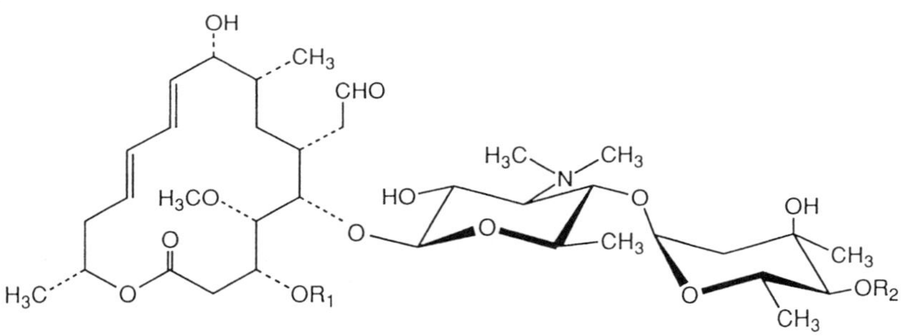

	R_1	R_2
Leucomycin A_1	-H	$-COCH_2CH(CH_3)_2$
Leucomycin A_3 (Josamycin)	$-COCH_3$	$-COCH_2CH(CH_3)_2$
Leucomycin A_4	$-COCH_3$	$-COCH_2CH_2CH_3$
Leucomycin A_5	-H	$-COCH_2CH_2CH_3$
Leucomycin A_6	$-COCH_3$	$-COCH_2CH_3$
Leucomycin A_7	-H	$-COCH_2CH_3$
Leucomycin A_8	$-COCH_3$	$-COCH_3$
Leucomycin A_9	-H	$-COCH_3$
Leucomycin A_{13}	-H	$-COCH_2CH_2CH_2CH_2CH_3$

Figure 2 Chemical structures of spiramycins and leucomycins

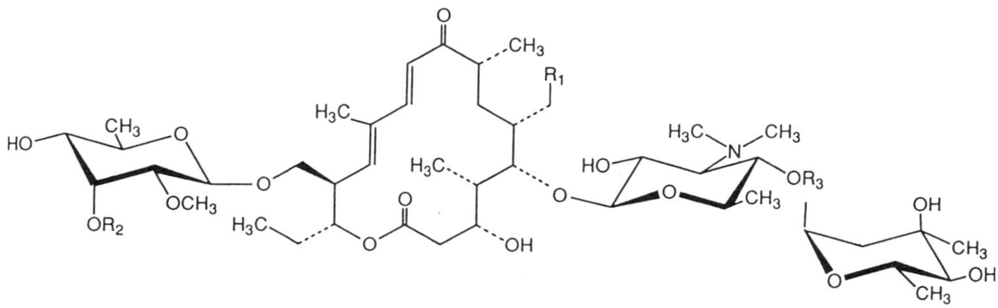

	R1	R2	R3
Tylosin A (Tylosin)	-CHO	$-CH_3$	-mycarosyl
Tylosin B (Desmycosin)	-CHO	$-CH_3$	-H
Tylosin C (Macrocin)	-CHO	-H	-mycarosyl
Tylosin D (Relomycin)	$-CH_2OH$	$-CH_3$	-mycarosyl

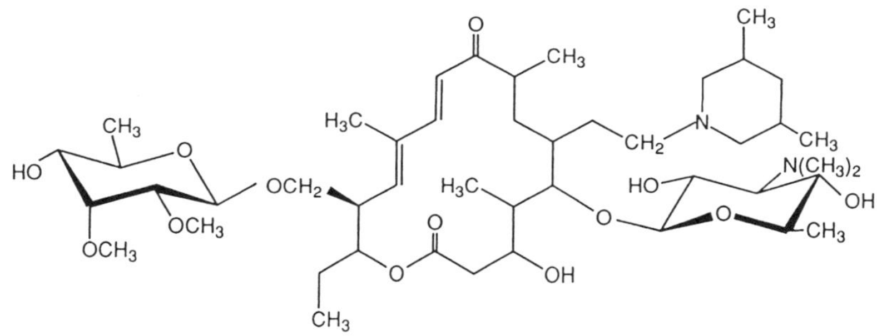

Tilmicosin

Figure 3 Chemical structures of tylosins and tilmicosin

antibiotics exhibiting a strong UV absorption peak at 282 nm. Aqueous solutions of tylosin exhibit a high degree of stability over pH 5.5–7.5. However, below pH 4 or above 9, degradation is observed. For example, in solution at pH <4, tylosin A is converted by acid degradation to tylosin B (desmycosin). An important feature of tylosin A is that the degradation product, tylosin B, exhibits physicochemical and biological properties that are similar to those of tylosin A.

Tilmicosin (Figure 3) is a new macrolide antibiotic recently approved for use in cattle in the United States and Canada. The drug is a new semisynthetic antibiotic prepared by reductive amination of the C-20 aldehyde group of desmycosin (*16*). Tilmicosin consists of approximately 85% *cis* isomer to 15% *trans* isomer. The in vitro antibacterial activity of tilmicosin is equivalent to desmycosin, but the in vivo activity is superior to desmycosin (*17*). In addition, the antibacterial activity of *cis* and *trans* isomers is not markedly different.

Mirosamicin (mycinamicin II, Figure 4), the principal of the five components of mycinamycins was isolated from the culture broth of *Micromonospora* in 1978 (*18*). This drug is highly active against a wide range of Gram-positive bacteria, certain Gram-negative bacteria, and *Mycoplasma*. The drug is well absorbed after oral administration and is distributed extensively in tissues, especially the lungs (*18, 19*). Mirosamicin has been used solely in livestock farming since 1988 in Japan. It exhibits a strong UV absorption peak at 218 nm in 30% acetonitrile.

Seventeen-Membered Macrolides

Sedecamycin (lankacidin A, Figure 4) is one of the lankacidin group of antibiotics produced by *Streptomyces rochei* (*20*). In general, the molecules of macrolide antibiotics have amino sugars as the basic moiety, however, sedecamycin does not contain amino sugar moieties. The drug is a neutral macrolide antibiotic. It possesses preeminent activity against *Treponema hyodysenteriae*, the pathogenic organism for swine dysentery, and received approval in Japan in 1985 for treating this disease. Sedecamycin exhibits a strong UV absorption peak at 226 nm.

Thin-Layer Chromatography

Thin-Layer Chromatography—Chemical Detection

The thin-layer chromatographic (TLC) technique is inherently simple, relatively inexpensive, and requires little time. It has a distinct advantage over other chromatographic methods, the possibility of analyzing many samples simultaneously rather than serially. Therefore, TLC methods are frequently used for separating and confirming antibiotics. The TLC methods (*21–31*) have been used for detecting macrolide antibiotics, mainly erythromycin, in bulk drug and pharmaceutical preparations. However, these methods have a limited application and are generally used only for screening or qualitative analysis (Table 2). The most widely used stationary phases are commercially available silica gel or cellulose plates, with various solvents or solvent mixtures used as the mobile phase.

The macrolide antibiotics are first separated by TLC and are then subsequently detected by visual color after spraying with a specific reagent, a number of which have been used for detection (*23*), the most useful being dilute sulfuric acid. TLC is rapid and simple, however, for residue analysis, the technique lacks accuracy, reproducibility, and sensitivity. Therefore, TLC is not considered suitable for most routine residue analyses, and it has rarely been applied to the analysis of macrolide antibiotic residues.

High-performance thin-layer chromatography (TLC) is occasionally used for the separation and determination of some macrolides. Generally, reversed-phase TLC plates (C8 or C18) have been used for this technique. The mobile phases used

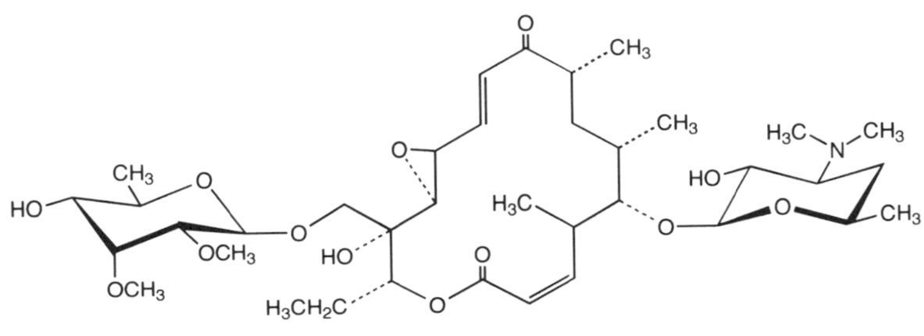

Mirosamicin

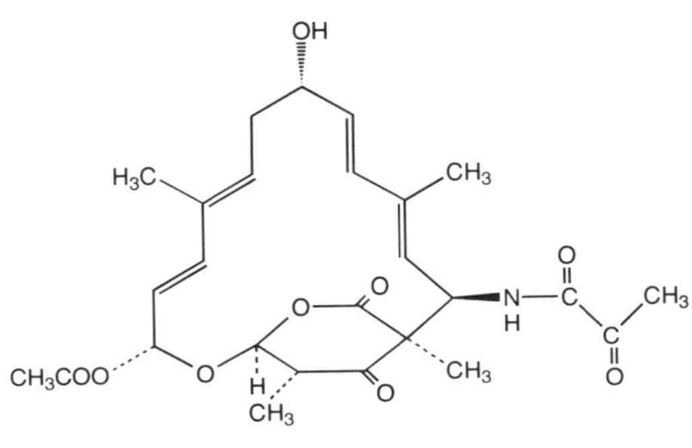

Sedecamycin

Figure 4 Chemical structures of mirosamicin and sedecamycin

Table 2 Analytical Thin-Layer Chromatography Methods of Macrolides

Macrolides	Sample	Extraction and cleanup	Analytical conditions	Detection	Limit	Ref.
Erythromycin, esters, salts	Drug	—	Silica gel $CHCl_3$-MeOH-CH_3COOH (90:10:1)	Spray reagent	50 μg/spot	*(21)*
Erythromycins	Drug	—	Silica gel 4 eluents	Spray reagent	10 μg/spot	*(22)*
EM, OM, SPM, JM, KT, others	Drug	—	Silica gel 8 eluents	Spray reagent		*(23)*
Erythromycins	Drug	CH_2Cl_2	Silica gel 4 eluents	Spray reagent	0.5 μg/spot	*(24)*
SPM, TS, others	Drug	—	Silica gel TLC, low layer of CH_2Cl_2-PrOH-NH_4OH (93.5:5:1.5)	Densitometry (282, 232 nm)		*(25)*
Tylsoin	Liver, milk	Deproteinized with $ZnSO_4$-$Ba(OH)_2$, liquid–liquid partition	Silica gel two-dimensional chromatography	UV (254 nm) spray reagent	1–2 ppm	*(26)*
Erythromycin, metabolites	Rat tissues	Acetonitrile, liquid–liquid partition	Silica gel 3 eluents	Spray reagent	—	*(27)*

Table 2 Analytical Thin-Layer Chromatography Methods of Macrolides—*continued*

Macrolides	Sample	Extraction and cleanup	Analytical conditions	Detection	Limit	Ref.
EM, TS, OM, SPM	Tissues, milk, eggs	Acetonitrile, liquid–liquid partition	Silica gel, *n*-BuOH-water-CH_3COOH (6:2:2)	Derivatization, densitometry (525nm)	0.02–0.5ppm	*(28)*
EM, OM, TS	Animal tissues	Methanol, liquid–liquid partition	Silica gel, $CHCl_3$-MeOH-acetone-glycerine (49:30:20:1)	Bioautography (*B. subtilis*)[a]	1–20 ppm	*(29)*
EM, OM, TS	Animal tissues	Methanol, liquid—liquid partition	Silica gel, $CHCl_3$-MeOH-acetone-glycerine (49:30:20:1)	Bioautography (*B. subtilis*)	0.1–0.5 ppm	*(30)*
EM, OM, TS, SPM, KT	Meat	Methanol, Amberlite XAD-2	Silica gel, cellulose, water saturated BuOH, EtOH-water (4:1)	Bioautography (*M. luteus*)[b]	0.05 ppm	*(31)*

EM: erythromycin; OM: oleandomycin; SPM: spiramycin; JM: josamycin; KT: kitasamycin; TS: tylosin
[a] Bacillus subtilis ATCC 6633.
[b] Micococcus luteus ATCC 9341.

with these plates are mainly two-component mixtures of water (or a buffer) and an organic modifier, such as methanol, acetonitrile, tetrahydrofuran, or dioxane.

It has been demonstrated (*25*) that the sixteen-membered ring macrolide antibiotics such as spiramycin and tylosin may be analyzed using either silica gel TLC plates or TLC reversed-phase C18 plates with detection by their UV maximum absorptions (282 nm for tylosin, 232 nm for spiramycin). Results obtained by this method for the quantitative determination of spiramycin components demonstrate that spiramycin preparations contain 50.4% spiramycin I, 22.6% spiramycin II, and 19.0% spiramycin III. These quantitative data agreed closely with those obtained by liquid chromatographic (LC) methods (*32*, *33*).

TLC is an excellent tool for analyzing macrolide antibiotics in fermentation broths and preparations. However, the sensitivity and the reproducibility of the method are inferior when compared with LC methods. Consequently, TLC methods are seldom applied in the residue analysis of macrolide antibiotics. Nevertheless, the possibility of developing LC systems based on thin-layer chromatographic results has been explored many times.

Thin-Layer Chromatography–Bioautography

Slaughter animals are tested for antibiotic residues in Japanese, U.S., and Canadian abattoirs by microbiological assays. Although the microbiological assays afford adequate sensitivity to the most commonly used macrolides, they lack specificity, and it is difficult to confirm what kinds of drugs remain in the animal tissues. In order to overcome these problems, thin-layer chromatographic–bioautographic (TLCBA) methods have been used to identify and determine macrolides (Table 2). TLCBA methods are comprised of a blend of physicochemical and biological techniques in which developed TLC plates are placed on a bacterial growth medium, and after incubation, the locations of the inhibition zones are used to identify antibiotics. The major drawbacks of this technique are that it is tedious and time-consuming, therefore, its application to residue analysis is limited.

Gas Chromatography

Gas chromatography (GC) possesses the requisite chromatographic resolution to separate closely related antibiotics. However, almost all antibiotics, including macrolide antibiotics, are not sufficiently volatile to be gas chromatographed without preliminary treatment such as derivatization. Therefore, a limited number of publications involving the use of GC for the analysis of macrolides have appeared.

For example, GC methodologies with very good accuracy, recovery, and precision for the analysis of erythromycin preparations have been reported (*36*, *37*). These methods involve the silylation of erythromycin with a combination of *N,O*-bis-(trimethylsilyl)-acetamide, *N*-trimethylsilylimazole, and trimethyl-chlorosilane in pyridine, followed by GC on a 3% OV-225 column at 275°C and using a flame-ionization detector (FID).

Recently a GC–MS method using single-ion monitoring (SIM) for determining erythromycin in beef and pork has been developed (*38*). The drug is extracted from meat with methanol and cleaned by *n*-hexane washing, followed by liquid-liquid partition. The drug is then acid-hydrolyzed and subsequently acetylated in an acetic anhydride–pyridine mixture. The recoveries for 1 and 0.1 ppm are 86 and 72%, respectively, and the detection limit is 0.01 ppm. A GC–MS method using SIM for determining of oleandomycin, kitasamycin, tylosin, and spiramycin has also been reported (*39*). In these studies, only the major component was analyzed for each antibiotic and the minor components were not mentioned.

The GC and GC–MS methods have excellent specificity, but the added steps necessary for derivatization are problematic for samples where a rapid turnaround time is required. Consequently, the GC and GC–MS methods are not always suitable for routine analysis. In recent years, methods based on LC have received much interest.

High-Performance Liquid Chromatography

LC is a powerful analytical tool that was developed during the 1960s and technologically improved during the 1970s and 1980s. The emergence of very small uniform particle packing materials with different physicochemical properties, sophisticated high-performance pumps, and various excellent flow detectors have made this technique extremely versatile. At present, LC has also become the most widely used technique for determining antibiotics, including macrolide antibiotics in drug preparations and biological samples.

The choice of adequate conditions for an LC procedure is governed by the ionizable groups, amino sugar(s), of the macrolide antibiotics. Silica-based C18 or C8 columns are overwhelmingly used with mixtures of buffer and acetonitrile (or methanol) as a mobile phase. The organic modifier content, the buffer concentration, and the pH of the buffer in the mobile phase are selected to provide sufficient retention of the ionizable compounds. In LC of macrolide antibiotics, the most popular detector is a UV spectrophotometer. Some macrolides such as tylosin, spiramycin, and josamycin have relatively strong UV absorption, so that UV detection is sufficiently sensitive in the determination of these macrolides. In the UV–LC analysis of chemicals, it is extremely important that an appropriate detection wavelength be selected. Without any interference from contaminants, the maximum absorption wavelength of the compound to be analyzed is generally employed as the wavelength of detection. However, there are exceptions such as erythromycin and oleandomycin. These compounds show extremely weak absorption in the ultraviolet region so therefore, UV detection cannot be used for determining therapeutic and residual concentrations in biological samples. More sensitive electrochemical methods and fluorometric methods have been described for determining erythromycin and other macrolides in pharmaceutical preparations and biological fluids. Some assays for these macrolide antibiotics are summarized in Table 3.

Spiramycin

Spiramycin (SPM) primarily consists of three closely related substances; spiramycin I, II, and III as shown in Figure 2, with three homologous subcomponents, neospiramycin (NSPM) I, II, and III, with other related, but minor, compounds also present. Therefore, a suitable LC method is needed to provide for the determination of individual components as a quality control measure in pharmaceutical preparations. Several LC methods have been developed to determine the components of spiramycin in bulk powders and pharmaceutical preparations (*32–35*). In this section, the method for setting the optimum LC conditions of spiramycin is described.

Isolation and determination of components. Spiramycin consists of several homologous components. It is, of course, possible to obtain high purity standards for antibiotics that are single substances, but in the case of SPM and other multicomponent biosynthetic substances it is difficult to procure pure standards for the individual components. This is where LC can be used to isolate and purify each component. LC has been applied to resolve SPM into its individual components with mass spectrometry (MS) used to positively identify them (*40*). Figure 5 shows the mass spectra.

Molecular ions for every component of SPM could not be obtained from either electron impact (EI) or chemical ionization (CI) methods, so secondary-ion mass spectrometry (SIMS) was employed as a mild and effective ionization method that yielded the molecular ions necessary for observing the molecular weights. The molecular weight values found for the three main components of SPM were 842, 884, and 898, thus confirming the identity of SPM-I, SPM-II, and SPM-III, respectively. NSPM-I, -II, and -III are a homologous series corresponding to SPM-I, -II, and -III, respectively, after removing the mycarose sugar moiety from SPM by acid hydrolysis. The identity of these components was confirmed also by MS, and the complete six component standard for SPM was obtained.

Chromatographic conditions. Mourot et al. (*33*) separated SPM-I, -II, and -III under reversed-phase conditions on a LiChrosorb RP-8 column, but they did not mention the NSPM sub-components. They employed a 74:26 mix of 1% sulfuric acid and acetonitrile as the mobile phase (pH of approximately 1.4), which,

Table 3 High-Performance Liquid Chromatographic Methods for the Analysis of Macrolide Antibiotics

Drugs	Sample	Extraction and cleanup	Analytical conditions	Limit (ppm)	Ref.
Spiramycin	Chicken	Extraction with MeCN, liquid–liquid partition	UV (231 nm), Zorbax BP-C8 (25 cm), MeOH-0.4% H_3PO_4 (7:3) containing 0.2% heptanesulfonate, 1.0 mL/min	0.05	*(82)*
Spiramycin	Meat	0.2% metaphosphoric acid-MeOH (7:3), liquid–liquid partition	UV (232 nm), Nucleosil 5C18 (25 cm), 0.05 M NaH_2PO_4 (pH 2.5)-CH_3CN (72:28)	0.05	*(40)*
Tylosin	Cattle tissues	dichloromethane, SPE with Baker Silica gel (500 mg)	UV (280 nm), Spherisorb C18 (25 cm), 5 mM $NH_4H_2PO_4$-MeCN-MeOH (8:72:20)	0.02	*(93)*
Tylosin	Meat	0.2% metaphosphoric acid-MeOH (7:3), liquid–liquid partition	UV (287 nm), Inertsil ODS (25 cm), 0.05 M NaH_2PO_4 (pH2.5)-CH_3CN (65:35)	0.05	*(42)*

Table 3 High-Performance Liquid Chromatographic Methods for the Analysis of Macrolide Antibiotics—*continued*

Drugs	Sample	Extraction and cleanup	Analytical conditions	Limit (ppm)	Ref.
Tylosin	Swine	MeCN, liquid–liquid partition	UV (280 nm), Varian Micropak MCH-10-N-Cap (30 cm), 5 mM $NH_4H_2PO_4$-MeCN-MeOH (1:6:3)	<0.1	*(94)*
Tilmicosin Tylosin	Cattle, swine tissues	Extraction with MeCN, SPE with C18 cartridge	UV (287 nm), Inertsil ODS (25 cm), 0.1 M NH_4COOH (pH 5)-CH_3CN-CH_3OH (60:30:10)	0.01 0.02	*(53)*
Mirosamicin	Animal tissues	0.3% metaphosphoric acid-MeOH (7:3), SPE with Bond Elut SCX	UV (230 nm), Puresil 5C18 (15 cm), 0.05 M NaH_2PO_4 (pH 2.5)-CH_3CN (70:30)	0.05	*(81)*
Sedecamycin	Swine tissues	Ethyl acetate, silica gel-Florisil double-layered column chromatography	UV (227 nm), µ-Porasil/*n*-hexane-*i*-PrOH (80:20), µ-Bondapak C_{18}/0.01 M Phosphate buffer (pH 8.2)-MeCN (60:40)	0.05	*(73, 74)*

Table 3 High-Performance Liquid Chromatographic Methods for the Analysis of Macrolide Antibiotics—*continued*

Drugs	Sample	Extraction and cleanup	Analytical conditions	Limit (ppm)	Ref.
Erythromycin, Oleandomycin	Serum urine	MeCN, SPE with Baker C_{18}	UV (200 nm), Novapak C_{18} (15 cm), 0.05 M phosphate buffer-MeCN (7:3), pH 6.3	0.25–1.0	*(84)*
Erythromycin	Serum	Ether, postcolumn derivatization (naphthotriazole disulfonate)	FL (360/440 nm), LiChrosorb RP-18 (25 cm) MeCN-0.2 M ammonium acetate-water (60:10:30), pH 7.0	0.01	*(68)*
Erythromycin	Plasma, urine	Ether (plasma, MeCN(urine)	ECD (0.7/0.9 V), μBondapak C18 (30 cm), MeCN-MeOH-0.2 M Na acetate (40:5:55), column temp. 70°C	0.01–0.03	*(69)*
Erythromycin	Plasma	Diethylether	ECD (0.9 V), Novapak C18, MeCN-MeOH-56 mM sodium acetate (50:4:56), pH 7.0	0.01	*(71)*

Table 3 High-Performance Liquid Chromatographic Methods for the Analysis of Macrolide Antibiotics—*continued*

Drugs	Sample	Extraction and cleanup	Analytical conditions	Limit (ppm)	Ref.
Erythromycin, Oleandomycin (IS)	Plasma	Methyl *t*-butyl ether	ECD (1.0–1.1 V), PRP-1 column (15 cm), 0.05 M phosphate buffer (pH 9.5)-MeCN (60:40)	0.1	*(85)*
Erythromycin, Oleandomycin (IS)	Serum	MeCN, SPE with Bond Elut C18	ECD (0.7/0.9 V), C18 column (25 cm), 0.05 M phosphate buffer (pH 6.3)-MeCN (35:65)	0.1	*(86)*
Erythromycin	Serum, plasma	Ether-isopentane (3:2), hexane-octanol (95:5, defat)	ECD (0.6/0.75 V), C8 column (25 cm), 15 mM phosphate buffer (pH 6.7)-MeCN (60:40)	0.01	*(87)*
Erythromycin, Oleandomycin (IS)	Plasma	Methyl *t*-butyl ether	ECD (0.65/0.85 V), μBondapak C18 (25 cm), 0.1 M acetate buffer (pH 5)-MeCN-MeOH (48:42:10)	<0.25	*(88)*

Table 3 High-Performance Liquid Chromatographic Methods for the Analysis of Macrolide Antibiotics—*continued*

Drugs	Sample	Extraction and cleanup	Analytical conditions	Limit (ppm)	Ref.
Erythromycin	Eggs, milk, tissue	Acetonitrile, dicloromethane partition, LC cleanup	FL (255/315), LiChrospher C18, end-capped (25 cm), phosphate buffer-acetonitrile	0.01	*(72)*
Oleandomycin, Erythromycin (IS)	Serum, urine	MeCN, SPE with Baker C18	UV (200 nm), Novapak C18 (15 cm), 0.05 M phosphate buffer (pH 6.3)-MeCN (70:30)	0.25-0.5	*(78)*
Josamycin	Plasma	Ion-pair (SDS) extraction with hexane-isoamyl alcohol (9:1), derivatization (dansylhydrazine)	FL (352/537 nm), Nucleosil C18 (15 cm), 0.05 M ammonium acetate-MeCN (72:28)	0.05	*(54)*
Josamycin	Serum, urine	Extraction with MeCN, solid-phase extraction with Bond Elut C18	UV (231 nm), C18 (10 µm, 25 cm), 0.015 M phosphate buffer (pH 6.0)-CH_3CN (2:5)	0.025	*(80, 89)*

Table 3 High-Performance Liquid Chromatographic Methods for the Analysis of Macrolide Antibiotics—*continued*

Drugs	Sample	Extraction and cleanup	Analytical conditions	Limit (ppm)	Ref.
Josamycin	Plasma	Column switching technique	UV (230 nm), Shandon Hypersil CPS (25 cm), 0.01 M Na acetate-CH_3CN (60:40)	0.025	*(79)*
Spiramycin I	Plasma	Column switching technique	UV (230 nm), Nucleosil 5C8 (15 cm), MeCN-0.2% perchloric acid (26:74)	0.05	*(90)*
Spiramycin	Plasma	SPE with Sep-Pak C18	UV (231 nm), RiChrosphere RP-18e (12.5 cm), 0.15 M NaH_2PO_4-MeCN (77:23)	0.05	*(91)*
Spiramycin	Bovine plasma	MeCN, C_2 cartridge (PrepStation)	UV (231 nm), LiChrosphere RP-8 (125 mm), 0.5% sulfuric acid-MeCN (79:21), Comparison between LC and bioassay	<0.1	*(92)*

however, considerably speeds column deterioration and would be impractical for conducting routine analyses.

Similar to a number of other macrolide antibiotics, SPM, which is a basic molecule containing an amino sugar in its structure, is strongly affected by silanol groups that remain in a column packing material. Therefore, an end-capped ODS column based on silica gel was selected. The choice of adequate conditions for an LC procedure for SPM is governed by the ionizable amino sugar group. Optimum conditions were sought by varying the mobile phase pH and the concentration ratio

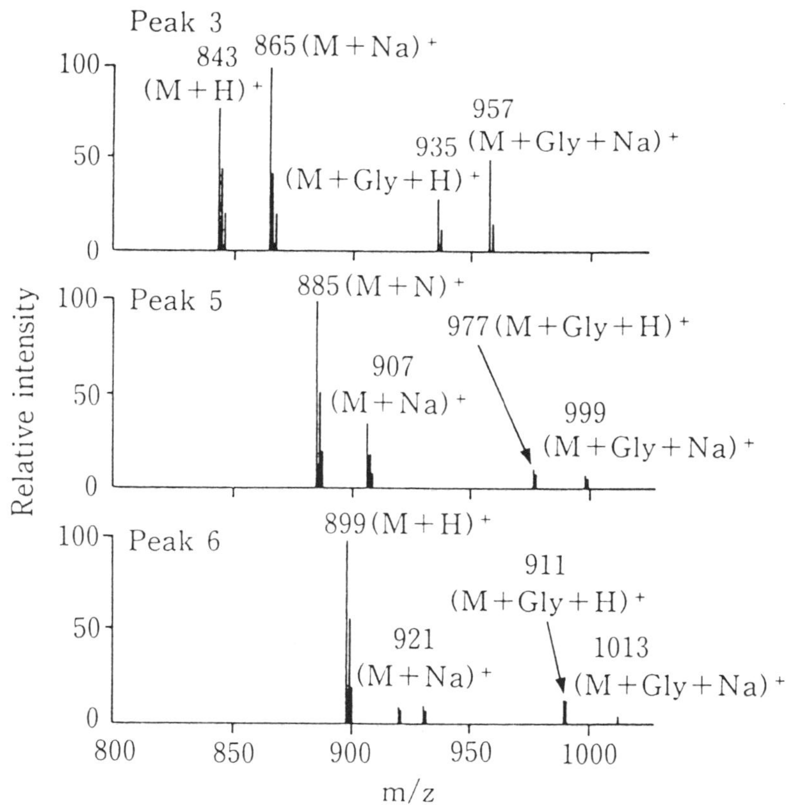

Figure 5 Secondary ion mass spectra of 3 peaks of authentic spiramycin separated by HPLC, as in Figure 8.

SOURCE: **Reproduced by permission of The Japan Society for Analytical Chemistry. See reference (*40*).**

of sodium phosphate buffer to acetonitrile in the mobile phase. As illustrated in Figure 6, the capacity factor k ′ decreased markedly for each component as the acetonitrile concentration rose. After considering the retention time and the resolution between components, an acetonitrile concentration of 27% was chosen (*34*).

Even at this mixture ratio, the simple aqueous acetonitrile mobile phase has a pH of approximately 7, which, it is thought, would dissociate any residual silanols to weakly cationic species that could strongly retain the basic SPMs. To prevent this, the mobile phase pH was adjusted to be acidic by the addition of 0.1 M phosphoric acid. However, even under these conditions, if the mobile phase did not contain a salt buffer, the compounds could still exhibit relatively high capacity factors, as shown in Figure 7 where the effect of salt concentration is plotted. As a result of the above investigations, 0.05 M phosphate buffer (pH 2.5)–acetonitrile (73:27, v/v) was chosen as the mobile phase (*34*). Figure 8 shows typical chromatograms of the spiramycin preparations obtained under these conditions.

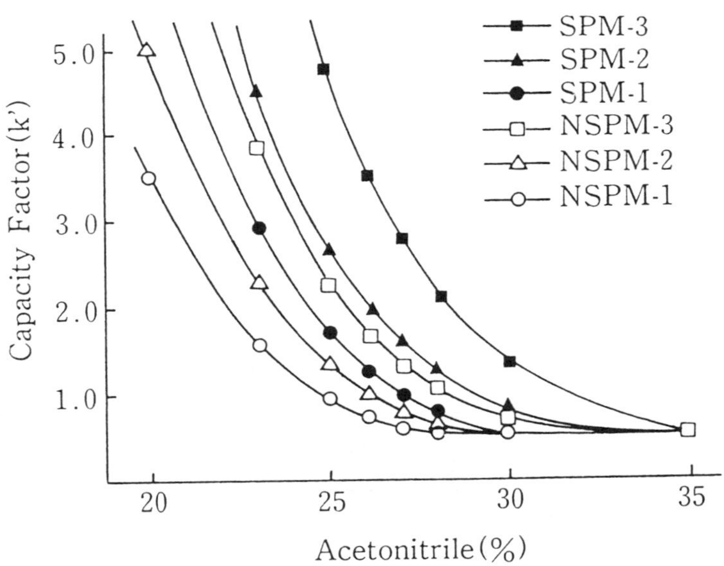

Figure 6 Effect of acetonitrile concentration in mobile phase on capacity factor (*k*′) of spiramycins. LC conditions: column Chemcosorb 5-ODS-H (250 × 4.6 mm i.d.); mobile phase, 0.05 M sodium phosphate buffer (pH 2.5)-acetonitrile (100-X:X); flow-rate, 0.5 mL/min; detection, 232 nm.

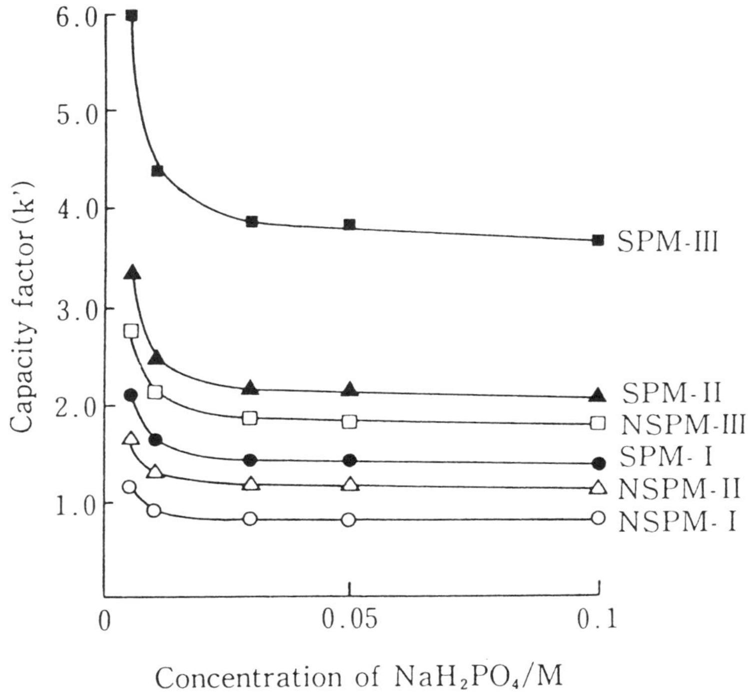

Figure 7 Effect of concentrations of sodium dihydrogenphosphate in mobile phase on capacity factor (k′) of spiramycins. LC conditions: column; Chemcosorb 5-ODS-H (250 × 4.6 mm), mobile phase; sodium dihydrogenphosphate (pH 4.0)-acetonitrile (73:27), flow rate; 0.5 mL/min, detection UV (232 nm). SOURCE: **Reproduced by permission of The Japan Society of Analytical Chemistry. See reference (*34*).**

Generally, macrolide antibiotics are not stable in acidic solutions. However, investigations of the stability of 5.0 µg/mL of spiramycin in the mobile phase at an ambient temperature (about 25°C) demonstrated that it is relatively stable under these conditions. Consequently, the use of an acidic LC mobile phase did not present a problem.

Quantitative methods. Omura and Nakagawa (*12*) reported on the spectrometric characteristics of SPM-I, -II, and -III, each of which had a maximum absorption at a wavelength of 232 nm, with similar molar extinction coefficients. The spectrometric characteristics of the three NSPM components, however, had not

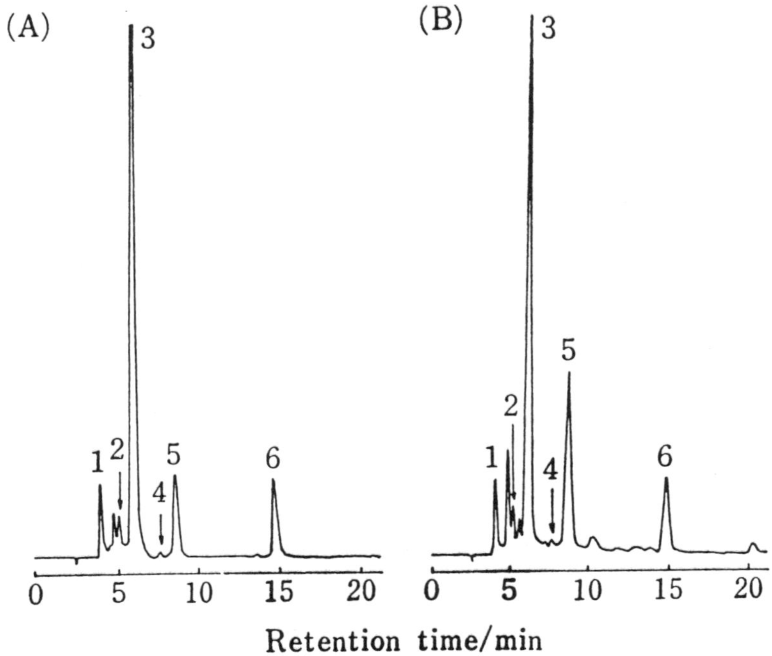

Figure 8 Typical chromatograms of (A) current reference preparations and (B) old reference preparations of spiramycin. LC conditions same as Figure 6, except 0.05M sodium dihydrogenphosphate (pH 2.5)-acetonitrile (73:27) in mobile phase.
Peaks: 1, neospiramycin I; 2, neospiramycin II; 3, spiramycin I; 4, neospiramycin III; 5, spiramycin II; 6, spiramycin III.
SOURCE: Reproduced by permission of the Japan Society for Analytical Chemistry. See reference (34).

been reported. The individual component standards were dissolved in a mobile phase and observed their spectra. The absorption maxima were 232 nm for all six components, and their molar absorptivities were nearly equal at about 28,000, thus confirming and extending the work of Omura and Nakagawa.

When the molecular weights of the six components are compared, the three NSPM components are about 15% smaller than the three corresponding parent compounds, but they account for only a very small fraction of the total. Therefore, it was decided that the differences in molecular weight could be disregarded and the six components were considered equal in molecular weight when a quantitative calibration method was chosen using the sum total of the individual peak areas on the chromatogram. The calibration was linear over 5–200 ng. The content of spiramycin I in several commercial and reference preparations by this quantitative method was about 70% (*34*).

By the old method of microbiological testing, it would have been difficult to measure the quantities of the individual components of such multicomponent antibiotics, but when chemical assay methods using LC are selected, each component can be quantitatively determined with speed, ease, and accuracy.

Tylosin and Tilmicosin

Chromatographic conditions. Tylosin consists of a mixture of several closely related compounds. It is necessary to separate these tylosins because they have different antibiotic activities (*41*, *42*). The British Veterinary Pharmacopoeia (*43*) specifies that LC be used in the analysis of the multicomponent antibiotic, tylosin, to provide for the determination of individual components as a quality control measure in pharmaceutical preparations. Therefore, suitable LC methods required tests that were capable of determining the content of tylosin A (the major component) and the levels of the other factors.

When selecting the optimum conditions for LC separation of the components of tylosin the considerations are similar to those for spiramycin. Moreover, the quantitative method for determining tylosin components is the same as that for spiramycin. That is to say, although tylosin consists of four major components (tylosin A, B, C, and D), their maximum molecular extinction coefficients of 287 nm are virtually identical. Therefore, calculating the contents of tylosin was done by a total peak area normalization of tylosin A, B, C, and D. This method, by which each component of tylosin can be separated and determined, makes it possible to evaluate the exact residue characteristics of tylosin when some part of tylosin A is decomposed by gastric acid and converted into tylosin B (*42*). This is an important consideration, because 90% or more of the pharmaceutical preparations of tylosin are administered orally.

A number of LC methods have been published for the quality control of tylosin. The first reversed-phase LC methods described did not provide separation of the different tylosins (*44*, *45*), while the methods reported by Kennedy (*46*, *47*) have poor selectivity for minor components that elute close to the main component. Fish and Carr (*48*) developed a method that is capable of separating and quantifying the factors present in tylosin raw material and dosage forms. The contents of tylosin A, B, C, and D are calculated by the total peak area normalization of all the peaks seen in the chromatogram, as discussed above.

Recently, Roets et al. (*49*) reported an excellent LC method for the quality control of tylosin A that shows good selectivity for the major components tylosin A, B, C, and D. The LC separation was carried out on different C18 or C8 columns using a mobile phase containing acetonitrile, 0.2 M tetra-*n*-butylammonium hydrogensulfate (TBA), 0.2 M phosphoric acid, and water. TBA masks the residual silanol activity and acts as a competing analyte, thereby improving both the separation efficiency and the peak symmetry. Roets et al. (*49*) found that tylosin

A comprises about 80–90% of commercial preparations. There are also several reports concerning the analysis of tylosin in drugs and fermentation broth (*50–52*).

Most recently, Chan et al. (*53*) developed a sensitive LC method for the simultaneous determination of tylosin and tilmicosin in muscle and kidney of slaughter animals. The LC separation was performed on an Inertsil C18 column (250 × 4.6 mm) with 0.1 M ammonium formate (pH 5.0)–acetonitrile–methanol (60:30:10) as the mobile phase. These drugs were detected at 287 nm, mean recovery from spiked tissues was 79.9% for tylosin and 92.2% for tilmicosin, and the detection limits were 0.02 ppm for tylosin and 0.01 ppm for tilmicosin. As the major component, tylosin A accounts for 80% or more of the total. Therefore, Chan et al. considered that the residue characteristics could be fully evaluated by monitoring the main component, tylosin A, alone.

In addition, because tilmicosin is a derivative of tylosin, the methods for tylosin discussed above could successfully be applied to tilmicosin.

Liquid chromatography and bioassay. Microbiological assays, which tend to lack specificity, are frequently found unsuitable in identifying residual antibiotics. However, these assays do excel as a qualitative screening method to cheaply and effectively detect any remaining amounts of such substances. Chicken muscle samples fortified with tylosin were analyzed by an LC method (*42*) and by a bioassay method in which *Micrococcus luteus* ATCC 9341 was used. (This test organism is frequently used to determine residual macrolide antibiotics in animal tissues.) The concentrations obtained with the two methods showed a linear correlation. The equation of the fitted curve was $y = 0.91x - 0.02$ ($n = 12$; $r = 0.99$) (*42*), demonstrating that it is reasonable to calculate the contents of tylosin by total peak area normalization of tylosin A, B, C and D's peaks observed on the chromatogram.

The LC method with UV detection has excellent selectivity and reproducibility. However, cross-referencing the method with a conventional bioassay method makes it possible to give more accurate analytical results.

Josamycin, Kitasamycin, and Mirosamicin

Several LC methods with UV detection have been reported for determining josamicin in serum and plasma (Table 3). There are also several reports on determinations using UV detection for kitasamycin (leucomycin) in drugs (*44*), but not in biological samples. No LC methods for determining mirosamicin in drugs and biological samples have been reported.

There is a recent report of a sensitive LC method with fluorescence detection (352/537 nm) in which josamycin is determined in plasma after ion-pair extraction into hexane-isoamyl alcohol with lauryl sulfate (SDS) and precolumn derivatization with dansyl hydrazine. Because dansyl hydrazine has previously been used to derivatize the aldehydes, the presence of an aldehyde group in josamycin led to the selection of dansyl hydrazine as the reagent for precolumn derivatization (*54*). The factors considered when selecting the optimum LC conditions for the analysis of josamycin, kitasamycin, and mirosamicin are almost the same as those for spiramycin.

Erythromycin and Oleandomycin.

Erythromycin base is rapidly degraded in gastric acid producing inactive anhydrous forms. To overcome this, erythromycin is usually administered as esters, which serve as prodrugs and are hydrolyzed gradually after absorption to the only antibacterially active form the free erythromycin base. Formerly, most pharmacokinetic studies have been employed using microbiological assays. However, the erythromycin esters are microbiologically inactive and are hydrolyzed to the free base during incubation, thus leading to an overestimation of the antimicrobial activity. Chemical methods are, therefore, required to simultaneously determine erythromycin's base and its esters. A number of LC methods using UV detection have been reported for determining erythromycin base and some erythromycin esters and their degradation products in fermentation broths, bulk powders, and pharmaceutical preparations (*55–67*), but applications to residue analysis have been limited. LC methods have demonstrated that pharmaceutical preparations contain 80–95% erythromycin A (*64*). A few reports have been published for the LC determination of oleandomycin in both drug and biological samples (Table 3).

Erythromycin and oleandomycin do not have a specific UV chromophore. In view of the low molar absorption of these antibiotics at conventional detection wavelengths (>215 nm), LC with UV detection cannot be considered sensitive enough for pharmacokinetic studies. I have not found any reference using UV–LC for determining erythromycin in animal tissues.

These antibiotics have been analyzed in biological samples using fluorescence or electrochemical detection (ECD) after derivatization. The method described by Tsuji (*68*) involves postcolumn derivatization with naphthotriazole disulfonate followed by an on-line extraction of the excess of reagent into chloroform and fluorescence detection. Naphthotriazole disulfonate forms an ion pair with a molecule protonated at a low pH. Erythromycin and its esters contain one tertiary amine in the desosamine moiety and can be readily protonated. The method has a potentially wide application for analyzing compounds containing various forms of amine groups, but requires complex instrumentation. The column is maintained at 70°C, which may cause the bonded phase material to dissolve. While fluorescence detection is more specific than UV detection, postcolumn derivatization requires a more complicated chromatographic system involving a pump for the derivatizing reagent and a reaction coil. Such methodology may only be suitable for routine applications if the LC system available includes current automated features.

In 1983, Chen and Chiou (*69*) developed a sensitive LC method with electrochemical detection (ECD) for determining erythromycin components and their potential metabolites or degradation products in both human plasma, where the detection limits for erythromycin A are 0.01–0.03 ppm, and urine. ECD has also been successful in analyzing trace elements of electroactive compounds. In erythromycin, the electrochemical response is probably due to the tertiary amine of the amino sugar in the molecule, which must be further deprotonated to be oxidized. Therefore, *N*-dimethyl-erythromycin proved to be undetectable (*70*, *71*). Electrochemical detection necessitates a mobile phase with a buffer in the low ionic range (10–100 mM) to provide adequate conductivity while keeping the contribution to background current to a minimum. Further, the use of an internal standard (e.g., oleandomycin) proved to be necessary because electrode sensitivity diminished slowly during sample analysis (*71*). The long-term stability of electrochemical detectors has, with earlier designs, proven to be a problem in their routine application to residues recovered from tissue samples.

Determining erythromycin residues in eggs, milk, and swine kidney, liver, and muscle has recently been reported, using precolumn derivatization with 9-fluorenylmethylchloroformate (FMOC) to produce fluorescence with excitation at 255 nm and emission at 315 nm (*72*). The residues are initially extracted with acetonitrile and cleaned up by liquid-liquid partition, followed by reversed-phase LC. Detection limits below 0.01 mg/kg for all matrices tested except liver were claimed, with recoveries ranging from 99% for egg fortified at 0.03 mg/kg to 38% for liver spiked at 0.06 mg/kg.

In summary, fluorescence or electrochemical detection are increasingly being used to develop LC methods for these antibiotics in animal tissues because of their high sensitivity and specificity.

Sedecamycin

Sedecamycin (lankacidin A), a neutral 17-membered macrolide antibiotic, does not contain amino sugar moieties and is extensively metabolized by swine (*73*). Its major metabolites, lankacidin C, lankacidinol A, and lankacidinol, exhibit antibacterial activity against *T. hyodysenteriae*. Okada et al. (*73*, *74*) developed an LC method to determine sedecamycin and its major metabolites, such as lankacidin C, lankacidinol A, and lankacidinol, in swine tissues. The sample was homogenized in a phosphate buffer containing sodium chloride and was then extracted with ethyl acetate. Further cleanup was done using silica gel-Florisil and double-layered column chromatography.

The LC analysis uses silica gel (μ-Porasil) and C18 (μ-Bondapak C18) as an analytical column and UV detection at 227 nm. The detection limits for sedecamycin and its metabolites are 0.05 μg/g. The LC method with UV detection is a useful approach to determine sedecamycin in animal tissues.

Extraction and Cleanup

The general procedure for cleanup includes extracting the macrolide antibiotic from the sample and then isolating it from coextracted materials using either additional solvent extraction or solid-phase extraction. Most reported methods for determining macrolide antibiotics using TLC, GC, and LC employ the same methodological approach as seen in Tables 2–3.

Plasma, Urine, and Serum

Biological fluids such as plasma, serum, and urine, are treated by chemical procedures intended to deplete the sample of proteins, fats, and other interfering substances. The extract from such samples contains, in addition to the possible traces of the target analytes, many diverse compounds that need to be excluded. There are a variety of techniques that can be employed to accomplish this.

A cleanup method for lipophilic compounds, such as macrolide antibiotics, that has been used for many years is the liquid-liquid partitioning method. These drugs are generally extracted from the biological matrix into organic solvents (typically employing chloroform, or dichloromethane as the immiscible organic) at a pH where the ionization of their basic function, amino sugar, is suppressed. Either the organic extract is evaporated to dryness and the residue is reconstituted, or the macrolides are back-extracted into an acidic buffer before analysis. Sometimes, several extraction steps are necessary to eliminate interfering substances, and pretreating becomes considerably time-consuming and complicated. In the liquid-liquid partitioning methods, it should be noted that macrolide antibiotics are not stable in acidic buffer solution. Conventional liquid-liquid partitioning methods are suitable methods for the cleanup of macrolides in biological samples, but they do have a drawback. They use chlorinated solvents that may pose both safety concerns and create expensive waste disposal problems.

Recently, a solid-phase extraction (SPE) method has been applied in the cleanup of biological materials prior to being quantitatively analyzed for trace amounts of antibacterial agents (*75*, *76*). Several SPE methods have also been reported for extracting and cleaning macrolide antibiotics (see Table 3) in biological samples. The most widely used solid-stationary phase for extraction and cleanup is octadecyl silane bonded to the silica gel, called C18 or ODS. In this SPE process, macrolides are isolated from a sample solution based on their relative solubility in the sample solution vs their solubility in a solid stationary phase (*77*).

Stubbs et al. (*78*) developed an LC method using erythromycin as an internal standard for determining oleandomycin in serum and urine. The success of this method is largely due to the efficiency of the solid-phase extraction procedure employed, which results in a clean sample extract suitable to be used at a detection wavelength of 200 nm. The lack of a specific UV chromophore in oleandomycin and erythromycin precludes the use of UV detection for these compounds at wavelengths above 215 nm, where most assays based on UV detection are conducted. For the cleanup procedure, the biological samples were deproteinized by adding acetonitrile and then centrifuged, after which the diluted sample was added to a disposable C18 cartridge. This was washed with 50% acetonitrile, and oleandomycin was then eluted with 0.05 M phosphate buffer (pH 6.3)–acetonitrile (1:1). Stubbs et al. reported that the elution behavior of oleandomycin from a C18 reversed-phase column depended upon the buffer concentration in the mobile phase. This phenomenon was used to develop the solid-phase extraction procedure, whereby oleandomycin could be selectively retained on the C18 extraction column, while interfering substances were washed off with a washing solvent. This behavior on the C18 cartridge seemingly relates to the basic and lipophilic nature of the macrolide antibiotics.

Column switching techniques have been applied for the determination of josamycin in plasma (*79*). This method has the important advantage that sample preparation is not necessary, because the plasma is directly injected into the system. However, this method does require a complex apparatus, and it is not suitable for cleaning animal tissue samples containing many contaminants.

Animal Tissues

The extract from a tissue sample contains many diverse compounds in addition to the possible traces of the target analytes. To exclude these physically- or chemically-interfering substances, a variety of techniques may be employed. Solid-phase extraction methods have been extensively used for the cleanup of biological material prior to the quantitative analysis for trace amounts of contaminants in the matrices. Based on a report by Skinner and Kanfer (*80*), who used a C18 (ODS) cartridge to recover josamycin in serum, the C18 cartridge was examined to see if it was applicable to the cleanup of spiramycin (*40*), tylosin (*42*), and mirosamicin (*81*) in animal tissues. Similar to other macrolides, these antibiotics are highly lipophilic and are strongly held by the C18 cartridge. In the case of liver and kidney samples, which contain many contaminants (e.g., proteins, lipids, and amino acids) as compared with the serum, the contaminants could not be completely eliminated by the use of the C18 cartridge.

For the cleanup of spiramycin and tylosin contained in meat, a liquid–liquid partition was employed (*40*, *42*). Figure 9 shows typical chromatograms of meat extracts obtained using this cleanup approach for tylosin. Similarly, Nagata and Saeki (*82*) employed a liquid–liquid partition for the cleanup of spiramycin in meat. Therefore, a cleanup procedure for mirosamicin was investigated by a liquid–liquid partition using dichloromethane. This procedure was found to be applicable for the cleanup of muscle samples, but was not suitable for liver and kidney samples due to extensive emulsion formation that prevented the elimination of interfering substances.

Another cleanup method was proposed (*81*), which used a cartridge packed with a silica-based cation exchange phase. Because mirosamicin is a basic and lipophilic compound, Bond Elut SCX packed with an ion-exchange phase (benzenesulfonylpropyl) and cationic and lipophilic properties was selected as the cartridge. To elute mirosamicin from the cartridge, the drug was first neutralized by washing the cartridge with 0.1 M dipotassium hydrogenphosphate (pH 8.9) and then eluted with methanol. By using this method, test solutions free from any effect of contaminants could be prepared, as Figure 10 shows. The recoveries of mirosamicin from tissue samples fortified at a level of 1.0 μg/g were greater than 80% with relative standard deviations of 5% or less. This cleanup method is suitable for other macrolide antibiotics, such as spiramycin, tylosin, josamycin, and kitasamycin. However, it is not applicable to sedecamycin, a neutral macrolide antibiotic. On the other hand, other cleanup methods (84–94) have been developed

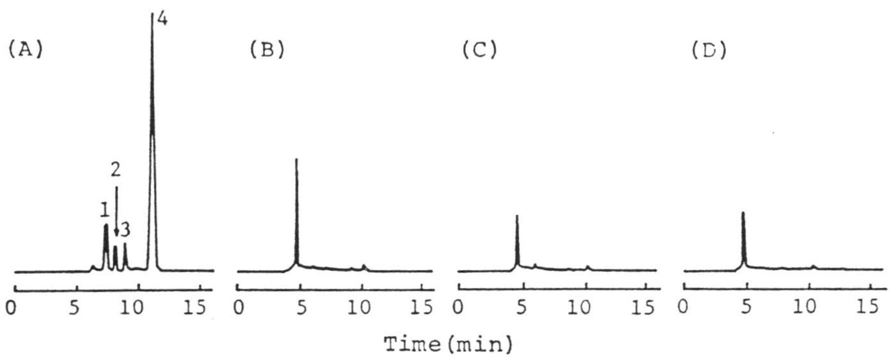

Figure 9 Typical chromatograms of extracts from meat
(A) Standard (220 ng). Peaks: 1, Tylosin C; 2, Tylosin B; 3, Tylosin D; 4, Tylosin A. (B) Chicken extract. (C) Pork extract. (D) Beef extract.
Conditions: column, Inertsil ODS (250 × 4.6 mm i.d.); mobile phase, 0.05 M sodium dihydrogenphosphate (pH 2.4)-acetonirtile (65:35); flow-rate, 0.5 mL/min; detector, UV 287 nm; sensitivity, 0.04 a.u.f.s.; injection volume, 20 μL
SOURCE: Reproduced by permission of The Pharmaceutical Society of Japan. See reference (42).

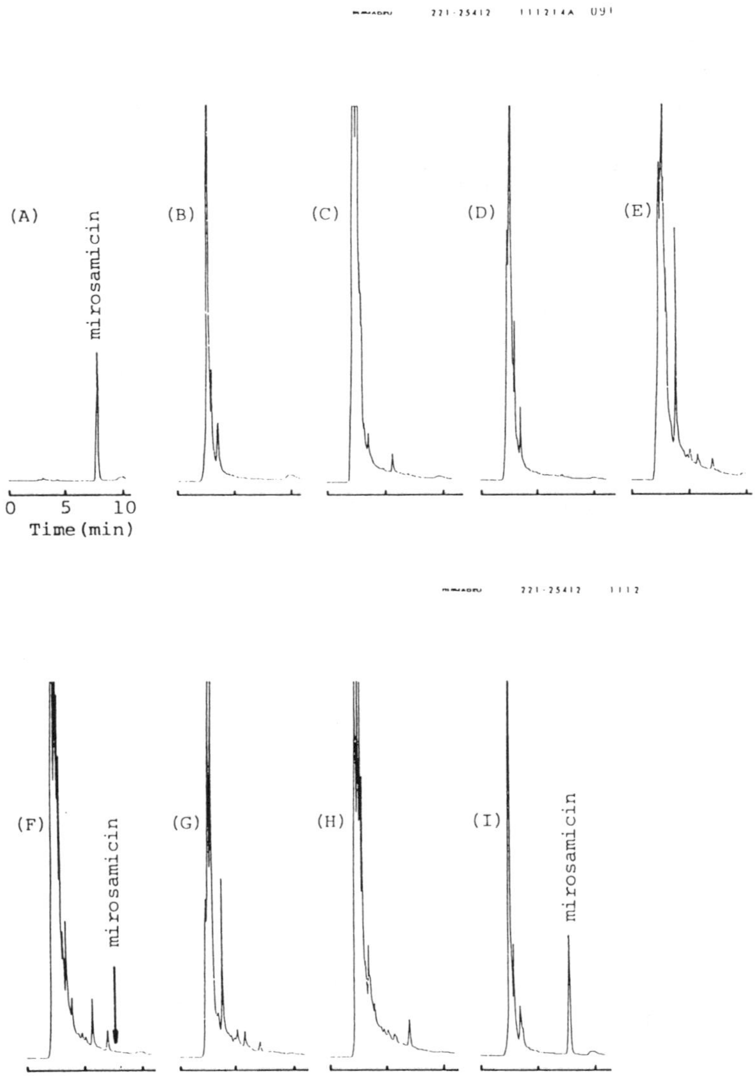

Figure 10 Typical chromatograms of (A) standard mirosamicin (50 ng), (B) chicken muscle extract, (C) swine muscle extract, (D) cattle muscle extract, (E) chicken liver extract, (F) swine liver extract, (G) cattle liver extract, (H) swine liver exrtract and (I) extract of chicken muscle fortified at 1.0 μg/g. LC conditions: column, Pruesil 5C18; mobile phase, 0.05 M phosphate buffer (pH 2.5)-acetonitrile (70:30); flow-rate, 0.5 mL/min; detection , UV (230 nm).
SOURCE: Reproduced by permission of Elsevier Science. See reference (*81*).

to determine macrolide antibiotics in livestock products and biological fluids (Table 3).

In Japan, guidelines have been set for developing residue analysis methods. For example, "detection limit: not more than 0.05 μg/g; recovery under addition of 1–2 μg/g: not less than 70%; relative standard deviation (RSD): not more than 10%" (*83*). In countries such as the United States a maximum residue limit (MRL) or tolerance has been established for a compound, it is required that the detection limit of a regulatory method should be low enough to enforce the MRL.

Liquid–liquid partition and solid-phase extraction are suitable procedures for the cleanup of serum, plasma, and urine. However, these methods are not always suitable for the cleanup of animal tissues such as liver and kidney. In my experience, cleanup using cation exchange cartridges is the most powerful approach for animal tissues.

Residues

Almost all macrolide antibiotics are weakly basic compounds that have good lipid solubility. With these characteristics they would be expected to accumulate in body tissues that have a lower pH than the plasma because of the ion-trapping mechanism. A distinguishing characteristic of macrolide antibiotics is that these drugs are highly concentrated in the lungs. As a result, adherence to the required withholding times prior to slaughter or use of eggs and milk is important, especially considering the high tissue/plasma ratios. An overview of the residues and metabolism of the macrolide antibiotics follows.

Tylosin

In general, macrolide antibiotics are unstable in acidic solutions. Mild acid hydrolysis of tylosin results in sugar mycarose and desmycosin production. Further acid hydrolysis results in producing two additional sugars and an inactive residue. After oral administration of radiolabeled tylosin to swine, almost all the radioactive components were excreted through the feces including tylosin A, tylosin D, and dihydrodesmycosin. In swine liver and kidney, only very small amounts of tylosin A, tylosin C, and dihydrodesmycosin could be found (*95*). Although tylosin is extensively metabolized, no single metabolite appears to be present in a greater concentration in tissue than the parent compound, suggesting that the latter should serve as the marker residue (*53*, *96*).

Residue-depletion studies in swine and cattle showed that the tissue where the highest residue levels occur depends on the method of administration. Injectable forms of tylosin, excluding the injection site, residue concentrations were highest and had the slowest decrease in the kidney. In contrast, with oral dosage forms, the liver contained the highest residue concentration, and therefore, might be the most appropriate target tissue when the drug is given orally (*95*).

No tylosin residues were found in swine fed 500 mg/kg of feed with zero withdrawal time (detection limit, 0.2 ppm) (*97*). At a dosage of 1000 mg/kg of feed with no withdrawal time, detectable residues were found only in the liver at slaughter. Residues were undetectable by LC in poultry tissues after oral dosing, even at zero withdrawal time (detection limit, 0.05 ppm).

Following oral dosing with tylosin, residue depletion appears to be rapid. After injecting tylosin, residues tend to be higher in both the kidney and the liver, and persist at the injection site. In general, oral dosing of animals resulted in lower residues in tissues than injection (*94, 98*); orally administered tylosin used for production applications requires a shorter withdrawal time than drug injected for therapeutic purposes.

Spiramycin

Residues of spiramycin were evaluated at the Thirty-eighth Joint FAO/WHO Expert Committee (*95*), as follows. In young cattle that received injections of spiramycin at 30 mg/kg, residues were highest in the liver and the kidney and lowest in muscle and fat. Detectable residues of spiramycin (<0.06 ppm) persisted in liver for 49 days after treatment, but quantifiable residues (0.14 ppm) persisted until day 28. A residue depletion study was carried out in calves fed spiramycin at 25 mg/kg of body weight for 7 days. Residue levels in muscle and fat decreased to below the detection limit (0.1 ppm) in approximately 3 days. In contrast, residues were more persistent in the liver and the kidney, at levels ranging from 15.9 ppm on day 3 to below the limit of detection on day 24. Residue studies of spiramycin following oral administration were also performed in young pigs. The highest levels of the drug were also found in the liver and the kidney after dietary administration. After 10 days, concentrations of the drug decreased below the limit of detection (<0.3 ppm). Spiramycin, at a dose of 300 mg/kg of feed, was administered to broilers for 10 days. Spiramycin residues in liver decreased to below the detection limit (<0.02 ppm) after 8 days.

After absorption, some portion of the spiramycin is desmycarosylated into neospiramycin by gastric acid. Methods are available for determining the concentrations of spiramycin and the microbiologically active metabolite neospiramycin in edible tissues of swine and cattle. Neospiramycin is similar to the parent compound in its potent antibiotic activity (*99*).

Spiramycin produces more persistent residues than the other macrolide antibiotics, and it is necessary to fully consider its metabolism and depletion in assessing livestock products.

Others

Sedecamycin. Swine were fed a diet containing 50–500 mg/kg of sedecamycin for 14 or 28 days starting at 6 or 10 weeks of age. As is the case with most macrolide antibiotics, sedecamycin is extensively metabolized by swine to approximately 20 metabolites (*73*, *74*). Lankacidin C, lankacidinol A, and lankacidinol are major metabolites of sedecamycin and are active against bacteria. Of the edible tissues, the liver retained the highest metabolite concentration at 2 hours after withdrawal. Blood, muscle, and fat showed no drug residue at any time, even at the 500 mg/kg level of medication. The small intestine showed a higher concentration of the unchanged drug (0.00–0.13 ppm) and metabolites (0.05–0.63 ppm) than the liver. All compounds disappeared rapidly from all the tissues and were not detected in any tissue after 1 day of withdrawal.

Erythromycin. The principal metabolic pathway is by *N*-desmethylation of the desosamine sugar. The des-*N*-methyl form of erythromycin has only 5% of the antimicrobial activity of the parent compound (*5*). Similar to almost all macrolide antibiotics, the amount of erythromycin eliminated in the urine is small.

Oleandomycin. Oleandomycin is used as phosphate and triacetylate in clinical practice. After absorption, triacetyloleandomycin is converted to several metabolites, but the oleandomycin base is not converted into metabolites (*100*).

Kitasamycin. The tissue concentrations in rats after oral administration of 200 mg/kg are highest in the lungs, followed by the liver, the kidney, the spleen, and the heart. Leucomycin A_5 is converted to the hydroxylated compound and some unknown metabolites (*100*). The major metabolite josamycin (leucomycin A_3) is a hydroxylated compound that has been detected in serum and urine (*80*).

Much of the above-mentioned information was obtained using microbiological methods. Therefore, uncertainties remain as to whether the administered macrolide antibiotics persist in their original state (the parent drug) or whether the activity measured includes the metabolites. Thus, a detailed examination will be necessary using chemical analysis methods such as LC and LC–MS to separate and characterize microbiologically active components present as residues in tissues.

Liquid Chromatography–Mass Spectrometry

In a residue analysis, identification of detected analytes is required to confirm a result. A gas chromatographic–mass spectrometric (GC–MS) method is generally a well-established confirmation procedure. The few GC–MS methods developed for determining and confirming macrolide antibiotics in animal tissues (*38*,

39) are complicated and time-consuming. The combination of liquid chromatographic separation and mass spectrometric (LC–MS) detection is an important analytical technique that is increasingly being used to identify analytes in complex mixtures, because it offers high sensitivity and selectivity in the analysis of a wide range of compounds. A few LC–MS methods for determining macrolides in plasma (*101*, *102*) and salmon (*103*) have also been reported.

Kokkonen et al., (*102*) described a specific method for determining erythromycin ethylsuccinate in plasma. The method involves a liquid–liquid extraction procedure followed by the analysis of extracts using a phase-system switching continuous–flow fast atom bombardment (CF–FAB) LC–MS. The advantage of this method is that sample preparation is not necessary, although the method does require complex instrumentation. Pleasance et al. (*103*) developed a reversed-phase LC–MS method, incorporating gradient elution, for characterizing residual erythromycin A and its metabolites in salmon tissue. The method uses ion spray, a mild atmospheric pressure ionization technique that provides an abundant protonated molecule well suited for selected ion monitoring experiments. The sample preparation was done as follows based on a procedure described by Takatsuki et al. (*38*). The drugs are extracted with acetonitrile from salmon and cleaned up by an *n*-hexane washing followed by a liquid–liquid partition. Using selected ion and selected reaction monitoring techniques, the LC–MS and LC–MS–MS methods provided detection limits of <0.01 and 0.05 ppm, respectively. Confirmatory full-scan LC–MS and LC–MS–MS spectra were obtained at the 0.5 and 1 ppm levels, respectively. Because this method was acceptable for a few sample injections, after which the guard column had to be replaced, application of the method to high-throughput quantitative analyses will require an improved cleanup procedure to be developed. LC–MS is the most promising approach to the confirmation of residues of macrolide antibiotics in animal tissues.

Conclusion

A general awareness of the greater need for food safety has been steadily increasing in many countries in recent years. The public has become notably concerned about contamination of food with potentially dangerous substances such as antibiotic residues in food-producing animals and certain aquacultured seafood. Moreover, the development of new veterinary drugs must be evaluated based on their efficacy, safety to the intended animal species, and safety to humans consuming products of animal origin. Therefore, simple and reliable analytical methods are required to monitor drug residues in animal tissues. Numerous methods have been proposed for the separation, detection, and determination of the macrolide antibiotics in pharmaceutical preparations and biological fluids.

However, only several methods for the determination of these drugs in animal tissues have been reported.

UV–LC assays, for example, can be affected by contaminants and, therefore, it can be difficult to establish a method for residue analysis based on this method of detection. It is expected that a convenient and widely applicable analytical method for residues of macrolide antibiotics will be developed in the future by overcoming the difficulties discussed in this chapter.

References

1. J. M. MacGuire, R. L. Bunch, R. C. Anderson, H. E. Boaz, E. H. Flynn, H. M. Powell, and J. W. Smith, *Antibiot. Chemother.*, **2**: 281, 1952.
2. R. B. Woodward, *Agnew. Chem.*, **69**: 50, 1957.
3. S. Omura, ed., *Macrolide Antibiotics: Chemistry, Biology, and Practice* (Orlando, FL: Academic Press, 1984), 26.
4. K. Ninomiya, *Antibiotics for Animals, (Tokyo: Youkendo Press, 1987), 307.*
5. S. Omura, ed., *Macrolide Antibiotics: Chemistry, Biology, and Practice* (Orlando, FL: Academic Press, 1984), 261.
6. Joint FAO/WHO Expert Committee on Food Additives, *WHO Tech. Rep. Ser., No. 430* (1968).
7. H. Nakazawa, *Eisei Kagaku*, **36**: 163, 1990.
8. Code of Federal Regulations Title 21, Part 556, 1993.
9. Association of Official Analytical Chemists, *Official Methods of Analysis*, 15th ed. (Arlington, VA: AOAC, 1990), sec. **988.08**.
10. Ministry of Health and Welfare, *Official Methods for Residual Substances in Livestock Products* (Tokyo, Japan: Ministry of Health and Welfare, 1982).
11. R. Hutter, W. S. Keller, and H. Zahner, *Arch. Mikrobiol.*, **39**: 158, 1961.
12. S. Omura and A. Nakagawa, *J. Antibiot.*, **28**: 401, 1975.
13. S. Omura, ed., *Macrolide Antibiotics: Chemistry, Biology, and Practice*, (Orlando, FL: Academic Press, 1984), 3.
14. S. Omura, Y. Hironaka, A. Nakagawa, I. Umezawa, and T. Hata, *J. Antibiot.*, **25**: 105, 1972.
15. S. Omura, Y. Hironaka, and T. Hata, *J. Antibiot.*, **23**: 511, 1970.
16. M. Debono, K. E. Willard, H. A. Kirst, J. A. Wind, G. D. Crouse, E. V. Tao, J. T. Vicenzi, F. T. Counter, J. L. Ott, E. E. Ose, and S. Omura, *J. Antibiot.*, **42**: 1253, 1989.

17. E. E. Ose, *J. Antibiot.*, **42**: 1253, 1987.
18. S. Satio, N. Muto, M. Hayashi, T. Fujii, and M. Otani, *J. Antibiot.*, **33**: 364, 1980.
19. N. Watanabe, *Proc. Jpn. Soc. Antimicrob. Anim.*, **12**: 29, 1991.
20. S. Harada, E. Higaside, and T. Kishi, *Tetrahedron Lett.*, **1969**: 2239, 1969.
21. K. C. Graham, W. L. Wilson, and A. Vilim, *J. Chromatogr.*, **125**: 447, 1976.
22. G. Kobrehel, Z. Tamburasev, and S. Djokic, *J. Chromatogr.*, **133**: 415, 1977.
23. H. Vanderhaeghe and L. Kerremans, *J. Chromatogr.*, **193**: 119, 1980.
24. Th. Cachet, E. Roets, J. Hoogmartens, and H. Vanderhaeghe, *J. Chromatogr.*, **403**: 343, 1987.
25. G. A. Bens, W. Van den Bossche, and P. De Moerloose, *J. High Resolut. Chromatogr. Commun.*, **3**: 433, 1980.
26. M. Debackere and K. Baeten, *J. Chromatogr.*, **61**: 125, 1971.
27. I. O. Kibwage, E. Roets, A. Verbruggen, J. Hoogmartens, and H. Vanderhaeghe, *J. Chromatogr.*, **434**: 177, 1988.
28. M. Petz, *J. Assoc. Off. Anal. Chem.*, **70**: 691, 1987.
29. C. D. C. Salisbury, C. E. Rigby, and W. Chan, *J. Agric. Food. Chem.*, **37**: 105, 1989.
30. E. Neidert, P. W. Saschenbrecker, and F. Tittiger, *J. Assoc. Off. Anal. Chem.*, **70**: 197, 1987.
31. B. Saito, S. Fujisaki, and H. Uchibe, *J. Food Hyg. Soc. Jpn.*, **24**: 130, 1983.
32. G. A. Bens, W. Van den Bossche, and P. De Moerloose, *Chromatographia*, **12**: 294, 1979.
33. D. Mourot, B. Delepine, J. Boissear, and G. Gayot, *J. Chromatogr.*, **161**: 386, 1978.
34. M. Horie, K. Saito, N. Nose, Y. Sida, and H. Nakazawa, *Bunseki-kagaku*, **37**: 580, 1988.
35. G. A. Bens, E. Crombez, and P. De Moerloose, *J. Liq. Chromatogr.*, **5**: 1449, 1982.
36. K. Tsuji and J. H. Robertson, *Anal. Chem.*, **43**: 818, 1971.
37. J. H. Robertson and K. Tsuji, *J. Pharm. Sci.*, **61**: 1633, 1972.
38. K. Takatsuki, S. Suzuki, N. Sato, I. Ushizawa, and T. Shoji, *J. Assoc. Off. Anal. Chem.*, **70**: 708, 1987.
39. K. Takatsuki, I. Ushizawa, and T. Shoji, *J. Chromatogr.*, **391**: 207, 1987.
40. M. Horie, K. Saito, N. Nose, Y. Sida, and H. Nakazawa, *Bunseki-kagaku*, **35**: 219, 1986.
41. H. A. Kirst, G. M. Wild, R. H. Baltz, R. L. Hamill, J. L. Ott, F. T. Counter, and E. E. Ose, *J. Antibiot.*, **35**: 1675, 1982.
42. M. Horie, K. Saito, Y. Hoshino, N. Nose, and H. Nakazawa, *Eisei Kagaku*, **34**: 128, 1988.

43. H. M. Stationery Office, *British Veterinary Pharmacopoeia 1985* (London: H.M. Stationery Office, 1985).
44. S. Omura, Y. Suzuki, A. Nakagawa, and T. Hata, *J. Antibiot.*, **26**: 794, 1973.
45. S. Bhuwapathanapun and P. Gray, *J. Antibiot.*, **30**: 673, 1977.
46. J. H. Kennedy, *J. Chromatogr.*, **281**: 288, 1983.
47. J. H. Kennedy, *J. Chromatogr. Sci.*, **16**: 492, 1978.
48. B. J. Fish and G. P. R. Carr, *J. Chromatogr.*, **353**: 39, 1986.
49. E. Roets, P. Beirinckx, I. Quintens, and J. Hoogmartens, *J. Chromatogr.*, **630**: 159, 1993.
50. F. L. Neely, *Chromatographia*, **34**: 51, 1992.
51. W. A. Moats, *J. Chromatogr.*, **366**: 69, 1986.
52. W. Yeh, N. J. Bauer, and J. E. Dotzlaf, *J. Chromatogr.*, **288**: 157, 1984.
53. W. Chan, G. C. Gerhardt, and C. D. C. Salisbury, *J. AOAC Int.*, **77**: 331, 1994.
54. M. Tod, O. Biarez, P. Nicolas, and O. Petitjean, *J. Chromatogr.*, **575**: 171, 1992.
55. K. Tsuji and J. F. Goetz, *J. Chromatogr.*, **147**: 359, 1978.
56. K. Tsuji and J. F. Goetz, *J. Antibiot.*, (Tokyo), **31**: 302, 1978.
57. K. Tsuji and J. F. Goetz, *J. Chromatogr.*, **157**: 185, 1978.
58. K. Tsuji and M. P. Kane, *J. Pharm. Sci.*, **71**: 1160, 1982.
59. G. Pellegatta, G. Carugati, and G. Coppi, *J. Chromatogr.*, **269**: 33, 1983.
60. I. O. Kibwage, E. Roets, J. Hoogmartens, and H. Vanderhaeghe, *J. Chromatogr.*, **330**: 275, 1985.
61. I. O. Kibwage, G. Janssen, E. Roets, J. Hoogmartens, and H. Vanderhaeghe, *J. Chromatogr.*, **346**: 309, 1985.
62. Th. Cachet, I. O. Kibwage, E. Roets, J. Hoogmartens, and H. Vanderhaeghe, *J. Chromatogr.*, **409**: 91, 1987.
63. Th. Cachet, G. Haest, R. Busson, G. Janssen, and J. Hoogmartens, *J. Chromatogr.*, **445**: 290, 1988.
64. J. Paesen, E. Roets, and J. Hoogmartens, *Chromatographia*, **32**: 162, 1991.
65. Th. Cachet, I. Quintens, E. Roets, and J. Hoogmartens, *J. Liq. Chromatogr.*, **14**: 1203, 1991.
66. Th. Cachet, P. Lannoo, J. Paesen, G. Janssen, and J. Hoogmartens, *J. Chromatogr.*, **600**: 99, 1992.
67. J. Paesen, P. Claeys, E. Roets, and J. Hoogmartens, *J. Chromatogr.*, **630**: 177, 1993.
68. K. Tsuji, *J. Chromatogr.*, **158**: 337, 1978.
69. M. Chen and W. L. Chiou, *J. Chromatogr.*, **278**: 91, 1983.
70. G. S. Duthu, *J. Liquid Chromatogr.*, **7**: 1023, 1984.
71. D. Croteau, F. Vallee, M. G. Bergeron, and M. Lebel, *J. Chromatogr.*, **419**: 205, 1987.
72. G. Zierfels and M. Petz, *Z. Lebensm. Unters. Forsch.*, **198**: 307, 1994.

73. J. Okada and S. Kondo, *J. Assoc. Off. Anal. Chem.*, **70**: 818, 1987.
74. J. Okada, S. Yamamoto, H. Yamamoto, and S. Kondo, *J. Agric. Food Chem.*, **36**: 104, 1988.
75. V. K. Agarwal, *J. Chromatogr.*, **624**: 411, 1992.
76. C. C. Walker, H. M. Lott, and S. A. Barker, *J. Chromatogr.*, **642**: 225, 1993.
77. K. C. Van Horne, ed., *Sorbent Extraction Technology*, (Harbor City, CA: Analytichem International, Inc., 1985), 18.
78. C. Stubbs, J. M. Haigh, and I. Kanfer, *J. Chromatogr.*, **353**: 33, 1986.
79. K. Rader, A. Wildfeuer, A. Schwedass, and H. Laufen, *J. Chromatogr.*, **344**: 416, 1985.
80. M. Skinner and I. Kanfer, *J. Chromatogr.*, **459**: 261, 1988.
81. M. Horie, K. Saito, N. Nose, H. Oka, and H. Nakazawa, *J. Chromatogr.*, **655**: 47, 1994.
82. T. Nagata and M. Saeki, *J. Assoc. Off. Anal. Chem.*, **69**: 644, 1986.
83. Corporation of Japan Veterinary Pharmaceutical Association, *GLP Standard of the Veterinary Drugs*, Livestock Agency, Ministry of Agriculture, Forestry and Fisheries, ed. (Tokyo, Japan: Corp. Japan Veterinary Pharmaceutical Association, 1988), 106.
84. C. Stubbs, J. M. Haigh, and I. Kanfer, *J. Pharm. Sci.*, **74**: 1126, 1985.
85. L. Nilsson, B. Walldorf, and O. Paulsen, *J. Chromatogr.*, **423**: 189, 1987.
86. C. Stubbs and I. Kanfer, *J. Chromatogr.*, **427**: 93, 1988.
87. N. Grgurinovich and A. Matthews, *J. Chromatogr.*, **433**: 298, 1988.
88. S. Laakso, M. Scheinin, and M. Anttila, *J. Chromatogr.*, **526**: 475, 1990.
89. C. Stubbs, M. F. Skinner, and I. Kanfer, *Chromatographia*, **29**: 31, 1990.
90. J. Dow, M. Lemar, A. Frydman, and J. Gaillot, *J. Chromatogr.*, **344**: 275, 1985.
91. D. Carlhant, M. A. Le Bot, Y. Guedes, C. Riche, F. Mimouri, J. Colin, and F. Berthou, *Biomed. Chromatogr.*, **3**: 1, 1989.
92. P. Sanders, G. Moulin, C. Gaudiche, B. Delepine, and D. Mourot, *J. Assoc. Off. Anal. Chem.*, **74**: 912, 1991.
93. L. J. Keng and J. O. Boison, *J. Liq. Chromatogr.*, **15**: 2025, 1992.
94. W. A. Moats, E. W. Harris, and N. C. Steele, *J. Assoc. Off. Anal. Chem.*, **68**: 413, 1985.
95. *WHO Tech. Rep. Ser.*, No. 815, 1991.
96. *FAO Food and Nutrition Paper 41/4, "Residues of Some Veterinary Drugs in Animals and Foods"* (1991), 109.
97. R. M. Kline and W. P. Waitt, *J. Assoc. Off. Anal. Chem.*, **54**: 112, 1971.
98. M. G. Lauridsen, C. Lund, and M. Jacobsen, *J. Assoc. Off. Anal. Chem.*, **71**: 921, 1988.
99. H. Takahira, *Jap. J. Antibiot.*, **23**: 424, 1970.
100. S. Omura, ed., *Macrolide Antibiotics: Chemistry, Biology, and Practice*, (Orlando, FL: Academic Press, 1984), 301.

101. P. Kokkonen, E. Schroder, W. M. A. Niessen, U. R. Tjaden, and J. van der Greef, *J. Chromatogr.*, **511**: 35, 1990.
102. P. S. Kokkonen, W. M. A. Niessen, U. R. Tjaden, and J. van der Greef, *J. Chromatogr.*, **565**: 265, 1991.
103. S. Pleasance, J. Kelly, M. D. LeBlanc, M. A. Quillian, R. K. Boyd, D. D. Kitts, K. McErlane, and D. H. North, *Bio. Mass Spectrom.*, **21**: 675, 1992.

Chapter 7

Chemical Analysis of Chloramphenicols

Tomoko Nagata

The Public Health Laboratory of Chiba Prefecture, Chiba, Japan

Contents

Introduction

Chloramphenicol (CAP) (D-(−)-threo-2-(dichloroacetamido)-1 (*p*-nitrophenyl)-1,3-propanediol), thiamphenicol (TAP) (D(+)-threo-2-dichloroacetamido-1-(4-methylsulfonylphenyl)propane-1,3-diol), and florfenicol (FF) (2,2-dichloro-*N*-(αS,βR)-α-(fluoromethyl)-β-hydroxy-*p*-(methylsulfonyl)-phe nyl acetamide) are synthetic broad-spectrum antibiotics with similar chemical structures, as shown in Figure 1 (*1*). In TAP, the *p*-nitro group on the benzene ring of CAP is replaced with a methyl sulfonyl group. In FF, the hydroxy group on the side chain of TAP is replaced with a fluorine. The plasma half-lives of CAP and TAP in humans are 1.5–3.5 h and 2.6–3.5 h, respectively. CAP, TAP, and FF are light stable, and 100 parts per million (μg/g) standard solutions of these compounds are stable for more than four months.

CAP, TAP and FF have a broad spectrum of activity, and they are suitable for treatment of infections caused by *Staphylococcus epidermidis*, *Staphylococcus*

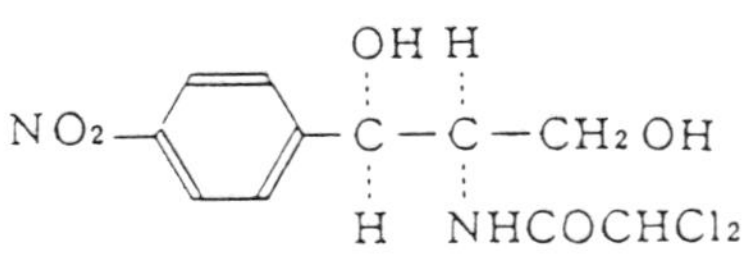

Chloramphenicol (CAP)

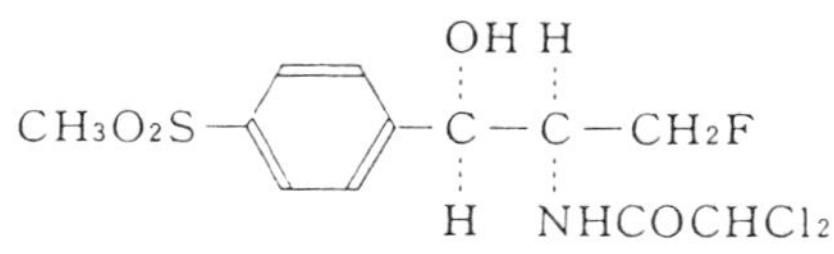

Florfenicol (FF)

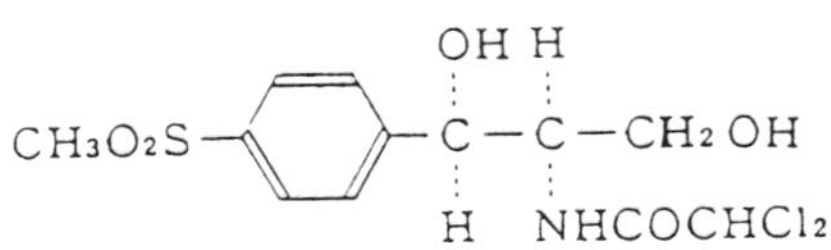

Thiamphenicol (TAP)

Figure 1 Chemical structures of chloramphenicol, thiamphenicol, and florfenicol

aureus, *Streptococcus pneumonide*, *Neisseria meningitidis*, *Hemophilus influenza*, *Leptospira*, *Mycoplasma*, *Treponema pallidum*, *Rickettsia*, *Chlamydia* and other agents by interacting with ribosomes to inhibit the organisms from synthesizing protein (*2*). CAP is an effective antibiotic, but it has a side effect. It causes aplastic anemia, an irreversible bone marrow insufficiency, in a small percentage of humans exposed to the drug. The absence of the *p*-nitro group on the benzene ring, however, results in TAP having a low toxicity relative to that of CAP, which in turn causes a weak dose-related reversible inhibition of erythropoiesis. FF was developed in the United States as a derivative of TAP and recently, in Japan, has been approved in the aquaculture industry for preventing yellowtail disease. FF, like CAP, has a potential for producing aplastic anaemia (*2*).

To determine CAP and TAP, bioassays can be used, but the sensitivity of these assays is limited to 0.5–1 μg/g, and is poor in samples containing other antibacterial agents. To overcome these disadvantages, chromatographic determinations of CAP, TAP, and FF have been developed. In gas chromatography (GC) and gas chromatography–mass spectrometry (GC–MS) analyses, CAP and TAP, which are nonvolatile, require derivatization prior to GC determination and GC–MS identification. GC determination of CAP has good sensitivity, with limits of detection 0.1–5 ng/g, while those for TAP are 50–200 ng/g. Because CAP, TAP, and FF all have a strong UV absorption, they can also be determined by high-performance liquid chromatography (LC) without derivatization. Furthermore, CAP has been identified by LC–MS. In LC determinations, detection limits are 1–5 ng/g for CAP and 10 ng/g for TAP and FF.

A review of chromatographic methods for determining residual CAP in foods of animal origin was published in *J. AOAC* in 1985, which included eight GC methods, nine LC methods and seven methods using thin-layer chromatography (TLC) (*3*).

Gas-Liquid Chromatography Determination of CAP and TAP

At present, CAP and TAP are usually determined by LC, although in most laboratories they were previously determined by GC, and then were confirmed by GC–MS. These molecules, whose structures are shown in Figure 1, including hydroxy, amino, sulfuryl, or imino groups, are highly polar and are difficult to volatilize. They can only be effectively analyzed by GC or GC–MS after esterification, silylation, or acylation of their polar functional groups to produce heat-stable, volatile derivatives. For effective use in a regulatory method, the derivatization reactions selected require a simple, rapid completion with no side reactions (*4–13*). Compounds containing hydroxyl groups have been derivatized quantitatively by trimethylsilylation (TMS) to a volatile and heat-stable compound (*10*). Because CAP and TAP have hydroxyl groups, TMS methods have generally been employed in forming CAP and TAP derivatives as shown in Table 1 (*13–24*). Selecting the TMS reagent and the conditions of derivatization are important in obtaining the desired product, as described below.

The choice of TMS reagents, the solvent used, the temperature, and the reaction time all depend on the substituted position of the functional groups of CAP and TAP (*13*). To determine CAP quantitatively, silylation with a mixture of hexamethyldisilazone (HMDS) and trimethylchlorosilane in pyridine, which gives only the bis-TMS derivative, has been recommended (*13*). In a method to determine and confirm CAP and TAP and their metabolites in human urine, for example, the methylation of CAP and TAP's carboxyl group of glucuronides was performed by adding an ethyl ether solution of diazomethane (*16*). The glucuronides of CAP and TAP contain both carboxyl and hydroxy groups, both of which are derivatized by the TMS reagent.

To perform the derivatization quantitatively and precisely, carboxy groups of glucuronic acid may be methylated before the silylation. In the above derivatization, bis-TMS derivatives of CAP and TAP were observed; that is, only two hydroxy groups of CAP and TAP were substituted with the TMS reagent. However, when the Sil-Prep reagent was used at room temperature for 15 min, the tri-TMS derivative of TAP was observed; that is, an amido nitrogen atom and two hydroxy groups of TAP were substituted with the TMS reagent.

Table 1 Determination methods of CAP and TAP by GC

Drug	Matrix	Sample preparation	Derivatization	GC column	Detection	Recovery (%)	Detection limit (ppm or ppb)	Ref.
TAP	Plasma, urine	Ethyl acetate extraction	TMS	1.5% DEGS Chromosorbw W, 1.5% OV-17 Shimalite W	ECD	—	1	*(14)*
TAP	Tissues, Body fluid	Ethyl acetate extn	TMS	6.6% OV-3 Chromosorb, 3.3% OV-17 W AW DMCS	FID GC–MS	—	0.2	*(15)*
TAP, CAP	Urine (Metabolites)	Ethyl acetate extn	diazomethane, TMS	1.5% OV-17 Chromosorb, W AW DMCS	FID GC–MS	—	50	*(16)*
TAP	Plasma, body fluid	Ethyl acetate extn	TMS	3% OV-17 GasChrom Q	ECD GC–MS	97	0.1	*(17)*
CAP	Tissues	isopentyl acetate extn	TMS	3% OV-1 Chromosorb, W AW DMCS	ECD	—	10	*(18)*
CAP	Milk	H_2SO_4-Na tungstate ethyl acetate extn	TMS	5% Se-30 Gas Chrom Q	FID	—	10	*(27)*

Table 1 Determination methods of CAP and TAP by GC—*continued*

Drug	Matrix	Sample preparation	Derivatization	GC column	Detection	Recovery (%)	Detection limit (ppm or ppb)	Ref.
CAP	Milk	MeCN extn, iosoctane partition, ethyl acetate extn	HFBA	3% Dexsil 300 Supelcoport AW DMCS	ECD	81	50	*(19)*
CAP	Tissues, milk, body fluid	ethyl ether-petroleum ether, K oxalate-MeOH partition, ethyl ether extn, residue in *n*-hexane, MeOH extn	TMS	3% DC 200 Chromosorb W DMCS	ECD	59–73	10–20	*(25)*
CAP	Tissues, milk, eggs	MeCN extn, ethyl acetate extn, Sephadex HL, MeOH elute	diazomethane, HFBA	capillary column coated with Se-30/52 (1 + 1)	TCD	90–94	10	*(20)*

Table 1 Determination methods of CAP and TAP by GC—*continued*

Drug	Matrix	Sample preparation	Derivatization	GC column	Detection	Recovery (%)	Detection limit (ppm or ppb)	Ref.
CAP	Tissues	ethyl acetate extn, petroleum ether partition, Sep-pak C_{18}	TMS	3% OV-1 Gas Chrom Q	ECD	51.9–68.0	5	*(21)*
CAP	Tissues, Milk	MeCN extn, *n*-hexane partition, NaCl-CH_2Cl_2-ethyl acetate-buffer partition	TMS	capillary column coated with Se-30/52 (1 + 1)	ECD	—	1	*(22)*
CAP	Muscle, Milk, Eggs	MeCN-NaCl/MeCN/$CHCl_3$-isooctane extn, *n*-hexane-H_2O partition, ethyl acetate extn	TMS	3% OV-101 Gas Chrom Q	ECD	91.3	1–5	*(23)*

Table 1 Determination methods of CAP and TAP by GC—*continued*

Drug	Matrix	Sample preparation	Derivatization	GC column	Detection	Recovery (%)	Detection limit (ppm or ppb)	Ref.
CAP	Milk	H_2O extn, Sep-pak C_{18}, MeCN-H_2O elute, ethyl acetate extn	TMS	capillary column coated with OV-17, CB	ECD	75–99	0.1	*(24)*
CAP	Tissues, milk	PBS extn, collect HPLC eluate, ethyl acetate ext	TMS	capillary column coated with CP Sil-5	GC–MS	—	0.2–5	*(30)*

TAP: thiamphenicol; TMS: trimethylsilation; ECD: electron-capture detector; FID: flame-ionization detector; CAP: chloramphenicol; extn: extraction

As shown in Figure 1, the CAP molecule contains two chlorine atoms, while TAP contains both two chlorine atoms and an aromatic methysulfonyl group, all of which have a high electron affinity. Therefore, an electron capture detector (ECD) has frequently been used to detect CAP (*18*, *19*, *21–25*) and TAP (*14*, *17*) with good sensitivity and specificity (*26*). FF, also shown in Figure 1, contains not only an aromatic methylsulfonyl group but also a fluorine in its molecular structure, so derivatized FF should be detected by ECD with good sensitivity, although this has not been published. Flame ionization detection (FID) has also been used to detect CAP and TAP (*15*, *16*, *27*), but it has the disadvantages of limited sensitivity and a lack of specificity (*28*, *29*). To overcome the lack of specificity, gas chromatography–mass spectrometry (GC–MS) has been used to determine and confirm TAP (*15–17*) and CAP (*16*, *30*), including their metabolites (*16*).

Chromatograms of CAP, TAP deacylated CAP and TAP metabolites, and the glucuronides of CAP and TAP are observed in Figures 2 and 3. As seen in Figure

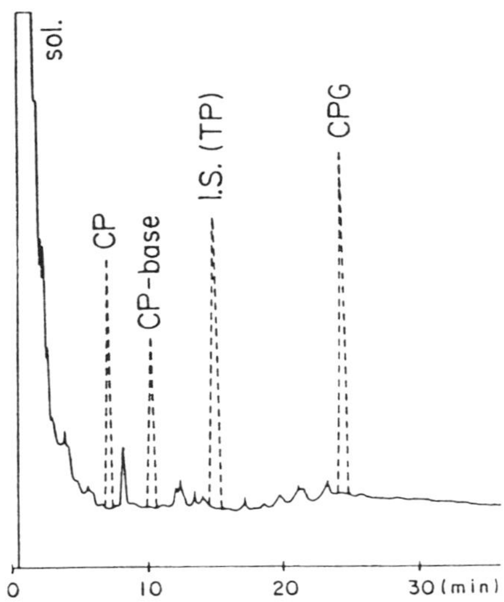

Figure 2 GC chromatogram of CAP
CP: CAP, CP-base: deacylated CAP, CPG: glucuronide of CAP
Column: 1 m × 3 mm i.d., packed with 1.5% OV-17 on Chromosorb W AW DMCS, 60-80 mesh, column temperature: at 180°C for the first 6 min, then increased to 260°C at the rate of 5°C/min, injection port temperature: 280°C, carrier gas: nitrogen, flow-rate: 65 mL/min
SOURCE: Reproduced by permission of J. Chromatogr. See reference (*16*).

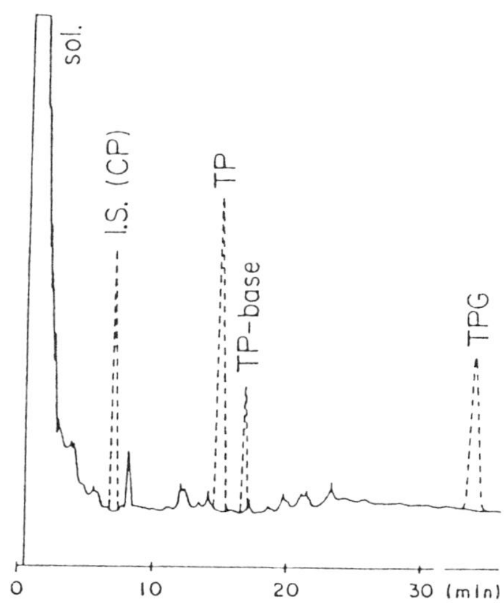

Figure 3 GC chromatogram of TAP.
TP: TAP, TP-base: deacylated TAP, TPG: glucuronide of TAP. The conditions are shown in Figure 2
SOURCE: **Reproduced by permission of J. Chromatogr. See reference (*16*).**

1, the chemical structures of CAP and TAP are quite similar, so they can be determined or identified simultaneously (*14*, *15*, *21*, *30*).

To improve the accuracy of methods in determining CAP or TAP by correcting for variation in the extraction and injection procedure, CAP or TAP analog chemicals, such as TAP (*16*, *21*, *30*), CAP (*16*), D(+)-threo-1-(*p*-methylsulfonyl-phenyl-2-acetamide-1,3-propanediol (*15*, *17*), CAP-*m*-isomer (*23*) and PCB-138 (*30*) have been used as internal standards based on their similar extractability and chromatographic behavior in GC.

The detection limit of CAP, 1–20 ng/g, is acceptable, but that of TAP, 50 ng/g, is not appropriate to detect residual TAP in animal-derived food samples. However, using GC methods to improve the detection limit of TAP should not be difficult.

When trimethylsilylation of the hydroxy groups is performed by adding acetonitrile and TMS reagent *N,O*-Bis-(trimethylsilyl)acetamide (BSA), the major fragment ions in the mass spectrum of derivatized TAP (molecular ion (M) = 499) are 486 and 484 (M-CH_3(15)), which are characteristic of TMS derivatives, 394 and 396, (M-15-$(CH_3)_3SiOH$(90)), representing the elimination of trimethylsilanol. Those of CAP (M = 466) are 453 and 451 (M-CH_3), 361 and 363 (M-CH_3-$(CH_3)_3SiOH$), 225, 208, and 178, as shown in Figures 4 and 5.

The presence of CAP was confirmed by using two diagnostic ions in selected-ion monitoring (SIM), *m/z* 225, the base peak, and *m/z* 208, produced by the elimination of a hydroxy group from the m/z 225 ion. The TMS reagent (BSTFA

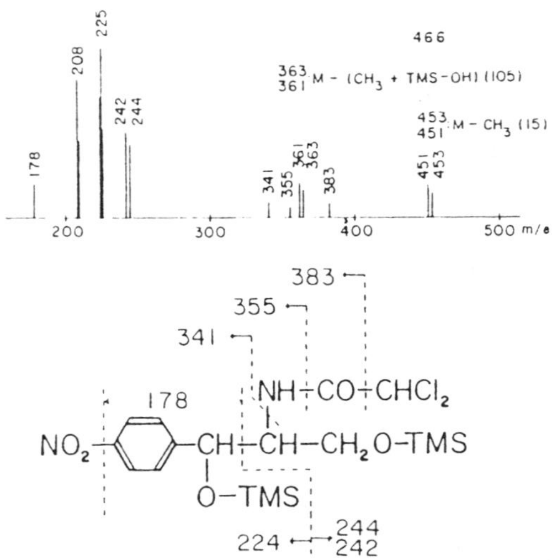

Figure 4 GC-MS Specturm of CAP.
Separation temperature: 290°C, ionization temperature: 270°C, ionization energy: 70 eV, acceleration voltage: 3.5 kV, trap current: 60 μA

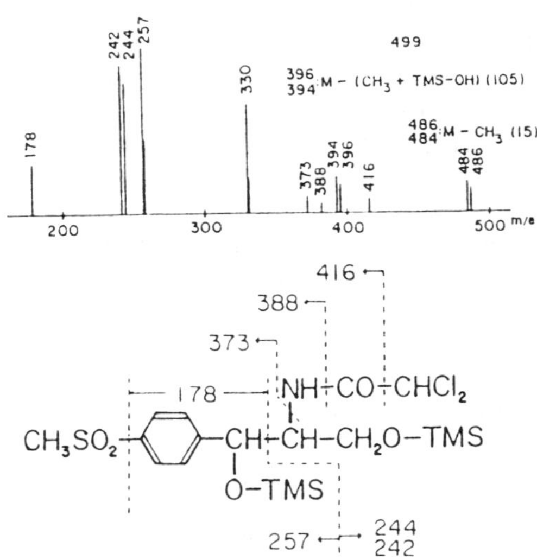

Figure 5 GC-MS Spectrum of TAP
GC-MS conditions are shown in Figure 4.
SOURCE: Reproduced by permission of J. Chromatogr. See reference (16).

+ TMCB, 99+1) was chosen after comparing the performance of several silylation reagents, including MSTFA with acetonitrile, HMDS–TMCS–pyridine, BSTFA–TMCS. BSTFA + TMCB, 99:1, yielded only the disilylated product and provided the strongest intensity of the *m/z* 225 base peak (*30*). This procedure was applied as the confirmatory step in a combination of three determinative methods, i.e., an immunochemical card test with a detection limit of 2 ng/g (screening test), an LC with a detection limit of 10 ng/g (quantitative determination), followed by a GC–MS with a limit of identification of 5 ng/g (confirmation). To perform the immunochemical card test, CAP is extracted with a filtered phosphate buffer (pH 5). After applying the filtrate to the card, the presence of CAP s indicated within 5–15 min (*31*, *32*). Muscle samples containing more than 4 ng/g CAP gave positive results. No false negative and no false positive results have been observed at this level (*30*). The second step in the testing scheme, the LC method, is described in the following section. This combination of several methodologies to detect, quantify, and confirm CAP residues in tissues has been reported by Keukens et al. (*30*). Such a combination of detection methods based on different principles is an effective way to approach the analysis of regulatory samples for residues of CAP.

GC analysis of CAP and TAP is less common because it requires derivatization, which makes the analysis somewhat laborious. However, confirmation by GC–MS methods, which provide molecular weight and characteristic fragmentation patterns, is considered to be necessary to eliminate false-positive results obtained by GC or LC methods.

High-Performance Liquid Chromatographic Determination of CAP, TAP, and FF

CAP, TAP, and FF have strong ultraviolet (UV) absorption and are therefore ideal for direct determination without derivatization by high-performance liquid chromatography (LC) with UV detection. Residues of these drugs have been determined by LC with visible UV detection, after cleanup by liquid–liquid/solid–liquid partition, as summarized in Table 2. In most reports, C18 columns were selected as the analytical column for LC, however, the use of a C8 analytical column has also been described (*32*, *33*). In general, C8 columns are used to analyze a compound that is not well retained on a C18 column in the usual mobile phases, such as methanol/acetonitrile–water. Because CAP and TAP are satisfactorily retained by C18, it has been selected as the initial choice when analyzing each drug in many of the reported methods, usually with commonly used mobile phases, such as mixtures of methanol or acetonitrile with phosphate buffer, acetate buffer or water.

The detection wavelengths of CAP have been set at 214 nm, 254 nm, or near 278 nm, and are presented in Table 2. Using these three wavelengths, CAP was detected with greatest sensitivity at 214 nm (Figure 6). However, at this short

Table 2 Determination methods of CAP, TAP, and FF by Liquid Chromatography

Drug	Matrix	Sample preparation	LC column	Mobile phase	Detection (nm)	Recovery (%)	Detection limit	Ref.
CAP	Muscle	MeCN extn, *n*-hexane partition	Nucleosil 10-NO_2	diiso-propyl ether-MeOH	254	71	10 ppb	*(45)*
CAP	Milk	ethyl acetate extn, iso-octane-MeCN partition, ethyl acetate extn, residue in propylene glycol, H_2O partition	Spherisorb ODS	MeOH-H_2O	280	72–99.5	5 ppb	*(55)*
CAP	Tissues	saline extn, ethyl ether partition, residue in ethyl ether, H_2O partition	MOS-Hypersil C8 spherical	MeOH-perchrolic acid containing pentane sulfonate	278/350	85	66 ppb	*(56)*
CAP	Muscle	Tris buffer extn, ethyl ether extn, residue in MeOH	Nucleosil C_{18}	MeOH-H_2O	278/350	—	2 ppb	*(46)*
CAP	Plasma	MeOH-Na phosphate extn	LiChrosorb RP18	MeOH-PBS (pH 4.9)	278	96.3	3 ppm	*(47)*
CAP	Tissues	ethyl acetate extn, iso-octane-MeCN partition, $CHCl_3$-glycol-H_2O partition	Spherib ODS	MeOH-H_2O	280	77	5 ppb	*(57)*
CAP	Milk	CH_3Cl-isopropanol extn, H_2O partition residue in mobile phase	Bonded ODS	MeCN–acetate buffer (pH 4)	280	—	50 ppb	*(58)*

Table 2 Determination methods of CAP, TAP, and FF by Liquid Chromatography—*continued*

Drug	Matrix	Sample preparation	LC column	Mobile phase	Detection (nm)	Recovery (%)	Detection limit	Ref.
CAP	Muscle, milk, eggs	MeCN extn, NaCl-CH_2Cl_2 partition residue in mobile phase *n*-hexane partition	MOS-Hypersil	MeCN–Na acetate (pH 4.6)	275	88–104	20 ppb	*(48)*
CAP	Tissues, urine, serum, body fluids	ethyl acetate extn, NaCl partition, Baker C_{18} column, MeOH-H_2O elute, CH_2Cl_2 partition	LiChrosorb RP18	MeOH–K phosphate	278	49	10 ppb	*(44)*
CAP	Muscle, milk	ethyl acetate extn, H_2O MeCN-petroleum ether partition, H_2O-MeCN pentane partition	Hypersil H5 ODS	MeCN–acetate buffer	276	62–98	10 ppb	*(59)*
CAP	Muscle	H_2O extn, Extrelut, CH_2Cl_2 elute residue in H_2O toluene partition	CpTM Spher C_{18}	MeCN–Na acetate buffer (pH 4.3)	278	58	5 ppb	*(31)*
CAP	muscle	ethyl acetate extn residue in CH_2Cl_2-petroleum ether silica Sep-Pak ethyl acetate-hexane elute residue in Tris buffer ethyl ehter extn residue in H_2O toluene partition	CpTM Sher C_{18}	MeCN-N a acetate buffer (pH 4.3)	diode array	85	10 ppb	*(31)*
CAP	Serum	MeOH/ethyl acetate extn residue in MeOH	Spherisorb C_{18}, Nucleosil C_{18}	MeCN-NA acetate (pH 6.0)	254	—	15 ppb	*(34)*

Table 2 Determination methods of CAP, TAP, and FF by Liquid Chromatography—*continued*

Drug	Matrix	Sample preparation	LC column	Mobile phase	Detection (nm)	Recovery (%)	Detection limit	Ref.
CAP	Tissues, serum, blood, urine	MeOH/$HClO_4$/H_2O extn, extrelut, ethyl acetate elute	Hypersil ODS	MeCN–NA acetate buffer (pH 6.1)	278	96.5	1–10 ppm	*(35)*
CAP	Serum	MeCN-zinc sulfate extn	Nucleosil C_{18}	MeCN-NH_3 phosphate	280	—	—	*(49)*
CAP	Plasma	Dichloromethane–isopropanol extn	LiCrosorb RP18	MeOH-H_2O	254	—	—	*(54)*
CAP	Tissues, milk	Acetone extn, NaCl-CH_2Cl_2 extn, MeOH-H_2O-*n*-hexane partition, Wofatit Y29, MeOH elute	Separon SIX C_{18}	MeOH-H_2O	254	80	10 ppb	*(50)*
CAP	Tissues, plasma, bile	PBS (pH 7.8) extn, Sep-pak C_{18}, ethyl ether elute	LiCrosorb RP18	MeCN-H_2O	280	90.6–91.4	5–50 ppb	*(64)*
CAP	Muscle	H_2O extn Extrelut, CH_2Cl_2 elute residue in H_2O toluene partition	Chromospher C_{18}, Chromospher C8	MeCN–NA acetate buffer (pH 4.3)	285	55.1	1.5 ppb	*(32)*
CAP	Muscle	MeCN extn residue in *n*—hexane–CH_2Cl_2–MeCN–PSB partition	Supelco LC_{18}	MeCN-PBS (pH 4.6)	275	—	50 ppb	*(51)*
CAP	Serum	MeOH/$HClO_4$ extn	LiChrosorb C_{18}	MeOH-Na PBS (pH 6.0)	214	—	—	*(36)*

Table 2 Determination methods of CAP, TAP, and FF by Liquid Chromatography—*continued*

Drug	Matrix	Sample preparation	LC column	Mobile phase	Detection (nm)	Recovery (%)	Detection limit	Ref.
CAP	Milk	Bulk C_{18} ethyl acetate elute	Bulk C_{18}	MeCN–Phosphate	278	60.9	62.5 ppb	*(39)*
CAP	Muscle	ethyl acetate extn, *n*-hexane partition Silica gel MeOH elute	Chromspher C8	MeCN–Na acetate buffer (pH 4.3)	280	79	10 ppb	*(33)*
CAP	Serum	EtOH-MeCN-H_2O extn	Brownlee C_{18}	MeCN-ammonium acetate buffer	278	88.6–99.8	25 ppb	*(52)*
CAP	Kidney	ethyl ether extn residue in MeCN petroleum ether partition	Chromsep C_{18}	MeCN-H_2O	280	41.7	10 ppb	*(40)*
CAP	Musclemilk	ethyl acetate extn silica Sep-Pak	C_{18} Sherisorb 5	MeOH–H_2O 0.1% TFA	278	50–67%	500 ppb	*(37)*
CAP	Serum	ethyl acetate extn	Perkin-Elmer C_{18}	isopropanol–Na acetate buffer (pH 5.0)	278	89.2	—	*(43)*
CAP	Tissues	ethyl acetate extn *n*-hexane-$CHCl_3$-H_2O partition	Novapack C_{18}	MeCN–ammonium phosphate (pH 7.9)	278	75.2–78.5	1 ppb	*(60)*
CAP	Muscle	ethyl acetate extn *n*-hexane–CH_2Cl_2–H_2O partition	LiChrospher RP18	MeOH–0.2% formic acid	LC–MS	100.4	2 ppb	*(42)*

Table 2 Determination methods of CAP, TAP, and FF by Liquid Chromatography—*continued*

Drug	Matrix	Sample preparation	LC column	Mobile phase	Detection (nm)	Recovery (%)	Detection limit	Ref.
TAP	Serum	MeOH extn	μ-Bondapak C_{18}	MeOH–H_2O	254	97	0.5 ppm	*(53)*
TAP	Plasma	ethyl acetate extn	C_{18} Ultrasphere	MeOH–H_2O	224	102	0.1 ppm	*(38)*
TAP	Muscle	ethyl acetate extn residue in NaCl, *n*-hexane partition, ethyl acetate extn residue in 95% MeCN aluminum column, 95% MeCN elute	Nucleosil C_{18}	MeCN–H_2O	230	90.0	50 ppb	*(61)*
FF, CAP, TAP	Muscle	ethyl acetate extn residue in NaCl *n*-hexane partition, ethyl acetate extn residue in *n*-hexane-ethyl ether, Florisil, ethyl ether-methanol elute	Chromatorex ODS	MeOH–H_2O	225/270	67.6–82.6	10 ppb	*(62)*
TAP	Muscle	ethyl acetate extn, 4% NaCl partition residue in *n*-hexane $CHCl_3$, H_2O partition	Nucleosil C_{18}	MeOH–H_2O	224	75.5	10 ppb	*(63)*
CAP	Muscle	CH_3CN extn, *n*-hexane partition silica Sep-Pak ethyl acetate extn	Novapack C_{18}	MeCN–Na acetate buffer (pH 4.8)	278	74.7–81.4	10 ppb	*(41)*

CAP: chloramphenicol; TAP: thiamphenicol; FF: florfenicol

wavelength, many interfering peaks of endogenous compounds are observed in the chromatograms of sample extracts. CAP's maximum UV absorption is 278 nm, and at this wavelength CAP would be detected with fewer interfering peaks. Because TAP and FF have a stronger UV absorption at shorter wavelengths, as shown in Figure 6, detection has been performed at 224 nm, 225 nm or 230 nm as a compromise to provide good sensitivity, fewer interfering peaks, and less background in the chromatograms.

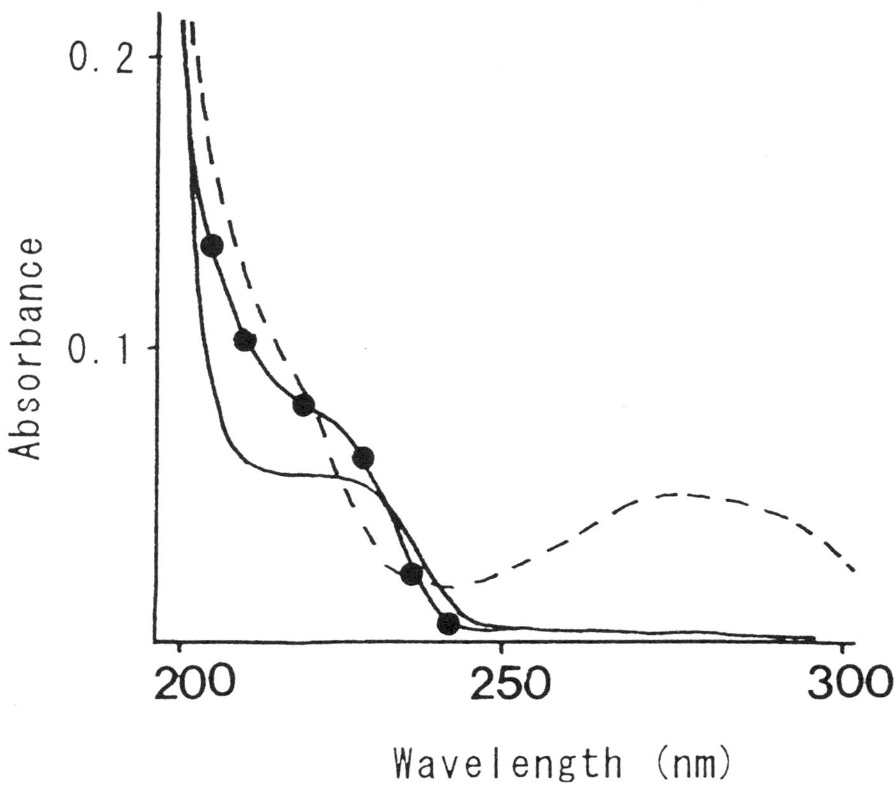

Figure 6 Absorption spectra of chloramphenicol, thiamphenicol and fluorfenicol in LC mobile phase
- - - - : Chloramphenicol
—— : Thiamphenicol
. —. — . : Florfenicol
(each 1.5 μg/mL in methanol-water (15:85) solution.)

In LC, 5-ethyl-*p*-tolybarbituric acid (*34*), 3-isobutyl-1-methylxanthine (*35*), tinidazole (*36*), and 4-nitroacetanilide (*37*) have been used as internal standards in analyzing for CAP, while *N*-acetyl-*p*-aminophenol was used for TAP (*38*).

To confirm CAP in LC-based methodologies, the photodiode array detector, which collects continuous spectral data during the analysis to check for interfering substances by comparing the spectrum of sample with that of the standard (*30, 31, 39, 40, 41*), and high-performance liquid chromatography–mass spectrometry (LC–MS) have been used. Although confirmation with a photo-diode array detector is simple, specificity and the limit of detection are not sufficient to determine or identify trace levels of residual CAP in samples.

During confirmation of CAP by LC–MS (*42*), CAP residue in muscle was extracted, and after cleanup by liquid–liquid partition, it was applied to an LC connected to a mass spectrometer via a particle beam interface. The MS system was operated in the negative-ion chemical ionization mode using methane as the reagent gas. CAP was determined in SIM mode to measure the fragments *m/z* 324, due to the chlorine-37 isotope, 322, M, 288, 286, (M–HCl) and 250 (M–$(HCl)_2$), as shown in Figure 7. The detection limit was 2 ng/g and the recoveries were more

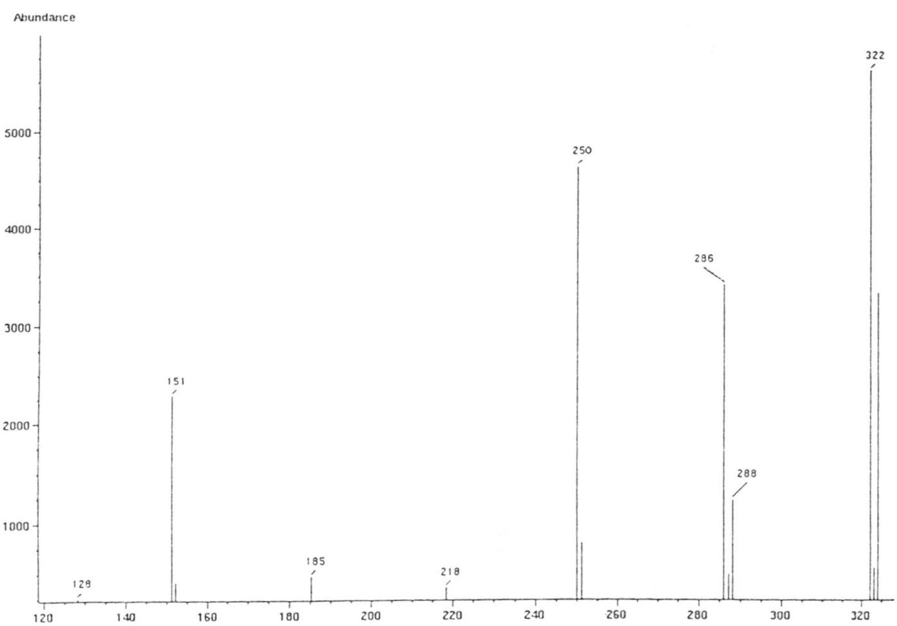

Figure 7 LC–MS Spectra of CAP.
Negative-ion chemical ionizatoin mode, electron energy: 230 eV, emission current: 300 μA, ion source temperature: 200°C
SOURCE: Reproduced by permission of J. Chromatogr. See reference (42).

than 99% with a variation coefficient of 1% at 20 ng/g levels in muscle. In the report of Ramsey et al. (*37*), LC–MS gave a complex background spectrum at the retention time of CAP, which was extracted from trout by Haagsma et al.'s method (*33*). In this method, ethyl acetate extract was applied to a silica Sep-Pak cartridge and CAP was eluted along with hexane–acetone, but LC tandem mass spectrometry (LC–MS–MS), was effective in producing interference-free daughter ion spectra. In this LC–MS–MS method (*37*), ammonia was used as a reagent gas in CI, and CAP was confirmed by collision-induced dissociation (CID), *m/z* 323, which corresponds to $(M + H)^+$, and *m/z* 305 and 275, which corresponds to the loss of H_2O and H_2O + CH_2O, respectively. Without an extensive cleanup procedure, this LC–MS–MS method gives unambiguous confirmation of residual CAP in trout and milk, but the limit of confirmation of 0.5 µg/g is not appropriate to identify low levels of residual CAP typically encountered in regulatory programs. Using these LC--MS methods, CAP is identified more rapidly than by GC–MS, which requires silylation of the residue. However, at present, the analytical performance of GC–MS is superior to LC–MS, based on sensitivity and the availability of suitable fragment ions for confirmation.

Usually, it is necessary to use different analytical methods to analyze different kinds of residual antibacterials in samples, making analysis very time-consuming and tedious. Therefore, it is desirable to determine residual antibacterials simultaneously. However, in order to accomplish this, the extraction and chromatographic characteristics of the drugs must be very similar as are CAP, TAP, and FF. As shown in Table 2, CAP and TAP have been individually analyzed by LC using C18 and C8 columns. To analyze the three drugs simultaneously, the retention times of CAP, TAP, and FF were investigated on 150 cm LC columns containing three different packings: Inertsil ODS-2, Inertsil C8 (Gaskuro Kogyo Co., Tokyo, Japan), and TSK-GEL ODS-80TM (Tosho Co., Tokyo) using methanol–water (20:80) as the mobile phase. The retention times of TAP, FF, and CAP were 5.11, 9.07, and 18.0 min, respectively, on the Inertsil ODS-2 column, 5.44, 9.76, and 19.56 min, on the Inertsil C8 column, and 5.86, 10.69, and 19.26 min on the TSK-GEL 80TM column. Peak shapes were symmetrical for the three compounds using each LC column.

To analyze the three drugs simultaneously, either C18 or C8 columns could be used. TAP, which has less retention on the analytical column and has a strong UV absorption at a shorter wavelength, is detected at 225 nm, followed by FF. The wavelength is then switched to 278 nm to detect CAP. This should create no problem for the analyst using current detector technology, because most UV detectors now have wavelength programming capability.

In LC methods, CAP, TAP, and FF are determined with sufficient sensitivity without derivatization and may be determined simultaneously. Therefore, I regard LC as a superior method to GC in performing such analyses effectively in a limited time. The retention time and the UV absorption characteristics of the peaks, however, are not sufficient to confirm each peak making it necessary to use

LC–MS or the more sensitive and selective GC–MS procedures to obtain final confirmation of each residue identity.

Extraction and Cleanup

CAP is rapidly metabolized to glucuronide conjugate in liver. In an investigation of the stability of tissue spiked with CAP, under storage at -20°C, CAP degraded in sliced liver and kidney, but remained stable in samples that had been cut into cubes (*43*). In swine liver, kidney, and serum, where more than one-half of the CAP present is in the conjugated form, the glucuronide conjugate of CAP was effectively determined after deglucuronidation by β-glucuronidase (*44*), but in muscles of chicken, calf, swine, and pork, conjugated CAP was not observed under various deglucuronidation or desulfatation conditions (*31*). In liver, kidney, and serum samples, residual CAP is accurately determined after a deglucuronidation step or by simultaneous determination with its metabolites. In this respect, the GC method (*16*) is useful in determining and identifying CAP, TAP, and their metabolites in tissues, provided that adequate cleanup procedures and an improved detection limit for TAP are added to the present determination.

Extracts of CAP and TAP from serum, plasma, and urine containing relatively high concentrations of drugs, are usually determined by GC or LC without cleanup. Those from milk and tissue samples, however, are purified with a liquid–liquid partition, either alone or in combination with a solid-phase extraction cartridge or a column packed with reversed-phase octadecylsilyl (C18) derivatized silica, diatomaceous earth (Extrelut), alumina, or Florisil.

In both GC and LC determinations, the drugs have been extracted from samples primarily with ethyl acetate, methanol, or acetonitrile, as summarized in Tables 1 and 2. The effect of the solvent on the extraction of these three drugs from cultured yellowtail (*Seriola lalandei*) was therefore investigated. Samples, 10 g each, of homogenized yellowtail muscle were spiked with 5 µg of CAP, TAP, and FF and refrigerated overnight. The drugs were then extracted with 50 mL of methanol, acetonitrile, or ethyl acetate, each extract was concentrated, and the resultant residue was dissolved with a 4% sodium chloride solution. After partitioning with *n*-hexane to remove any lipids, the drugs were back-extracted twice into ethyl acetate and then evaporated to dryness. The residue was then dissolved with methanol–water (1:1) and analyzed by LC.

As seen in Table 3, the recoveries of CAP, TAP, and FF were not as high for methanol or acetonitrile as they were for ethyl acetate, making it the most suitable solvent for extracting the three drugs. The results suggest that CAP is bound to muscle components. The ethyl acetate extract was the most oily of the three solvent extracts. However, almost all lipids were removed from the extracts by partition with *n*-hexane. Water was reported to extract CAP effectively without extensive cleanup and without significant formation of emulsion (*31*, *32*). In that extraction

Table 3 The Effect of Extract Solvent on the Recoveries of CAP, TAP, and FF

	Recovery (%)		
Solvent	CAP	TAP	FF
MeOH	74.5	89.8	83.4
MeCN	81.1	97.4	94.7
EtOAc	93.6	98.4	98.1

Yellowtail muscles spiked with 5 μ g of each drug were extracted with each solvent two times. Recoveries are the average of two trials.

of CAP with water, lower recoveries were also observed when the amount of sample increased (*31*)

In some extraction procedures, sample extracts were purified by precipitating and centrifuging out the protein (*35, 36, 45–53*). Residual CAP and TAP in plasma (*47, 54*) and serum (*36, 49, 53*) have been determined after protein precipitation without any further cleanup. However, low concentrations of residual CAP and TAP could not be detected using the above procedures because of the many interfering chromatogram peaks.

A popular purification method of sample extracts is a combination of multiple liquid–liquid extractions (*19, 21–23, 25, 27, 31–33, 40–42, 44–46, 48, 50–51, 55–63*). These partitions are performed to extract the drugs from aqueous solution with an organic solvent or to wash the organic or aqueous solution to remove interfering substances. Table 4 shows that ethyl acetate is more effective in extracting the three drugs from aqueous solution than any other organic solvent tested. To back-extract the three drugs from aqueous solution efficiently with organic solvent, sodium chloride is added to the aqueous solution (*22, 23, 41, 44, 48, 50, 61–63*). As shown in Table 5, by adding sodium chloride to aqueous solution, TAP and FF, which are relatively polar substances, were effectively extracted with ethyl acetate. The addition of too much salt to aqueous solution, however, causes problems. Three to four percent of sodium chloride solution is considered to be optimal for the extraction of the drugs. To remove lipids, the drugs in aqueous solution have often been partitioned with *n*-hexane (*22,23,25,33,37,41,42,45,48,50,51,60,61–63*), petroleum ether (*21, 25, 31, 40, 59*), toluene (*31, 32*), or isooctane (*20, 25, 55, 57*). Of these solvents, *n*-hexane, isooctane, and petroleum ether are preferable, because CAP and FF can transfer into toluene to some extent, as shown in Table 4.

Liquid–liquid partitions are frequently applied as a cleanup step prior to solid–liquid partition using a column or a cartridge. Sephadex HL (*20*), Sep-Pak C18 (*21, 24, 64*), Baker C18 (*44*), Extrelute (*31, 32, 35*), Silica Sep-Pak (*31, 37,*

Table 4 The Effect of Extract Solvents on the Recoveries of CAP, TAP, and FF From 4% NaCl Solutions[a]

	Recovery (%)		
Drug	CAP	TAP	FF
n-hexane	0	0	0
toluene	2.5	0	3.9
isooctane	0	0	0
petroleum ether	0	0	0
carbon tetra chloride	0	0	0
dichloromethane	57.0	3.2	66.2
chloroform	37.0	1.5	50.7
ethyl ether	61.6	4.6	57.5
ethyl acetate	67.5	57.0	74.0

CAP: chloramphenicol; TAP: thiamphenicol; FF: florfenicol
[a]Nine 4% NaCl solutions containing 5 μg of each drug were extracted with each solvent two times.

41), Wofatit Y29 (*50*), silica gel (*33*), alumina (*61*), and Florisil (*62*) have all been used in such applications. An interlaboratory study with 13 collaborators was performed using Extrelut cartridges to determine residual CAP in meat, with a concentration range of 6.5–21 ng/g, after extraction with water (*32*). Recoveries

Table 5 The Effect of NaCl Concentrations on the Extraction of CAP, TAP, and FF[a]

NaCl concentra-	Recovery (%)		
tion (%)	CAP	TAP	FF
0	45.0	40.9	54.0
2	47.5	44.0	58.6
3	48.2	44.0	63.6
4	48.8	46.0	64.0
7	50.0	46.0	63.6
10	50.0	47.6	64.0

CAP: chloramphenicol; TAP: thiamphenicol; FF: florfenicol
[a]Each NaCl solution containing 5 μg of each drug was partitioned with ethylactate one time.

were approximately 55%, with loss of analyte considered to take place during cartridge cleanup (*31, 32*).

C18 solid-phase extraction cartridges are often used in cleanup procedures, where they provide high recovery of the analyte. However, in many cases, after cleanup on C18 cartridges, analysis by LC using C18 columns reveals a similar elution pattern to that observed for the C18 cartridges. In such cases, solid-phase extraction has not proven effective in removing interfering substances in the sample extract. Therefore, the effects of Florisil and alumina cartridges on the cleanup of sample extracts were investigated, as described below.

Yellowtail muscle was spiked with 5μg of each drug, and the samples were extracted with ethyl acetate twice. The ethyl acetate phases were concentrated and 4% sodium chloride solution was added to the oily residues and then partitioned with *n*-hexane. The drugs were then back-extracted with ethyl acetate. The ethyl acetate extracts were concentrated to dryness, the residues were dissolved in *n*-hexane or acetonitrile–water (95:5), and then were applied to Florisil (*62*) or alumina (*61*) cartridges, respectively. In the former procedure, the cartridge was washed with *n*-hexane followed by ethyl ether, and the drugs were eluted with ethyl ether–methanol (7:3). In the latter, the drugs were eluted with acetonitrile–water (95:5). When the cleanup effects were compared, the recoveries of CAP, TAP, and FF using Florisil were 82.6% (n = 4, CV = 4.6%), 75.4% (n = 4, CV = 5.2%), 79.1% (n = 4, CV = 2.6%), while results on alumina were 75.8% (n = 4, CV = 2.7%), 65.6% (n = 4, CV = 4.9%), 63.4% (n = 4, CV = 9.6%), respectively. The Florisil cartridge was superior at removing lipids after being washed with *n*-hexane followed by ethyl ether. Even if eluate from the alumina cartridge was partitioned with organic solvents, such as isooctane, lipids could not be removed effectively, because this kind of liquid–liquid partition, using 4% sodium chloride solution-*n*-hexane, had already been performed in the initial cleanup. Florisil cartridge cleanup is therefore recommended to obtain relatively high recovery rates and to effectively remove lipids. Chromatograms of yellowtail extract obtained after the Florisil cleanup are shown in Figure 8. The detection limits for CAP, FF, and TAP are each 10 ng/g.

In the report of Long et al. (*39*), CAP in milk was extracted using the matrix solid-phase dispersion technique, in which the milk is added to C18 particles. After grinding together with a mortar and pestle, the mixture was packed into a glass column to be purified. This batch process might be useful to avoid column clogging, but as relatively small amounts of C18 particles are used, only a small volume of sample may be applied.

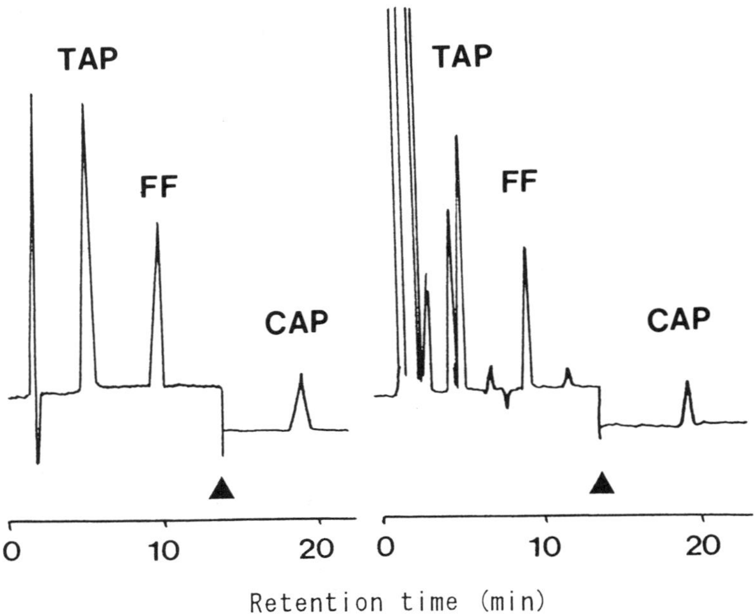

Figure 8 Chromatograms of standards and yellowtail extracts
1: Standards (each 10 ng)
2: Yellowtail extract (spiked with 1 μg of each drug to 10 g yellowtail muscle). wavelength was switched from 225 nm to 270 nm. Column: Intertil C8 150 × 4.6 mm i.d., mobile phase: methanol-water (20:80), flow rate: 1.0 mL/min, column temperature: 50°C

Conclusion

CAP and TAP have been successfully determined in samples of foods of animal origin by both GC and LC methods, while FF has been analyzed only by LC to date. To confirm CAP and TAP in tissues, GC–MS, LC–MS and LC–MS–MS methods have been described. Ethyl acetate is the most suitable solvent for the extraction of CAP, TAP, and FF residues from tissue samples. To remove interfering substances in serum and plasma extracts, partitioning with *n*-hexane, iso-octane, or petroleum ether is sufficient to obtain a satisfactory chromatogram without interfering peaks. In fatty tissues, however, in addition to the liquid–liquid partition a solid–liquid partition is necessary to permit the determination of trace concentrations of the residual drugs. To determine the three drugs without derivatization, LC is superior to GC. However, in both GC and LC methods, it is necessary to confirm the positive samples by GC–MS.

For routine monitoring, using a screening test with enzyme-linked immunosorbent or radioimmunoassay, which should have high sensitivity and specificity, is recommended as an effective approach. For chromatographic analysis, appropriate cleanup steps should be selected for the sample to be analyzed. For example, a liquid–liquid partition for serum and plasma, a liquid–liquid followed by a liquid–solid partition for relatively fatty tissues. It is quite reasonable to select the determination methods that have been collaboratively studied, such as methods of Aerts et al. (*32*), or Epstein (*65*).

In the future, the development of GC and LC determination methods for CAP, TAP, and FF, including their metabolites, in many kinds of tissues, will be important to obtain information on the pharmacokinetics of these three drugs in animal tissues. To confirm the drugs and their metabolites, further development of GC–MS and LC–MS methods is necessary. In LC–MS, along with the molecular ion, better fragmentation is required to provide sufficient ions to enable accurate confirmation of the analytical results. Because derivatization is not required in LC- and LC–MS, the development of LC–MS and LC–MS–MS is considered a more promising approach for the development of effective confirmatory methods for these drugs, if a solution can be found to the problem of lack of suitable fragmentation.

References

1. S. Budavari, ed., *The Merck Index*, 11th ed. (Rahway, N.J.: Merck & Co., Inc., 1989), 2068, 9232.
2. H. P. Lambert and F. W. O'Gray, ed., *Antibiotic and Chemotherapy*, 6th ed. (Edinburgh, UK: Churchill Livingstone, 1992), 136.
3. E. H. Allen, *J. Assoc. Off. Anal. Chem.*, **68**: 990, 1985.
4. R. Kuhn, H. Trischmann, and I. Low, *Angew. Chem. Int. Ed. Engl.*, **67**: 32, 1955.
5. H. Schlenk and J. L. Gellerman, *Anal. Chem.*, **32**: 1412, 1960.
6. E. Fischer, *Ber. Dtsch. Chem. Ge.*, **34**: 433, 1901; **39**: 530, 1906.
7. H. G. Jones, J. K. N. Jones, and M. B. Pervy, *Can. J. Chem.*, **40**: 1559, 1962.
8. W. J. A. VandenHeuvel, J. Sjovall, and E. C. Horning, *Biochim. Biophys. Acta*, **48**: 596, 1961.
9. S. H. Langer, P. Patages, and I. Wender, *Chem. & Ind.*, 1664, 1985.
10. T. Luukkainen, W. J. A. VandenHeuval, E. O. A. Haahit, and E. C. Horning, *Biochim. Biophys. Acta*, **52**: 599, 1965.

11. A. K. Ruhlmann and W. Giesecke, *Angew. Chem. Int. Ed. Engl.*, **73**: 113, 1961.
12. C. C. Sweeley, R. Bentley, M. Makita, and W. W. Welles, *J. Am. Chem. Soc.*, **85**: 2497, 1963.
13. G. Janssen and H. Vanderhaeghe, *J. Chromatogr.*, **82**: 297, 1969.
14. T. Aoyama and S. Iguchi, *J. Chromatogr.*, **43**: 253, 1969.
15. A. Gazzaniga, E. Pezzotti, and A. C. Ramusino., *J. Chromatogr.*, **81**: 71, 1973.
16. T. Nakagawa, M. Masada, and T. Uno, *J. Chromatogr.*, **111**: 355, 1975.
17. T. A. Plomp, R. A. A. Maes, and M. Thier, *J. Chromatogr.*, **121**: 243, 1976.
18. S. Fabiansson, T. Nilsson, and J. Backstrom, *J. Sci. Food Agric.*, **27**: 1156, 1976.
19. J. M. Wal, J. C. Peleran, and G. Bories, *J. Chromatogr.*, **168**: 179, 1979.
20. H. Holtmannspotter and H. P. Thier, *D. Tsch. Lebensm. Rundsch.*, **78**: 347, 1982.
21. J. R. Nelson, K. F. T. Copeland, R. J. Forster, D. J. Campbell, and W. D. Black, *J. Chromatogr.*, **276**: 438, 1983.
22. R. Malisch, U. Sandmeyer, and K. Kypke-Hutter, *Lebensmittelchem. Gerichl. Chem.*, **38**: 11, 1983.
23. D. Arnold and A. Somogyi, *J. Assoc. Off. Anal. Chem.*, **68**: 984, 1985.
24. V. B. B. Lang and M. Kachele, *D. Tsch. Lebensm. Rundsch.*, **81**: 278, 1985.
25. E. Hollstein, W. Laue, and G. Zapfe, *Nahrung*, **25**: 143, 1979.
26. J. E. Lovelock and S. R. Lipsky, *J. Am. Chem. Soc.*, **82**: 431, 1960.
27. K. Karlowski, *Chem. Abstr.*, **93**: 24564h, 1980.
28. J. Harley, W. Nel, and V. Pretorius, *Nature*, **181**: 177, 1958.
29. I. G. McWilliam and R. A. Dewar, *Gas Chromatography 1958*, D. H. Desty, ed. (New York, NY: Academic Press, 1958), 142.
30. H. J. Keukens, M. M. L. Aerts, W. A. Traag, J. F. M. Nouws, W. G. Ruig, W. M. J. Beek, and J. M. P. Hartog, *J. Assoc. Off. Anal. Chem.*, **75**: 245, 1992.
31. H. J. Keukens, W. M. Beek, and M. M. L. Aerts, *J. Chromatogr.*, **352**: 445, 1986.
32. R. M. L. Aerts, H. J. Keukens, and G. A. Werdmuller, *J. Assoc. Off. Anal. Chem.*, **72**: 570, 1989.
33. N. Haagsma, C. Schreuder, and E. R. A. Rensen, *J. Chromatogr.*, **363**: 353, 1986.
34. S. H. Y. Wong, B. Cudny, O. Aziz, N. Marzouk, and S. R. Sheehan, *J. Liq. Chromatogr.*, **11**: 1143, 1988.
35. M. Ashton, *J. Liq. Chromatogr.*, **12**: 1719, 1989.
36. D. E. Holt, J. Louvois, R. Hurley, and D. Harve, *J. Antimicrob. Chemother.*, **26**: 107, 1990.

37. E. D. Ramsey, D. E. Games, J. R. Startin, C. Crews, and J. Gilbert, *Biomed. Environ. Mass Spectrom.*, **18**: 5, 1989.
38. L. Felice, E. H. Abdennebi, and M. Ashraf, *J. Assoc. Off. Anal. Chem.*, **71**: 1156, 1988.
39. A. R. Long, L. C. Hsieh, A. C. Bello, M. S. Malbrough, C. R. Short, and S. A. Barker, *J. Agric. Food Chem.*, **38**: 427, 1990.
40. U. R. Tjaden, D. S. Stegehuis, B. J. E. M. Reeuwijk, H. Lingeman, and J. V. D. Greef, *Analyst*, **113**: 171, 1988.
41. M. Ramos, T. H. Reuvers, A. Aranda, and J. Gomez, *J. Liq. Chromatogr.*, **17**: 385, 1994.
42. B. Delepine and P. Sanders, *J. Chromatogr.*, **582**: 113, 1992.
43. R. S. Markin, M. C. Wadman, P. L. Boffjen, M. C. Haven, and J. A. Huth, *J. Chromatogr.*, **525**: 464, 1990.
44. B. Yohannes, K. H. Korfer, J. Schad, and I. Ulbrich, *Arch. Lebensmittel-hyg.*, **34**: 1, 1983.
45. H. A. Russel, *Chromatographia*, **11**: 341, 1978.
46. R. Kutter, D. Jalr, and H. Stritzinger, *Fleischwirtschaft*, **62**: 515, 1982.
47. B. Wiese and K. Martin, *Chromatographia*, **15**: 737, 1982.
48. M. Petz, *Z. Lebensm. Unters. Forsch.*, **176**: 289, 1983.
49. S. Lam and G. Malikin, *J. Liq. Chromatogr.*, **12**: 1851, 1989.
50. A. Plantikow and P. Lohs, *Nahrung*, **10**: 1021, 1989.
51. A. M. Rizzo, M. Amorena, and M. L. Cortesi, *Industrie Alimentari*, **18**: 710, 1989.
52. K. Tyczkowska, K. M. Hedeen, D. P. Aucoin, and A. L. Aronson, *J. Chromatogr. Sci.*, **26**: 533, 1988.
53. A. Meulemans, C. Manuel, J. Mohler, A. Roncoropni, and M. Vulpillat, *J. Liq. Chromatogr.*, **4**: 145, 1981.
54. H. P. Nissen and H. W. Kreysel, *Chromatographia*, **28**: 49, 1989.
55. J. M. Wal, J. C. Peleran, and G. F. Bories, *J. Assoc. Off. Anal. Chem.*, **63**: 1044, 1980.
56. V. L. Lee, J. F. M. Nouws, and F. W. R. Bloemendal, *J. Vet. Pharmacol. Ther.*, **5**: 160, 1982.
57. G. S. F. Bories, J. C. Peleran, and J. M. Wal, *J. Assoc. Off. Anal. Chem.*, **66**: 1521, 1983.
58. V. S. Kruyswijk, J. H. V. Gend, and R. Kommerij, *Tijdschr. Diergeneesk. De.*, **108**: 145, 1983.
59. M. Becheiraz, A. Haldemann, and R. Etter, *Mitt. Geb. Lebensmittelunters. Hyg.*, **74**: 147, 1983.
60. P. Sanders, P. Guillot, D. Michele, and J. M. Delmas, *J. Assoc. Off. Anal. Chem.*, **74**: 483, 1991.
61. T. Nagata and M. Saeki, *J. Chromatogr.*, **565**: 471, 1991.
62. T. Nagata and M. Saeki, *J. Liq. Chromatogr.*, **15**: 2045, 1992.
63. J. E. Psomas and E. G. Iosifidoy, *J. Liq. Chromatogr.*, **16**: 2653, 1993.

64. A. Simonella, L. Torreti, and C. Filipponi, *J. High Resolut. Chromatogr.*, **12**: 555, 1989.
65. R. Epstein, *J. AOAC Int.*, **77**: 570, 1994.

Chapter 8

Chemical Analysis of β-Lactam Antibiotics

Joe O. Boison
Food Animal Chemical Residues Section, Health of Animals Laboratory, Food Production & Inspection Branch, Agriculture and Agri-Food Canada, 116 Veterinary Road, Saskatoon, Saskatchewan S7N 5E3 Canada

Contents

Introduction

In 1929 Flemming, a bacteriologist, first observed the dissolution of colonies of *Staphylococcus aureus* in the zone surrounding the contaminating *Penicillium* colony. This observation culminated in the discovery of one of the most remarkable therapeutic agents, penicillin, an antibiotic that has captured the attention of basic and clinical microbiologists, chemists, biochemists, physicians, and veterinarians for over six decades. The history of penicillin and the extensive work that went into its isolation, characterization, and eventual synthesis have been dealt with in great detail elsewhere and will not be considered in this chapter. Benzylpenicillin or penicillin G (Pen G), 6-aminopenicillanic acid (6-APA), and phenoxymethylpenicillinic acid or penicillin V (Pen V) are three of the naturally occurring compounds belonging to the group of antibiotics collectively referred to as β-lactams. This group is comprised of the penicillins, cephalosporins, and cephamycins whose basic nuclear structures are shown in Figure 1. The penicillin nucleus is a β-lactam ring coupled to a thiazolidine (five-membered) ring to form the penam ring system. The cephalosporin nucleus (3-cephem ring system) differs from the penicillin nucleus by having a dihydrothiazine (six-membered) ring coupled to a β-lactam ring. The primary distinguishing structural difference between the cephalosporins and the cephamycins is the methoxyl group substituted for the α-hydrogen at the 7-position on the β-lactam ring. Table 1 shows the chemical structures of the group of naturally-occurring and semisynthetic penicillins. The semisynthetics have been grouped according to whether they are resistant to bacterial penicillinase (antistaphylococcal penicillins), whether they are broad spectrum, or whether they show increased activity against Gram-negative bacteria (antipseudomonal penicillins). The development of semisynthetic penicillins from Pen G and 6-APA is presented in Table 2.

1948 marked the beginning of another significant phase in the evolution of the β-lactam group of antibiotics. Brotzu (*1*) sent a culture of Sardinian *Cephalosporium* that he had isolated from sea water near a sewage outlet at Cagliari to Oxford University where Professor Abraham, Professor Newton, and their research associates were able to isolate and identify cephalosporin C from the culture and demonstrate that the isolate was an antibiotic (*2*). Several of the cephalosporins available today are semisynthetic derivatives of cephalosporin C (Table 3). These have been arranged into first, second, and third generation cephalosporins, showing only those that have been approved for veterinary use (Table 4).

The cephamycins represent the third group of naturally-occurring β-lactam antibiotics. In 1962 Miller et al. (*3*) reported that the actinomycetes, which might be considered filamentous bacteria, do not only produce β-lactam antibiotics, but are also responsible for a much greater variety of such antibiotics. Several naturally-occurring cephamycins (Table 5) have been discovered by various investigators (*4–10*). As in the case of the penicillins and cephalosporins, chemical manipulation of the natural cephamycins has resulted in the development of a variety of semisynthetic cephamycins (Table 6).

There are other structural variants of the basic β-lactam framework, including the amidinopenicillin, mecillinam, the penicillinic acid sulfone, sulbactam, and the oxa-β-lactam compound, moxalactam, which do not fit into any one of the three classified groups and have been synthesized by chemists from newly-discovered natural compounds (Table 7).

Text continues on page 249

Penicillin
(Product of Filamentous Fungi)

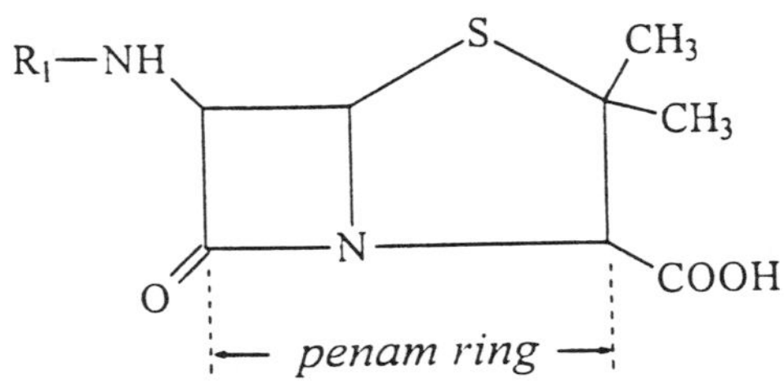

Cephalosporin
(Product of Filamentous Fungi)

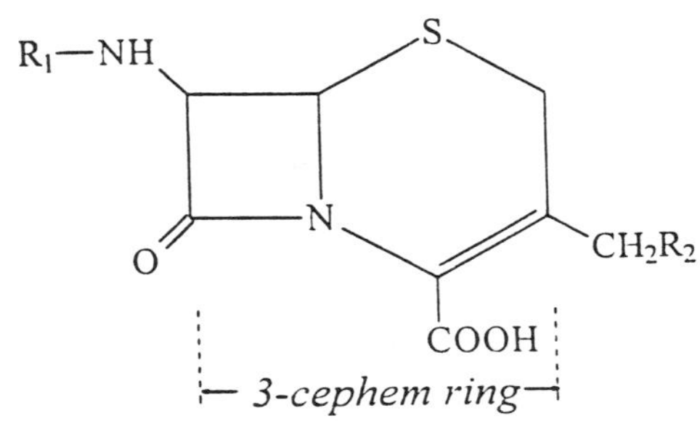

Cephamycin
(Product of Actinomycetes)

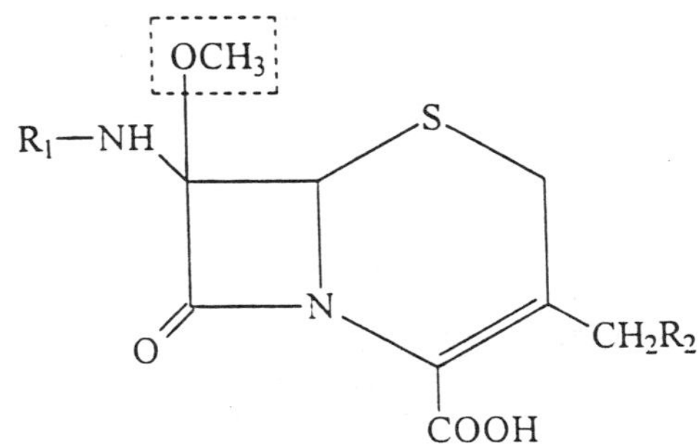

Figure 1 Structural features of the β-lactam antibiotics

Table 1 Natural and Semisynthetic Penicillins

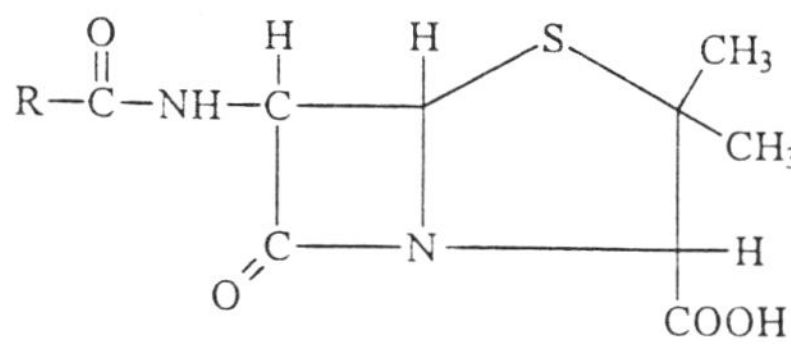

Penicillin	**Structure of side chain, R**
<u>Natural Penicillins</u>	
Benzylpenicllin (Penicillin G)	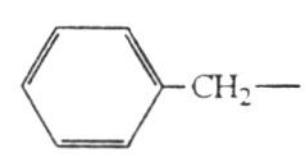
Phenoxymethylpenicillin (penicillin V)	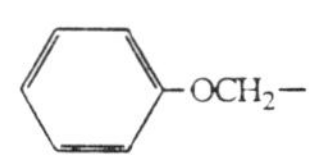
6-Aminopenicilloic acid (6-APA)	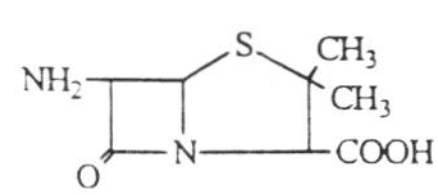 (Full Structure)
<u>Semi-synthetic Penicillins</u>	
<u>Penicillinase-resistant penicillins</u>	
Methicillin	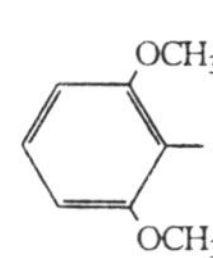

Table 1 Natural and Semisynthetic Penicillins—*continued*

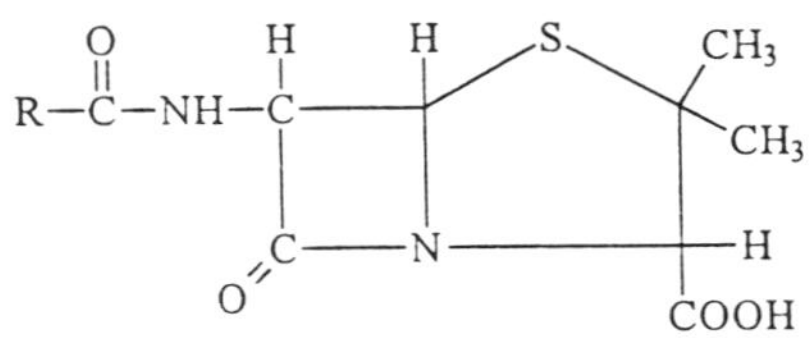

Penicillin	Structure of side chain, R
Antipseudomonal Penicillins	
Carbenicillin	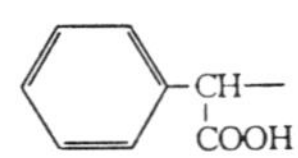
Ticarcillin	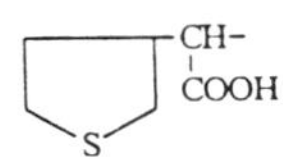
Piperacillin	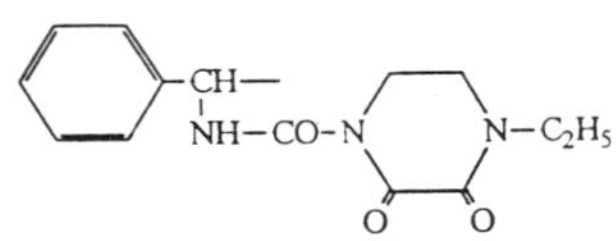
Broad Spectrum	
Ampicillin	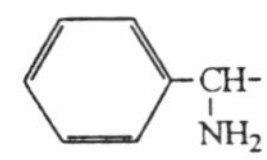
Amoxicillin	
Hetacillin	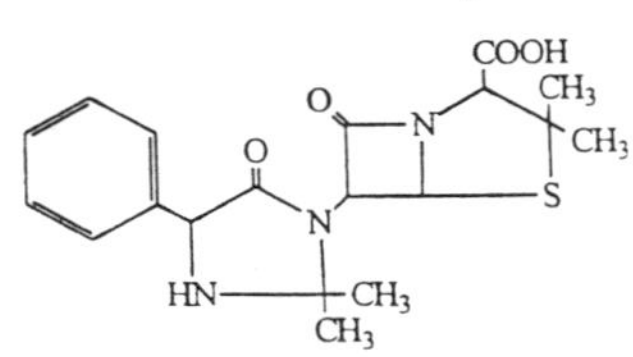 (Full Structure)

Table 1 Natural and Semisynthetic Penicillins—*continued*

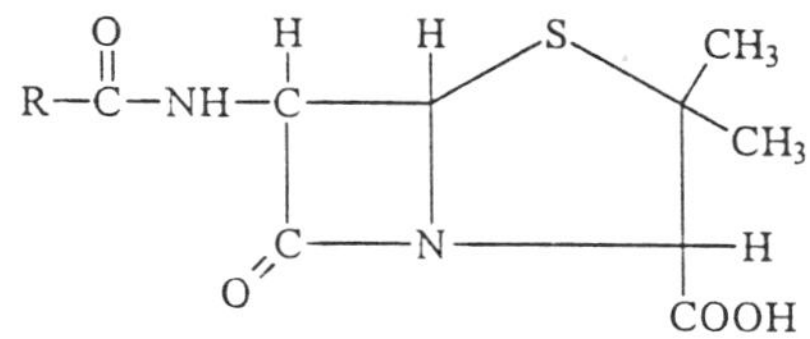

Penicillin	Structure of side chain, R
Isoxazolyl Penicillins	
Oxacillin	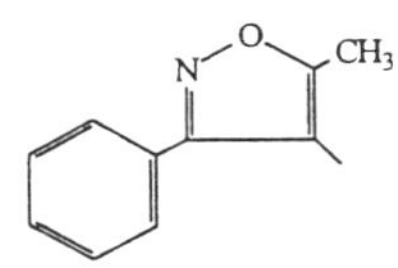
Cloxacillin	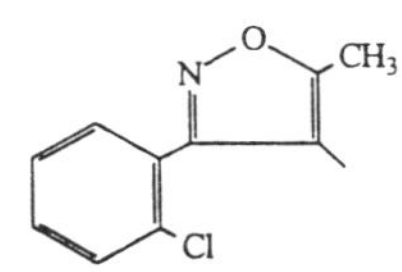
Dicloxacillin	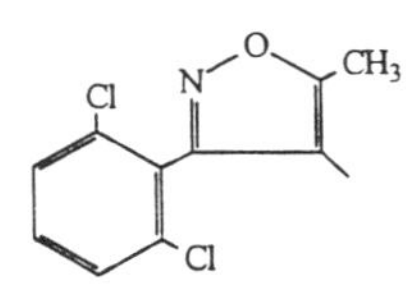
Nafcillin	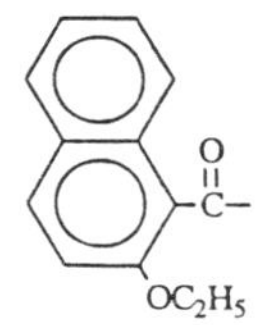

Table 2 The Penicillin Group of Antibiotics and Their Development From Benzylpenicillin and 6-aminopenicillanic Acid

Starting penicillin	Improved property	Resulting penicillin		Penicillin obtained after further improvement
	better absorption and acid stable	penicillin V		
	long acting	procaine Pen G, benzathine Pen G, benethamine Pen G		
	improved absorption	phenethicillin		
	stability to staphylococcal β-lactamase	methicillin →	improved oral absorption	dicloxacillin, cloxacillin, flucloxacillin, oxacillin, nafcillin
Pen G → 6-Aminopenicillanic acid				
	activity against enterobacteria	ampicillin, ciclacillin, hetacillin →	improved absorption	amoxicillin, pivampicillin
	activity against *Ps. aeruginosa*	carbenicillin, ticarcillin, sulbenicillin, piperacillin, apalcillin		

Table 3 The Cephalosporin Group of β-Lactams and Their Development From Cephalosporin C and 7-Aminocephlosporanic Acid

Starting cephalosporin	Mode of administration	First generation		Second generation		Third generation
		cephalexin				
	oral	cephadrine	more active	cefaclor		
	↑	cephaloglycine				
Cephalosporin C →	7-Aminocephalosporanic acid					
	↓	cephalothin				
	parenteral	cephaloridine		cefamondole		ceftiofur
		cefazolin	enzyme stability	cefoxitin	more active	cefotaxime
		cephapirin		cefanocid		ceftizoxime
		cephacetrile		cefuroxime		cefmenoxime
			active against *Ps. aeruginosa*	cefsulodin	more active	ceftazidime
				cefoperazone		

Table 4 Natural and Semisynthetic Cephalosporains

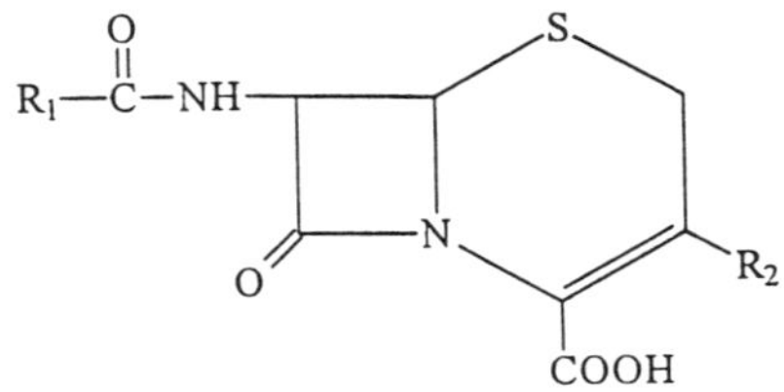

Cephalosporin	Structure of side chain R1	R2
Cephalothin	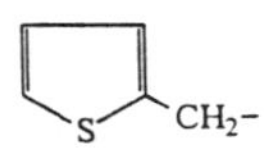	$-CH_2 \cdot OAc$
Cephazolin	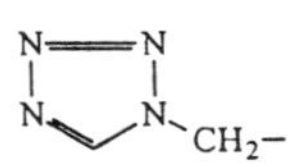	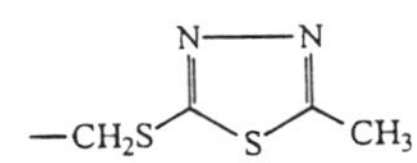
Cephacetrile	$N\equiv C-CH_2-$	$-CH_2 \cdot OAc$
Third Generation Cephalosporins		
Ceftiofur (Excenel)	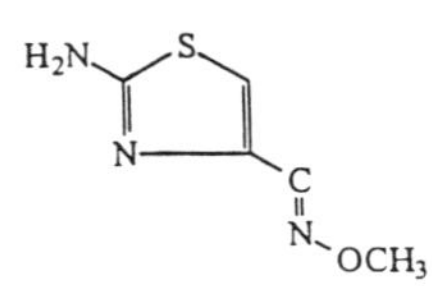	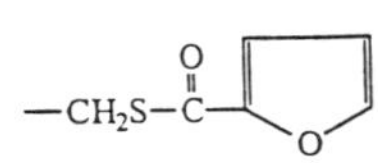

Table 4 Natural and Semisynthetic Cephalosporains—*continued*

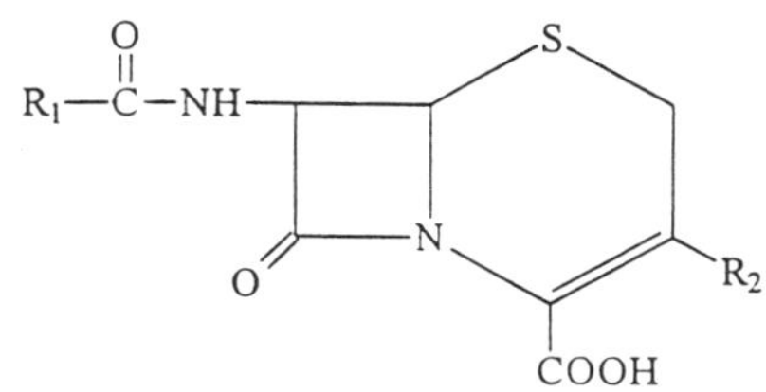

Cephalosporin	Structure of side chain R1	R2
Cephalosporin C	HOOC-CH(NH2)-CH2CH2CH2—	—O—C(=O)—CH3
7-Aminocephalosporanic acid (7-ACA)	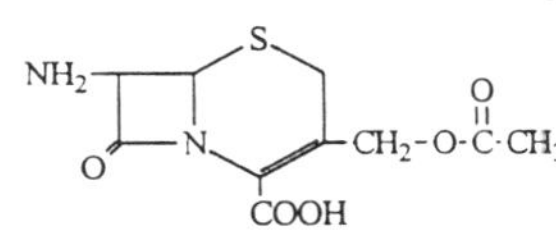 (Full Structure)	
<u>Semi-synthetics (First Generation)</u>		
Cephalexin	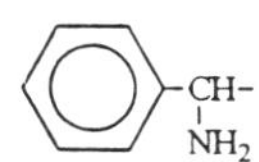	—CH3
Cephapirin	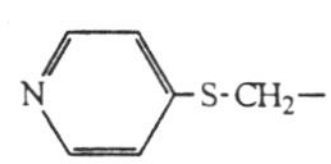	—CH2—O—Ac
Cephaloglycine		—CH2—O—Ac

Table 5 Natural and Semisynthetic Cephamycins

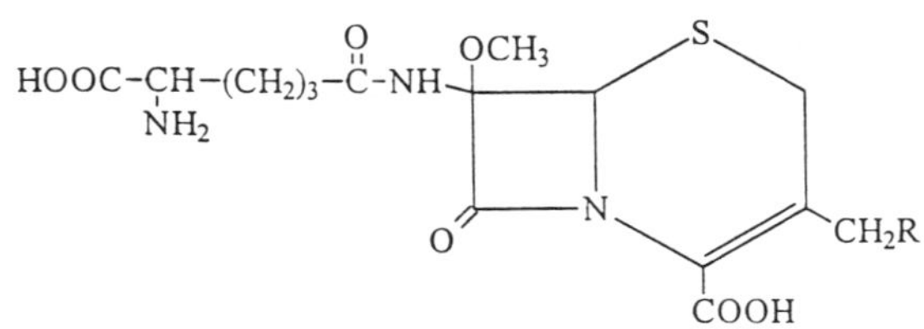

Cephamycin	Structure of side chain, R
Natural Cephamycins	
Cephamycin A	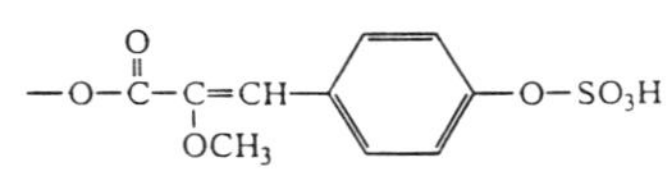
Cephamycin B	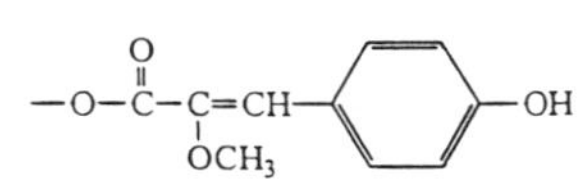
Cephamycin C	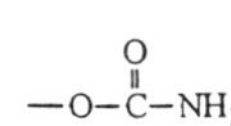

Table 5 Natural and Semisynthetic Cephamycins—*continued*

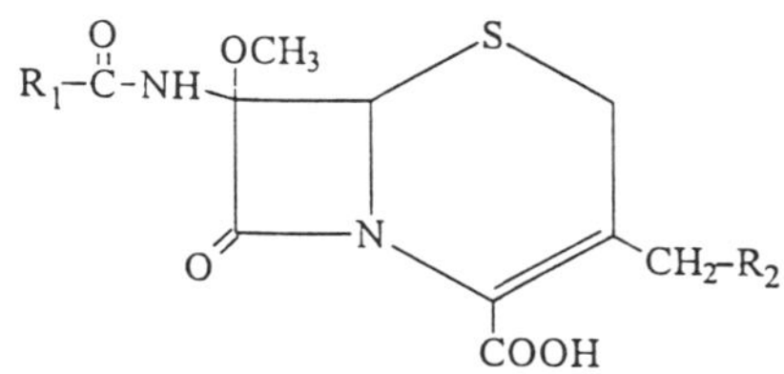

Cephamycin	Complete Structure
Semi-synthetic Cephamycins	
Cefoxitin	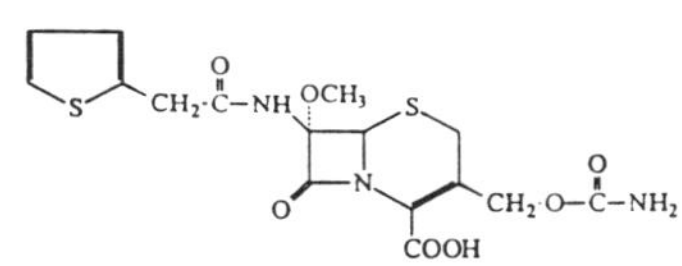
Cefmetazole	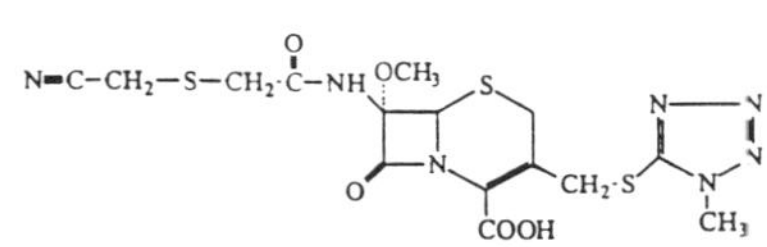
SK&F 73678	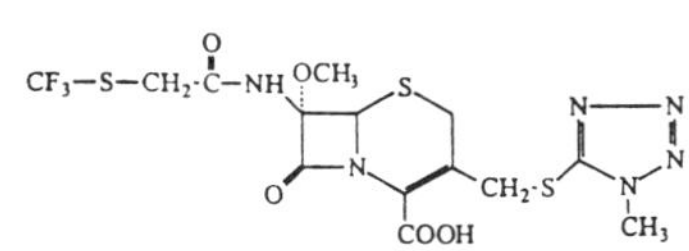
SQ14, 359	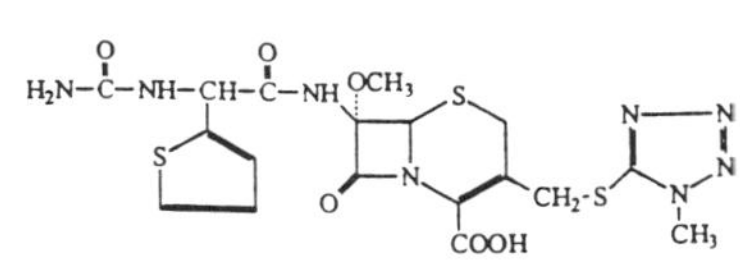

Table 6 The cephamycin group of β-lactam antibiotics and their development from natural cephamycins

Drug Name	Antibacterial property	Resulting cephamycin
Cephamycin (A, B, C)	β-Lactamase resistance	Cefoxitin, cefmetazole
	Broad spectrum	SK & F 73678
	Metabolism stability	SQ 14,359

Table 7 Other Compounds with a β-Lactam Ring Structure

Drug Name	Antibacterial Property	Additional Group Strucure/ Complete chemical structure
Clavulanic Acid	β-lactamase inhibitor	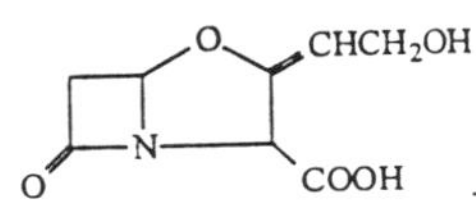 Fused β-lactam oxazolidine ring system
Thienamycin	β-lactamase stable	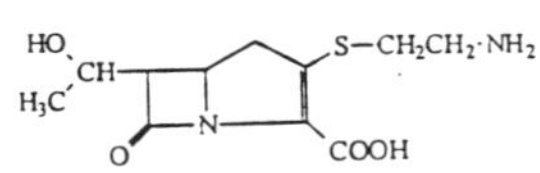 Carbon atom at position 1
Mecillinam	Improved activity against enterobacteria	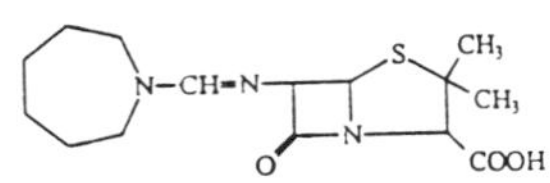 Side chain in amidino linkage
Moxalactam	Improved activity	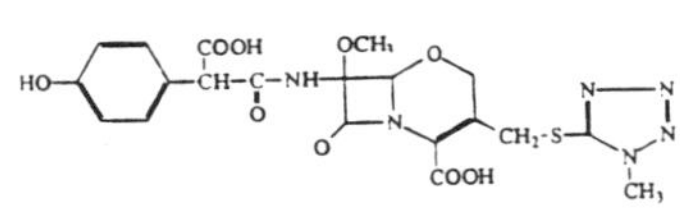 Fused β-lactam dihydrooxazine ring system
Sublbactam	β-lactamase inhibitor	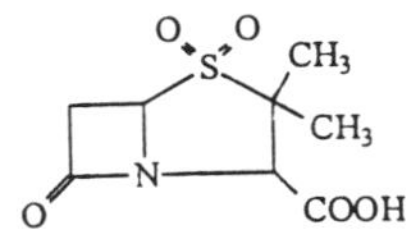 Penicillanic acid sulphone

Mechanism of β-Lactam Action on Bacteria

Scheme 1 illustrates, in general terms, the mechanism of antibiotic action. In an ideal situation, a given antibiotic is expected to reach its target site(s) unimpeded and intact, bind to the essential target(s) and, as a result, the cell should rapidly lose its viability and lyse. It is well known that Pen G and the other β-lactams prefer to bind to peptidoglycan transpeptidase and inhibit its normal cross-linking role in completing the cell wall synthesis in bacteria. It is this inhibition of the synthesis of peptidoglycan, the main contributor to the rigidity of the cell wall structure, which causes the bacterial cells to undergo lysis and death.

Resistance to β-Lactam Action by Bacteria

There are instances, however, when an antibiotic may be prevented by bacteria from reaching its target(s). Bacteria, which may be Gram-positive or Gram-negative, may do this in one of several ways. Either they react to exclude the drug from the cell preventing it from reaching its intended target (exclusion); they react to modify the target site(s) to reduce or prevent its interaction with, or affinity for, the antibiotic; or the bacteria possess enzymes capable of modifying or destroying the drug. When bacteria are exposed to antibiotics, they tend to liberate enzymes, typically penicillinase (β-lactamases), esterases, and amidases, which are capable of hydrolyzing specific bonds in β-lactams to produce antibiotically inactive products or products with very reduced antibiotic activity (Scheme 2). Hydrolysis of the penicillins by β-lactamases results in the production of the inactive penicilloic acid (Scheme 3); with cephalosporins and cephamycins, the corresponding acids are only transiently produced and spontaneously degrade into smaller fragments (Scheme 4). Penicillins are also susceptible to attack by enzymes (amidases) that cleave the CO–NH bond at the 6-amino position of the penam nucleus (Scheme 3). The corresponding cephalosporin amidase also exists, but is less commonly encountered. Those cephalosporins that have an acetoxymethyl group at the 3-position of the cephem ring (i.e., cephalosporin C, cephapirin, cephalothin, and cefotaxime), are susceptible to deacetylation and produce a corresponding hydroxymethyl derivative. This transformation can be accom-

Antibiotic	⇒	**Target(s)**	⇒	**Cell Response**
(Amount reaching the target)		**(Nature of target)**		**(Nature of response)**

Scheme 1 Mechanism of antibiotic drug action

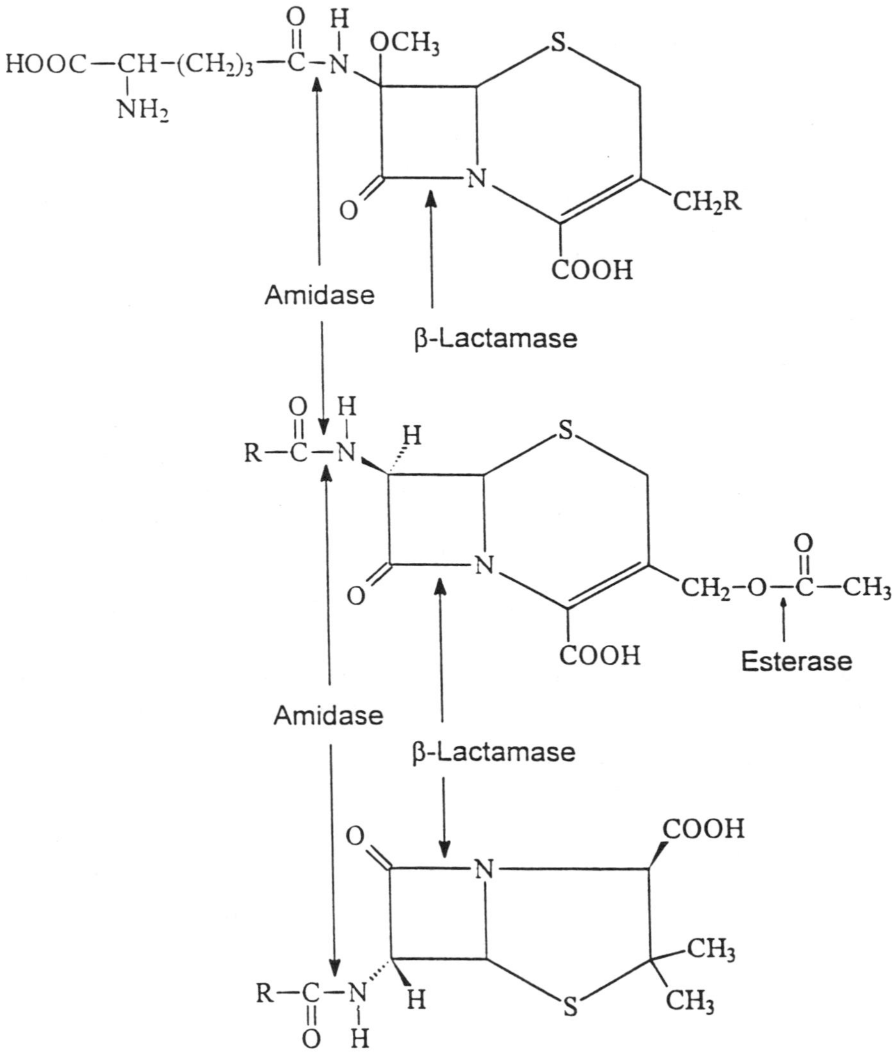

Scheme 2 Site of hydrolytic activity of bacterial enzymes in β-lactams

plished by esterases located in the liver and the kidneys. Further reaction of the hydroxymethyl group and the carboxylic acid group at C-2 may spontaneously occur to form the cyclic lactone (Scheme 4). Hydroxymethylcephalosporins and their corresponding lactones display reduced antibacterial activity when compared to the parent compound (*13*).

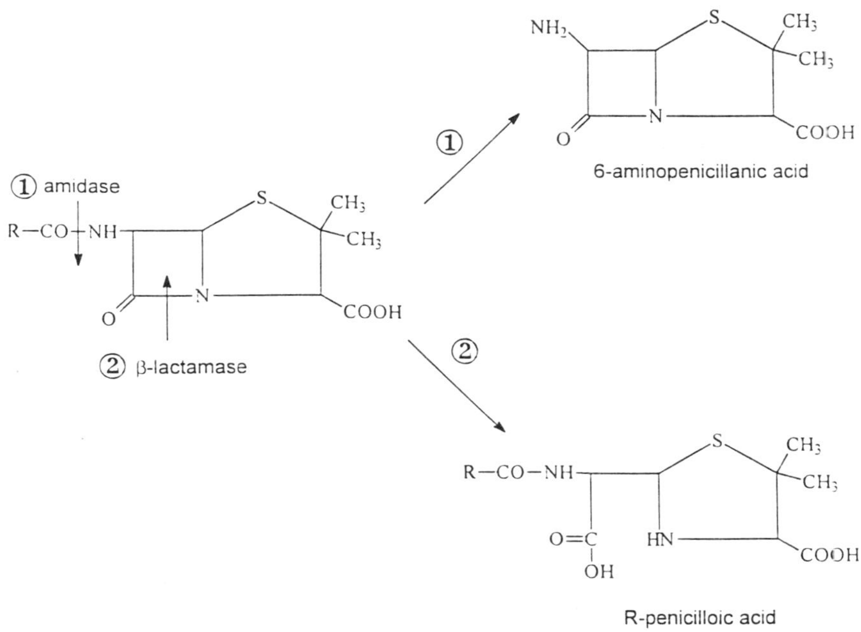

Scheme 3 Hydrolytic products of bacterial enzymes from penicillins

Only a small proportion of the β-lactams that have been synthesized have been approved for use in veterinary medicine. If this is the case, why is it that 65 years after the effect of the first β-lactam antibiotic on microorganisms was observed, scientists are still developing methods for detecting β-lactam residues in foods of animal origin? To answer this question properly and to be able to justify the continued need and interest in developing sensitive laboratory methods for the analyses of β-lactam drug residues in foods, let us first look at the reasons for β-lactam antibiotic use in veterinary medicine, and review some of the reasons why drug residues are still being found in foods.

The Beneficial Uses of β-Lactam Antibiotics in Food Production

β-Lactams are used at therapeutic levels in veterinary practice primarily to treat diseases and to prevent infection. They are also used at subtherapeutic levels to increase feed efficiency, promote growth, and prevent disease. Unlike the treatment of diseases in humans and companion animals, which is usually directed at the individual patient, treating diseases in food-producing animals is generally

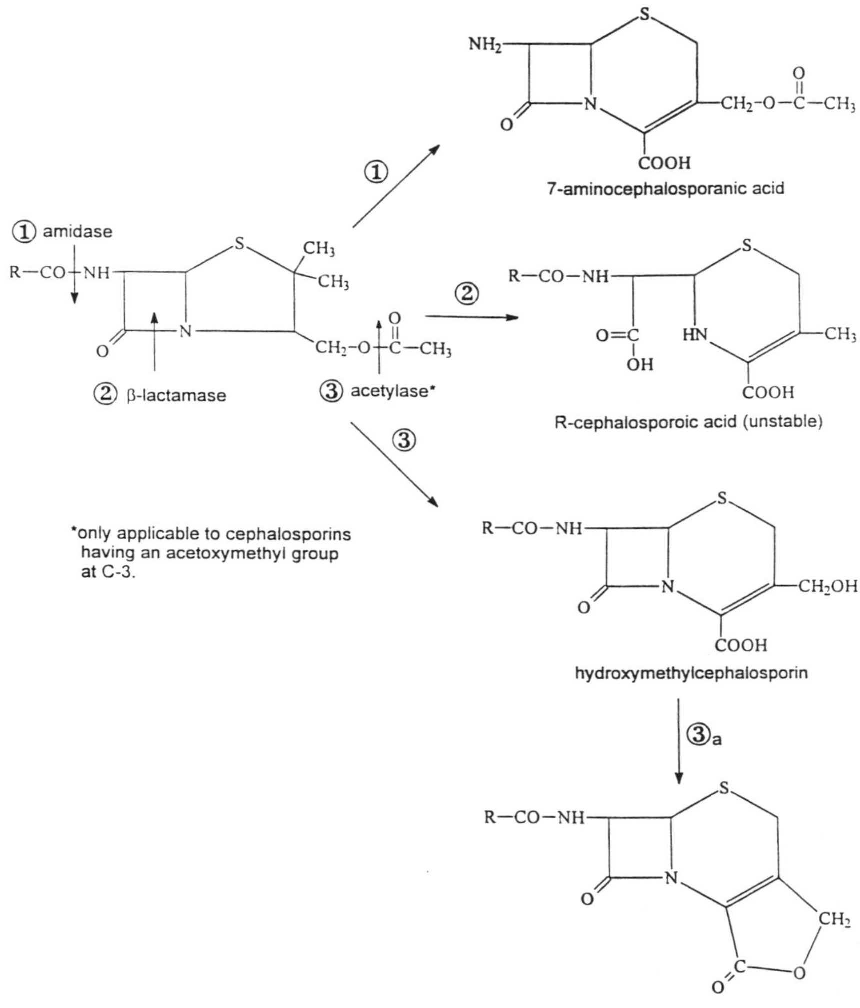

Scheme 4 Hydrolytic products of bacterial enzymes from cephalosporins

directed at the herd or flock to prevent the spread of infection to a large proportion of the animals (*14*). Economic considerations require that sick animals be returned to full health and productivity at the earliest opportunity, and that epidemics from infectious diseases be prevented from developing (*15*). Herd medicine, therefore, plays a very important role in food-animal production, especially in those areas of animal management where it is imperative to keep all the animals at a level of optimum productivity.

Advocates for the continued use of antibiotics in food-animal production have argued that large numbers of farm animals would die and chronic bacterial diseases would prevail causing a loss of animal life and a drastic reduction in milk and meat production, if there were no antibiotics available today for veterinary use. On the other hand, those who would rather see an end to veterinary drug use in food-animal production argue that (*a*) antibiotics are necessary only because of the stressful conditions under which the animals are managed, (*b*) that there would be an increase in the development of bacterial resistance to antibiotics, (*c*) that it is possible to transfer antibiotic-resistant bacteria from farm animals to human beings, (*d*) that there would be a prevalence of unwanted veterinary drug residues in milk and edible tissues of food-producing animals, and (*e*) that there is concern about the adverse hypersensitive reactions in some individuals from ingesting low levels of antibiotics in milk and animal tissues. Additionally, such residues may inhibit starter cultures used in the production of yogurt, cheese, and other solid milk products. Even though all of these issues have been actively and vigorously debated and challenged by advocates for banning the use of antibiotics in food-animal production, it is perhaps the last three which are of most interest to analytical chemists who provide the scientific analytical methodology to support some of the attempts by regulatory agencies to enforce legislation regulating the use of veterinary drugs in milk and tissues of food-producing animals.

Reasons for the Presence of β-Lactam Drug Residues in Foods

In a recent survey by Van Dressler and Wilcke (*16*) to determine the reasons for violative residues in animal tissues in the United States, Pen G was found to be among the most prevalent tissue residues in culled dairy cows, veal calves, and market hogs. In over 80% of the violative cases investigated, the producer was found to be the responsible party, and failing to observe the withdrawal time was one of the major management deficiencies noted. Analytical data gathered over several years at the Health of Animals Laboratory (HAL) are in agreement with the findings made by Van Dressler and Wilcke indicating that one of the most frequently detected antibiotics in violative tissue samples in Canadian slaughter animals is Pen G (*17*).

The Role of β-Lactam Antibiotics in the Induction of Allergic Reactions

Documented cases (*18–22*) of allergic reactions believed to have resulted from eating penicillin-contaminated foods indicate that Pen G is the only member of the β-lactam group of antibiotics that has been frequently implicated. At the 36th Joint FAO/WHO Expert Committee on Food Additives (JECFA) meeting held in

Rome 5–14 February, 1990 (*23*), it was concluded after reviewing documented cases of penicillin-implicated induced allergic reactions in humans that "the very small amount of any drug with allergenic potential likely to be encountered as a residue in food of animal origin was unlikely to lead to sensitization. Additionally, sensitization was only likely to occur as a result of exposure to large doses of an allergenic drug. In theory, however, small amounts of an allergenic drug present as a residue could elicit an allergenic response in a previously sensitized person, they maintained (*24*, *25*).

Physico-Chemical Properties of β-Lactams

Penicillins are relatively unstable in aqueous solutions. Their degradation is catalyzed both by acids and by bases (*26*). They are also extremely susceptible to nucleophilic attack in aqueous solutions. The optimum stability for the amphoteric penicillins occurs at a pH which coincides with the isoelectric point; for the monobasic penicillins, all of which have measured acid dissociation constants (pK_a) of less than 3 (*27*, *28*), this generally occurs at a pH between 6 and 7 (Table 8). The aqueous instability of the β-lactams has important consequences for extraction and chromatography conditions. It also has significant implications with regard to the stability of stock solutions and the storage of incurred biological samples. For example, Wiese and Martin (*29*) demonstrated that standard solutions of Pen G are rapidly degraded under conditions of low and high pH, while they remain fairly stable between pH 5 and 9 (see Figure 2). The stability of incurred Pen G residues in animal plasma (*30*, *31*), animal tissues (*31*), and milk (*32*, *33*) under various storage conditions has been studied. These studies show that Pen G is stable in these matrices if the samples are stored frozen at –76°C. The polar nature of the β-lactams, coupled with the fact that most of the β-lactam drugs are formulated as their sodium or potassium salts, makes them readily soluble in water.

Table 8 Acid Dissociation Constants (pK_as) for Common β-Lactams

β-Lactam	pK_{a1} (-COOH)	pK_{a2} ($-NH_2$)	pK_{a3} (-OH)
Amoxicillin	2.4[a]	7.4	9.6
Ampicillin	2.66[a]	7.24	—
Penicillin V	2.74[b]	—	—
Penicillin G	2.73[b]	—	—
Cloxacillin	2.73[b]	—	—

[a] Adapted from reference (*27*).
[b] Adapted from reference (*28*).

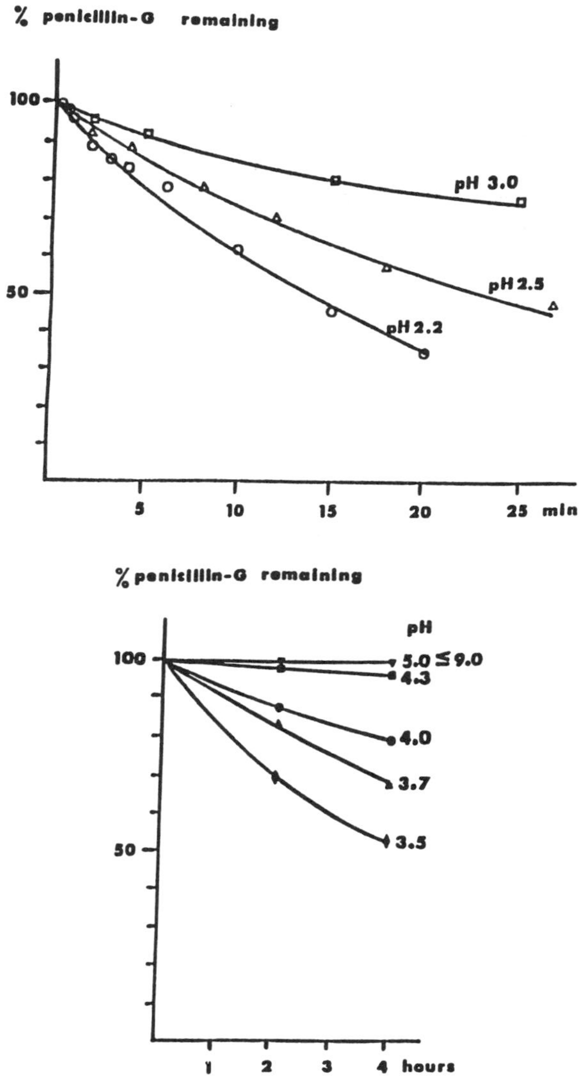

Figure 2 The stability of penicillin G at room temperature in aqueous buffer at different pH. Samples 100 μg/mL. See reference (*29*).

Others, such as amoxicillin (solubility 4.2 to 9 mg/mL at pH 4.8 (*34*)) and benzathine Pen G (solubility 0.15 mg/mL (*35*)), show decreased solubility in water.

Because most methods developed for the analysis of β-lactams use various organic solvents, especially during the extraction and isolation and chromatographic analysis stages, it is useful to review the effect such solvents have on the stability of β-lactam drugs. It is well known that gastric juices rapidly degrade Pen G. It is also well known that methanol rapidly degrades penicillins into their corresponding penicilloic acid esters. In a study of the stability of cloxacillin in water and aqueous organic solutions by thermospray liquid chromatography–mass spectrometry (LC–MS), Tyczkowska and co-workers (*36, 37*) found that cloxacillin in methanol results in the most rapid degradation of the drug, while 25% acetonitrile, 25% ethanol in water results in the least degradation. The methyl or ethyl esters of cloxacillin were the predominant products found in methanolic or ethanolic solutions of cloxacillin stored at –20°C for 2 weeks.

Animal tissues and biological fluids are typically complex mixtures composed of a wide variety of proteins, carbohydrates, phospholipids, etc. In addition to the deleterious effects these compounds have on the components of the chromatographic system (i.e., the column packing material and the pumps), their presence often interferes chromatographically with the analyte(s) of interest. These must, therefore, be selectively removed from any extracts containing the analyte(s) of interest prior to chromatographic analysis. An effective approach to developing suitable methods for analyzing β-lactams requires an understanding of the distribution and binding (reversible or irreversible) characteristics of these drugs in plasma and animal tissues.

The concentration-time profile for a drug in the bloodstream and in tissues is the result of the drug moving from the intravascular space into tissues. It is also related to the quantity of drug administered, the absorption rate of the drug from the site of administration, the rate of metabolism of the compound, the degree of urinary and hepatic excretion of the antibiotic, and the fractional distribution of the antibiotic between the extracellular fluid (ECF) and intracellular fluid (ICF) spaces. For the penicillins, a relatively important correlation exists between tissue and serum protein binding of drugs when the degree of serum protein binding is contrasted with the appearance of the antibiotics in interstitial fluid (*38*, *39*).

The physicochemical properties of the β-lactams briefly reviewed above will, to a large extent, dictate the experimental approaches that should be considered by analysts in their attempts to develop analytical methods for these analytes and their metabolites in biological matrices. These matrices have traditionally included milk, urine, lymph, interstitial fluid, liver, kidney, injection-site, and noninjection-site muscle tissue samples.

A number of highly sensitive rapid tests with detection sensitivities for most β-lactam antibiotics in the low µg/kg (ppb) range, are commercially available for screening antibiotic residues in milk. Because rapid tests for the analysis of β-lactam residues in food cannot discriminate among different β-lactams, alternative confirmatory methods with equivalent or better detection sensitivity are needed to confirm the identity or identities of the specific β-lactam drug(s) eliciting presumptive positive responses in the rapid test and to determine their residual concentrations.

Analytical methods developed for determining β-lactams over the 65 years since penicillin's antibiotic effect was first observed may be classified into two groups: those that were developed and published prior to the 1990 JECFA meeting, and those that were developed and published after the meeting.

Residue Methods Available by 1990

Table 9 lists the numerous chemical methods that had been developed for the analyses of β-lactams in milk, biological fluids, and animal tissues by 1990 (*40–96*). More methods had been developed for milk analyses than for tissues, which had detection sensitivities comparable to those of the microbiological assays being used for screening these drugs. For example, Wiese and Martin (*32*) had developed a sensitive liquid chromatographic–ultraviolet (LC–UV) method with a detection limit of 0.2 µg/kg for benzylpenicillin in milk, while Meetschen and Petz (*89*) had developed a gas chromatographic (GC) method with a detection limit of <0.4 µg/kg for the determination of seven penicillins in milk (see Figure 3). For tissue residues, the methods available had detection limits of about 50 µg/kg. Most of the chromatographic methods that had been published at that time generally included cumbersome and labor-intensive procedures that were sometimes difficult to reproduce for routine use.

Observations of JECFA and Recommendations on Residue Methods and Residue Data for β-Lactams

The Committee observed that analytical bioassay methods with detection limits between 1–10 µg/L (ppb) used for screening antibiotic residues in milk were well established and provided practical and suitable means for screening benzylpenicillin residue in milk at its maximum residue limit (MRL) of 4 µg/L. These screening methods have been extensively discussed in Chapter 4 of this book. It had become apparent to this Committee in February 1990 that bioassay methods were still being widely used to screen for residues of benzylpenicillin. The Committee also noted that there were several chemical methods available with detection limits of 50–100 µg/kg in meat and 10–50 µg/L in milk. Only very limited data on residues in meat were available from studies in cattle. There were no meaningful tissue depletion studies available, except for two reported studies conducted in which single cows were injected intramuscularly (IM) with a single

Table 9 Determinative Methods for Analyzing β-Lactam Residues in Animal Tissues and Fluids

Matrix and β-lactams analyzed	Analytical method	Detection limits (μg/g or μg/mL)	Year of pub. (ref.)
Standard solutions of Pen G and Pen V	FID of methyl esters	ND	1964 (*40*)
Tissue fluids and plasma in dogs, horses, oxen, and sheep; benzathine penicillin G	Tissue and plasma depletion study	ND	1965 (*41*)
Standard solutions of 8 penicillins	FID of silyl derivatives	ND	1971 (*42*)
Standard solutions; benzylpenicillin	UV (325 nm) spectrophotometric study	ND	1971 (*43*)
Standard solutions; benzylpenicillin	UV (325 nm) spectrophotometric study	ND	1972 (*44*)
Standard solutions; methicillin, carbenicillin, penicillins G and V, propicillin, oxacillin, cloxacillin, phenethicillin, and dicloxacillin	UV (325 nm) spectrophotometric study	ND	1972 (*45*)
Standard solutions; oxacillin, dicloxacillin, penicillins G and V, phenethicillin, propicillin, carbenicillin and ampicillin	UV (325–445 nm) spectrophotomretric study	ND	1972 (*46*)
Standard solutions; ampicillin	UV (325 nm) spectrophotometric study	ND	1974 (*47*)
Standard solution of 8 penicillins	ECD	ND	1975 (*48*)
Standard solutions of penicillins	ECD	ND	1975 (*49*)
Freeze-dried hamburger and chicken; penicillin G and its decomposition products after cooking	LC–UV (254 nm); MS, IR, TLC-autoradio-chromatography, NMR	ND	1977 (*50*)
Standard solutions of ampicillin	FID of trimethylsilyl derivatives	ND	1977 (*51*)
Standard solutions of 8 penicillins	FID of silyl derivatives	ND	1977 (*52*)
Standard solution of cephalexin	FID of trimethylsilyl derivative	ND	1978 (*53*)
Plasma of Dutch-Friesian calves; amoxicillin	A pharmacokinetic study	ND	1978 (*54*)

Table 9 Determinative Methods for Analyzing β-Lactam Residues in Animal Tissues and Fluids–*continued*

Matrix and β-lactams analyzed	Analytical method	Detection limits (μg/g or μg/mL)	Year of pub. (ref.)
Muscle and kidney tissues of dairy cows; amoxicillin trihydrate, sodium penicillin G, procaine penicillin G, sodium ampicillin, sodium cephacetrile, sodium cephapirin	Depletion study	ND	1978 (*55*)
Urine and plasma; amoxicillin	LC–UV (310 nm)	0.025 (plasma), 0.2 (urine)	1979 (*56*)
Fermentation media; penicillins K, N, V, G, X, and methicillin	Gradient LC–UV (325 nm)	1.0	1983 (*57*)
Milk; penicillin G, penicillin V, and cloxacillin	LC–UV (220 nm)	Cloxacillin (0.002), penicillins G and V (0.005)	1983 (*58*)
Beef and pork muscle, kidney, liver, and serum; penicillin G and cloxacillin	LC–UV (220 nm)	0.05	1984 (*59*)
Fermentation broth; penicillins V, G, X, N, 6-APA, cephamycin C, cephalosporin C, cephalexin, ampicillin, and cephadrine	LC–UV (325 nm)	Cephamycin C and penicillin N (0.5–1.0), ampicillin (26), cephalexin (14), cephadrin (8)	1984 (*60*)
Drug formulations; benzathine Pen V, benzathine cloxacillin, procaine penicillin, benethamine penicillin, benzathine diacetate, FID of trimethylsilyl derivatives	LC–UV (258 nm)	ND	1984 (*61*)
Drug formulations; ampicillin trihydrate, amoxicillin trihydrate, potassium Pen V, Potassium propicillin, sodiun Pen G, oxacillin sodium, fluoxacillin sodium, and cloxacillin sodium	LC–UV (254 nm)	ND	1984 (*62*)
Milk; penicillins G and V, phenethicillin, nafcillin, methacillin, oxacillin, cloxacillin, and dicloxacillin	LC–Fluoresxence (254 nm excit, 500 nm emission)	Detection limits not given but recoveries were calculated	1985 (*63*)
Urine and plasma; benzylpenicillin	LC–UV (214 nm)	Urine (10), plasma (0.2)	1985 (*64*)

Table 9 Determinative Methods for Analyzing β-Lactam Residues in Animal Tissues and Fluids–*continued*

Matrix and β-lactams analyzed	Analytical method	Detection limits (μg/g or μg/mL)	Year of pub. (ref.)
Urine and serum; sulbenicillin (SBPC), carbenicillin (CBPC), and ticarcillin (TIPC)	LC–UV (328 nm for CBPC, 327 nm for TIPC, 331 nm for SBPC)	Sulbenicillin (0.05), carbenicillin and ticarcillin (0.1)	1985 (*65*)
Milk; penicillins G and V, ampicillin	LC–UV (210 nm)	0.03	1985 (*66*)
Cattle muscle, kidney, and liver; Pen G	LC–UV (325 nm)	0.05	1985 (*67*)
Standard solutions; pen G, ampicillin, amoxicillin, cloxacillin, sulbenicillin, carbenicillin, ticarcillin	UV (250–400 nm) spectrophotometric study	ND	1985 (*68*)
Biological fluids and pharmaceutical preparations; penicillins	A review paper	Not applicable	1985 (*69*)
Plasma, amoxicillin	LC–UV (229 nm)	0.10	1985 (*70*)
Canine and equine serum; ampicillin	LC–UV (230 nm)	<0.5	1986 (*71*)
Pharmaceutical preparations; Pen G	UV–vis analysis at 650 nm	1.555	1986 (*72*)
Swine muscle, kidney, liver, and serum; penicillin G	LC–UV (220 nm)	ND	1986 (*73*)
Fish tissue; ampicillin	LC–UV (222 nm)	0.03	1986 (*74*)
Pharmaceutical preparations; pen G and V, amoxicillin, cloxacillin, dicloxacillin, methicillin, nafcillin, oxacillin	Review paper	ND	1986 (*75*)
Standard solutions and urine; pen G, ampicillin, ciclacillin, carbenicillin, ticarcillin and metabolites	LC-UV (postcolumn reaction; 270-280 nm)	0.1–0.2	1986 (*76*)
Standard solutions; cefixime, ampicillin, penicillin V and their thermal decomposition products	Thermospray LC–MS	ND	1987 (*77*)
Canine serum and tissues; cefazolin	LC–UV (230 nm)	0.01	1987 (*78*)
Urine, amoxicillin, and metabolites	LC–UV (300 nm)	0.5	1987 (*79*)

Table 9 Determinative Methods for Analyzing β-Lactam Residues in Animal Tissues and Fluids–*continued*

Matrix and β-lactams analyzed	Analytical method	Detection limits (μg/g or μg/mL)	Year of pub. (ref.)
Standard solutions; pen G and acid-base degradation products	LC–UV (230nm)	ND	1987 (*80*)
Rat plasma, bile, and urine; ampicillin and degradation products	LC-UV (postcolumn 270 nm)	Detection limit of 25 ng indicated	1987 (*81*)
Standard solutions; ampicillin, amoxicillin, benzylpenicillin, ampicillin piperazine-2,5-dione, and amoxicillin piperazine-2,5-dione	MS–MS	ND	1988 (*82*)
Fermentation media; 6-APA, isopenicillin N, δ-(L-aminoadipyl)-L–α-cysteinyl-D-valine (ACV) dimer, cephalosporin C, ampicillin, deacetylcephalosporin C, and deacetoxycephalosporin C	Gradient LC–Flourescence [260 nm (excit), 313 nm (emission)]	Isopenicillin N (0.5), 6-APA (0.01), cephalosporins (0.25), ACV (0.5)	1988 (*83*)
Fermentation media; 6-APA, penicillins G, V, X, K and penicillin G penicilloic acid	Gradient LC–UV (328 nm)	PenicillinsG, V, and X (0.001), 6-APA (0.010), penicillin K (0.100), penicllin G penicilloic acid (0.100)	1988 (*84*)
Canine skin biopsy fluid; cephalexin	LC–UV (230-320 nm; 260 nm); LC–MS	0.100	1988 (*85*)
Standard solutions; pen G, pen V, ampicillin, amoxicillin, piperacillin, cloxacillin	LC–UV (229 nm)	ND	1988 (*86*)
Standard solutions; 25 penicillins and 20 cephalosporins	FAB–MS	ND	1989 (*87*)
Plasma and lymph from Swedish Red and White cows; pen G	LC–UV (325 nm)	Plasma (0.0008), lymph (0.0016)	1989 (*30*)
Milk; benzylpenicillin	LC–UV (325 nm)	0.000214	1989 (*32*)

Table 9 Determinative Methods for Analyzing β-Lactam Residues in Animal Tissues and Fluids–*continued*

Matrix andβ-lactams analyzed	Analytical method	Detection limits (µg/g or µg/mL)	Year of pub. (ref.)
Milk; benzylpenicillin	LC–UV (210 nm); LC–MS	LC–UV (0.010), LC–MS (0.100)	1989 (*88*)
Bovine milk; penicillin G and V, methicillin, oxacillin, cloxacillin, dicloxacillin, and nafcillin	GC–NPD	<0.0004 for all β-lactams	1990 (*89*)
Milk; ampicillin	LC–UV (210 nm)	0.005	1990 (*90*)
Milk; penicillin G, penicillin V, oxacillin, methicillin, nafcillin, cloxacillin, and dicloxacillin	LC–Fluorescence [320 nm (excit), 400 nm (emission)]	200 pg	1990 (*91*)
Cattle plasma; desfuroyl-ceftiofur	LC–UV (254 nm)	0.100	1990 (*92*)
Standard solutions; amoxicillin, ampicillin, 6-APA, oxacillin, cloxacillin, ticarcillin, nafcillin, and dicloxacillin	CZE	ND	1990 (*93*)
Bovine tissues and milk; cloxacillin, penicillin G, cephapirin and other veterinary drugs	Particle beam LC–MS	0.010	1990 (*95*)
Bovine tissues and milk; penicillin G, cloxacillin, and ampicillin	Thermospray LC–MS	0.010	1990 (*95*)
Bovine kidneys; desfuroyl ceftiofur	LC-UV (254 nm) and ^{14}C-radioactivity counting	—	1988 (*96*)

FID: flame-ionization detector; ND: not determined; ECD: electron-capture detector; MS–MS: tandem mass spectometric analysis; FAB: Fast Atom Bombardment; NPD: nitrogen–phosphorus detection; CZE: capillary zone electrophoresis

dose of benzylpenicillin at 6 and 6.7 mg/kg body weight, respectively, and killed 2 hours later (*59*, *100*). The residue data measured using different assays (microbial inhibition, high-performance liquid chromatography (LC), and total residue) differed significantly from one another. Benzylpenicillin residues in calves injected IM with either one dose of Pen G at 10 mg/kg body weight or two doses of 3 mg/kg body weight 24 hours apart, and killed 24 hours later, had benzylpenicillin

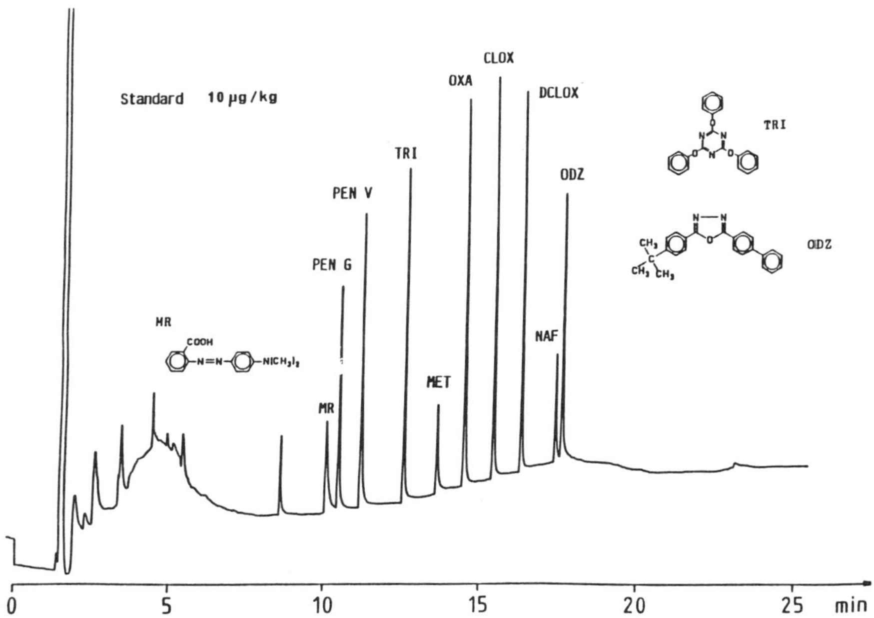

Figure 3 Capillary GC separation of methylated esters of monobasic penicillins on methyl silicone fused silica column (Ultra-1 Hewlett Packard 25 × 0.2 mm i.d., 0.4 mm o.d., 0.11 µm film thickness) obtained on a Varian Model 3700 GC equipped with a cooled injection system KAS 2 for cold sample injection and thermionic nitrogen/phosphorus detector (NPD).
Operating conditions: Carrier gas: nitrogen 1.4 bar inlet pressure; carrier gas venting (front pressure control KAS 2) 20 mL/min; solvent venting: 30 mL/min; venting at split mode: 6 mL/min. Purge gas: nitrogen at 25 mL/min; Detector gases: Hydrogen 1.4 bar inlet pressure; air 180 mL/min.
Injection program for KAS 2: split/splitless mode with solvent venting; injection volume 3 µL; injection temperature 48°C; first heat-up phase 2°C/s; first retention phase 50°C/10 s; second heat-up phase 12°C/s; second retention phase 300°C/180 s. Detector 310°C; bead current 3 pA, bias voltage 4V. Oven temperature program: 1 min at 70°C, then maximum heating to 220°C; increase from 220–295°C at 6°C/min. Electrometer 10–12 amp/mV; attenuation 4–16.
Peak identification: MR: methyl red; PEN G: benzylpenicillin; TRI: "Triazine" (internal standard); MET: methicillin; OXA: oxacillin; CLOX: cloxacillin; DCLOX: dicloxacillin; NAF: nafcillin; ODZ: "oxadiazole" (internal standard). See reference (*89*).

residues in excess of 40 µg/kg in muscle, liver, and kidney when assayed by a microbial assay method (*101*). With a recommended MRL for total residues of benzylpenicillin of 50 µg/kg for liver, kidney, and muscle (all species) and an MRL of 4 µg/L for milk, the Committee made the following recommendations:

1. the provision of results of new studies on the depletion of benzylpenicillin residues from edible tissues of food-producing animals,
2. investigation of the accuracy and precision of the assays used to measure residues of benzylpenicillin, and
3. the development of more sensitive chemical assays for benzylpenicillin.

The recommendations made by JECFA, coupled with the definition of MRLs for most of the β-lactam antibiotics in foods, have provided the impetus needed in today's regulatory community for the continued interest in developing more suitable chemical methods for determining these drug residues in food. This need has even been exacerbated with the recent certification of commercially available rapid test kits by the AOAC Research Institute (AOAC RI) in the United States for screening β-lactam residues in milk. In the European Union, Directive 86/469 (*102*) requires member states to provide confirmatory methods for veterinary drug residues, including β-lactams, detected in milk and blood plasma using rapid tests. Additionally, the recommendations made by JECFA provide guidelines for the characteristics that chemical methods to be developed for the purposes of addressing these issues should have. Such methods must have operational characteristics that must not only be specific for the individual β-lactam, but must also demonstrate detection sensitivities below MRLs established for the β-lactam drug(s) of interest.

Analytical Methods Developed for β-Lactam Analysis in Foods Subsequent to the JECFA Meeting

Table 10 catalogs the chemical methods (*33, 36, 37, 103–132*) that have been developed for analyzing β-lactam drug residues subsequent to the 1990 JECFA meeting. These methods will now be reviewed to ascertain whether or not they are addressing the needs identified in the JECFA recommendations. Very few GC methods have been developed for analyzing β-lactams in milk and animal tissues using mass-flow rate type detectors such as the flame ionization (FID) and thermionic or nitrogen-phosphorus sensitive (NPD) detectors, or concentration type detectors such as the electron capture detector (ECD). With the exception of two methods that have been recently developed (*89, 114*), most of the GC methods were developed in the 1970s (*40, 42, 48, 49, 51–53*). Using an NPD, Meetschen and Petz (*89*) have developed a sensitive GC method for determining seven monobasic penicillins in bovine milk and tissues. Unfortunately, this method cannot be used to determine the amphoteric penicillins such as ampicillin and

Table 10 Determinative Methods Developed or Published After 1990 for the Analyses of β-Lactam Residues in Animal Tissues and Biological Fluids

Matrix and β-lactams analyzed	Analytical methods	Detection limits (μg/g or μg/mL)	Year of pub. (Ref.)
Standard solutions and pharmaceutical preparations; pen G, ampicillin, oxacillin, amoxicillin, methicillin, benzypenicillenic acid, penicillic acid, 6-aminopenicillanic acid, penicillamine	UV–vis spectrophotometric analyses (460 and 842 nm)	ND	1991 (*103*)
Standard solutions; cloxacillin	Thermospray LC–MS	ND	1991 (*36*)
Fish tissue; pen G and other antibiotics	LC–MS	ND	1991 (*104*)
Calf muscle, liver, and kidney; pen G	LC–UV (325 nm)	0.005	1991 (*105*)
Fermentation broth; pen G, pen V, isopenicillin N, penicillin K, penicillin X, 6-APA, benzylpenicilloic acid, tripeptide δ-(L-aminoadipyl)-L-cysteinyl-D-valine (ACV)	LC–Fluorescence [324 nm (excit), 520 nm (emission)]	Isopen N (0.045), pen G (0.050), ACV (0.100), 6-APA (0.150)	1991 (*106*)
Standard solutions; cefazolin, cefaclor, and cefaloridine	Microbore LC–UV (262 nm), DLI–MS	Cefaloridine and cefazolin (1 ng), ceftioxime (5 ng)	1991 (*107*)
Bovine muscle, kidney and liver; pen G	LC–UV (325 nm)	0.005	1991 (*108*)
Bovine plasma; pen G	LC–UV (325 nm)	0.005	1992 (*109*)
Milk, plasma, and animal tissue; Pen G	Thermospray LC–MS	<0.005	1992 (*110*)
Standard solutions; cloxacillin	Thermospray LC–MS	ND	1992 (37)

Table 10 Determinative Methods Developed or Published After 1990 for the Analyses of β-Lactam Residues in Animal Tissues and Biological Fluids—*continued*

Matrix and β-lactams analyzed	Analytical methods	Detection limits (μg/g or μg/mL)	Year of pub. (Ref.)
Chinook salmon muscle tissue; pen G	LC–UV (214 nm)	0.05	1992 (*111*)
Milk; cloxacillin and pen V	LC–UV (210 nm)	Recoveries were calculated	1992 (*112*)
Milk; monobasic penicillins	LC–UV (210 nm)	<0.002	1992 (*113*)
Beef, pork and chicken meat; pen G, pen V, ampicillin, amoxicillin, oxacillin, cloxacillin, dicloxacillin	GC–FID	0.005–0.010	1992 (*114*)
Ceftiofur drug formulations; high molecular weight impurities	Size exclusion chromatography with UV (254 nm) and refractive index detection	Impurities >0.03%	1992 (*115*)
Standard solution; pen G and thermal isomerization products	Plasma spray LC–MS and Plasma spray LC–MS–MS	ND	1992 (*116*)
Food-animal tissues; penicillins	A review paper	—	1992 (*117*)
Milk and animal tissues; β-lactam antibiotics	A review paper	—	1992 (*118*)
Milk and tissues; penicillins	A review paper	—	1992 (*119*)
Urine; amoxicillin	LC–UV (229 nm)	7.5	1993 (*120*)
Pharmaceutical preparations; cefadroxil	LC–UV (230 nm)	ND	1993 (*121*)
Serum and urine; cephalosporin, SCE-2787	LC–UV (235 nm)	Serum (0.6), urine (3.5)	1993 (*122*)
Tissues and plasma of yearling beef; procaine pen G	LC–UV (325 nm)	0.005	1993 (*123*)
Plasma of feedlot steers; procaine penicillin G	LC–UV (325 nm)	0.005	1993 (*124*)

Table 10 Determinative Methods Developed or Published After 1990 for the Analyses of β-Lactam Residues in Animal Tissues and Biological Fluids—*continued*

Matrix and β-lactams analyzed	Analytical methods	Detection limits (μg/g or μg/mL)	Year of pub. (Ref.)
Serum; 17 penicillins and cephamycins	*n*-LC–ESI–MS	ND	1993 (*125*)
Milk; pen G, ampicillin, amoxicillin, cloxacillin and cephapirin	LC–ESI–MS	0.200	1993 (*126*)
Serum and milk; ceftiofur and its metabolite desfuroylceftiofur	LC–UV (289.6 nm for ceftiofur, 265.8 nm for desfuroylceftiofur, 271.4 nm for dimer of desfuroylceftiofur)	0.05	1993 (*127*)
Tissues of yearling beef steers; procaine and benzathine pen G	LC–UV (325 nm)	0.005	1993 (*128*)
Eggs; pen G	LC–UV (325 nm)	ND	1993 (*129*)
Animal tissues; penicillins and other antibiotics	Strategic residue testing plan	Not applicable	1993 (*130*)
Injection sites of yearling beef steers; benzathine pen G alone or combined with procaine pen G	LC–UV (325 nm)	0.005	1994 (*131*)
Milk; ampicillin and amoxicillin	LC–UV (210 nm)	0.002–0.005	1994 (*132*)
Milk; penicillin G	LC–UV (325 nm)	0.003	1994 (*33*)

ND: not determined; FID: flame-ionization detector

amoxicillin. Mineo et al. (*114*) have published a GC method using FID to determine the concentrations of seven penicillins, namely ampicillin, amoxicillin, Pen G, Pen V, oxacillin, cloxacillin, and dicloxacillin, in addition to 13 other antibiotics in pork, beef, and chicken meat as their trimethylsilyl derivatives. Even though these two GC methods appear to have a lot of potential in residue analyses, other residue analysts have not readily adapted or used them for determining β-lactam residues in foods. This is because the procedures include several tedious and cumbersome manipulative steps. In addition, many residue analysts prefer to use LC instead of GC techniques for the analysis of these polar, thermally labile, nonvolatile compounds. In fact, almost all of the methods developed to date for the analyses of β-lactam residues have relied on LC techniques. It may also be that, unlike LC techniques where the native drugs can be directly detected using some of their inherent physicochemical properties, it is imperative in GC analysis to resort to either precolumn or on-column chemical reaction schemes to form suitable volatile and thermally stable derivatives of the otherwise polar, nonvolatile and thermally unstable β-lactams before the analytes can be detected. The formation of suitable GC derivatives, however, may not always be readily controlled under experimental conditions.

Chromatographic methods developed for analyzing β-lactam drug residues in milk and milk products, biological fluids, and animal tissues, may be classified into two groups: single-analyte (single-residue) or multi-analyte (multi-residue) methods. Single-analyte residue methods developed for a specific β-lactam drug residue should be capable of detecting that residue at a concentration below the defined MRL, if any has been established. While no limits for minimum concentration levels have been set so far, it is practical to have the limit of determination at least a factor of two below the established MRL. Where no MRL has been defined, as is the case for veterinary drugs with a zero residue designation, every effort must be made to develop a method that is as sensitive as the prevailing analytical equipment will allow. In Canada, a zero residue designation for the tolerance of a veterinary drug implies that no drug residue that can be detected by the best analytical method available will be permitted. For a multiresidue method, some compromises are made when a single method is developed to analyze several components, even when the compounds of interest belong to the same family.

The method development process can be partitioned into three major stages: stage 1, isolation or extraction where the analyte(s) of interest are extracted from the biological matrix; stage 2, cleanup where the extracted analytes are purified and concentrated; and stage 3, chromatographic separation and detection. Boison (*117*), Moats (*118*), and Petz (*119*) have recently reviewed the chromatographic methods for analyzing penicillins and cephalosporins in animal-derived foods. Instead of duplicating what has been discussed quite well in those review papers, I will focus primarily on the third stage of the method development process by extensively discussing detection enhancement techniques that have been used to detect and quantify these residues. For completeness, a critical examination of the techniques that have been used to extract these veterinary drugs from biological fluids and animal tissues will be made in the light of the physicochemical properties of these drugs. The discussion will conclude with the introduction of new and emerging concepts or techniques being incorporated into the method development process to increase automation in sample analyses and to reduce sample preparation time by analysts while maximizing analytical repeatability and sample throughput.

Sample Extraction or Isolation (Stage 1)

Sampling of Tissue and Biological Fluid Samples for Antibiotic Analysis

Appropriate sampling, sample handling, and tissue preparation are the keys to successful, accurate, and reproducible antibiotic residue analysis. Unlike biological fluids and organ tissues where the drugs are usually evenly distributed, antibiotic residues in muscle tissues may not be evenly distributed. Because an analysis of the antibiotic content in a small mass of tissue is intended to reflect the general distribution in the large animal, it is imperative that the sample taken for analysis be homogeneous and be as representative as possible of the bulk matrix. In this regard, it has been determined from earlier experiments at Health of Animals Laboratory (HAL) that in order to obtain a representative sample for the analysis of Pen G from a noninjection-site muscle or organ tissue, a minimum sample size of 40 g is required for homogenization. Smaller subsamples from this homogenized sample can then be used for analysis in order for the variability (the coefficient of variation) in analytical results to be less than 15% (*117*).

Once a suitable representative sample size for homogenization has been defined, when to homogenize the tissue sample for analysis is considered. Homogenization of tissue samples provides the most favorable medium for accelerated hydrolysis or degradation of β-lactams by enzymes, even under frozen storage conditions. This is probably because of the large surface-to-volume ratio presented to the enzymes to carry out their metabolic activity. It is, therefore, advisable to store the minimum tissue sample size required for analysis whole and unhomogenized. Experiences gained over the last few years from Pen G residue analysis have prompted scientists at HAL to adopt a protocol whereby 40 g aliquots of tissue samples are stored unblended at –76°C. When the sample is needed for analysis, a 40 g sample is partially thawed, homogenized, and then analyzed all on the same day to prevent accelerated degradation of the penicillins. If the analysis needs to be repeated, it is important to use a different 40 g sub-set of an unhomogenized frozen sample from the same animal and not the rest of the previously analyzed, refrozen homogenized sample.

Protein Precipitation

Protein precipitation with organic solvents, inorganic acids, and ultrafiltration have frequently been used to free noncovalently bound β-lactams from macromolecules. Organic solvents including methanol (*17, 37, 50, 74, 88, 95*), acetonitrile (*30, 32, 36, 37, 58, 59, 89, 112*), and 2-propanol (*36, 88*) have been used.

Sulfuric acid and sodium tungstate (*66, 67, 108–110*), as well as acidified acetonitrile (pH 4.0–4.4) (*119*), or molecular mass cut-off filters (*71, 85, 88*) to remove proteinaceous material have also been used. While most of these solvents degrade penicillins to some extent, methanol, which is known to degrade the penicillins the most rapidly and would be considered the most unsuitable organic solvent to be used for extraction purposes, has been extensively used for protein precipitation.

Typically, after homogenizing a tissue sample with a deproteinizing agent to increase the contact surface area, protein precipitation is performed by shaking with an extracting solvent. With biological fluids, protein precipitation can be achieved by shaking the sample in the presence of an extracting solvent and deproteinizing agent. Once deproteinization is completed, the sample is centrifuged to separate the precipitated from the extracted solution. Moats (*18*) has successfully used organic solvents to extract and deproteinize penicillins from animal tissues and milk. This approach, which is unsuitable for extracting amphoteric penicillins, extracts many more coextractives, including lipids, from tissue or biological fluids than an aqueous extraction or isolation technique. As a result, it has been necessary to use several solvent–solvent partitioning steps to reduce the level of contamination in the residual extracts when organic solvents are used. By extracting penicillins into water in the presence of sulfuric acid and sodium tungstate, Boison et al. (*108*) have been able to selectively eliminate the extraction of several nonpolar endogenous compounds and have obtained reasonably clean aqueous sample extracts.

Barker and his research group (*133*) have developed a matrix solid-phase dispersion (MSPD) technique for extracting veterinary drug residues, including β-lactams, from biological fluids and animal tissues. Briefly, this technique consists of blending the sample matrix with a solid support (silica) chemically treated with a lipid-solubilizing hydrocarbon (typically C18) in an agate or glass mortar. The blended material is then packed into a column and the analytes of interest eluted with suitable solvents. This technique has been successfully used to analyze penicillin G, ampicillin, and cephapirin in bovine tissues (*134*).

The aqueous or organic extract obtained at this point may be a very dilute solution of the analyte(s) of interest. It may also contain coextractives, which if allowed in the final extract, will not only have deleterious effects on the chromatographic components, but also increase the chemical background noise in the detector making it impossible to determine trace level concentrations of the analyte(s) of interest. To reduce the background chemical interference noise and to concentrate the analyte(s), ion exchange (IE), solid-phase extraction (SPE), and immunoaffinity (IA) cleanup procedures have been used.

Cleanup and Concentration of Sample Extracts (Stage 2)

Almost without exception, one or a combination of solvent–solvent partitioning, SPE, IE, and IA have been used to further cleanup sample extracts just prior to chromatographic analysis. Moats (*90, 112, 132*) has also reported using an LC system to collect fractions from an LC column which concentrates the dilute fractions for analysis on the same LC system. Meetschen and Petz (*89*) use a combination of anion exchange and diol cartridges to cleanup and concentrate methylated penicillins in extracts from liver and kidney tissues. At HAL, C18 SPE cartridges have been used to cleanup and concentrate penicillin extracts. This same approach has successfully been used in a multiresidue β-lactam method recently developed by Boison and Keng (unpublished data) for the cleanup of tissue and milk extracts that contain amoxicillin, ampicillin, penicillins G and V, and cloxacillin.

Automation in Sample Preparation Techniques

With the exception of the instrumental analysis stage, which has been automated by automated injectors, computer controlled GC, capillary zone electrophoresis (CZE), supercritical fluid extractors (SFE), and LC systems, most methods that analyze β-lactams require extensive commitments of the analyst's time from the moment the samples are received until they are ready for LC analysis.

Over the last decade, the consumption and disposal of wastes from chemical solvents used in analytical methods has become an issue of considerable concern for analytical laboratories. Even though there have always been safety concerns associated with using chemical solvents, the depletion of our ozone layer has raised a lot of questions that may impact the future use of chemical solvents. To address this problem, in January 1989 an international agreement, the Montreal Protocol On Substances That Deplete The Ozone Layer was signed by over 50 countries, including Canada, the United States, and the European Union (*135*). The agreement is specifically directed at reducing or banning the production of ozone-depleting compounds such as carbon tetrachloride and 1,1,1-trichloroethane. Analytical laboratories too are contributing to a solution by reducing their solvent consumption and solvent waste through a process that enables used solvents to be reclaimed and recycled as well as increasing their use of SPE and SFE systems. Virtually all analytical methods developed and used at HAL use SPE technology. SPE technology, in addition to helping to reduce solvent waste and solvent consumption, lends itself to unattended, automated sample cleanup and purification of isolated material. Because the whole sample cleanup process can be combined with the sample injection and chromatographic analyses stages to be fully automated, analyst interaction with the sample preparation stages of the analytical procedure is limited to stage 1 (the extraction or isolation step) where the analyte(s) of interest are isolated or extracted from the biological matrix. Using this automated sample cleanup and analysis concept, Boison et al. (unpublished data) have succeeded in fully automating the method for the analysis of Pen G residues from animal tissues. Analytical laboratories are consistently expected to demonstrate innovative ways to increase the testing loads at the laboratory support staff's expense, causing laboratory method automation to be an asset. Thus, at HAL, instead of spending 2 h isolating and cleaning up tissue samples for LC analysis, an analyst may spend only 20–30 min a day extracting Pen G from animal tissues. The rest of the analysis from C18 sample cleanup to LC analysis is performed automatically and unattended. This concept has been used to automate the cleanup and LC analysis for tetracyclines in animal tissues, the LC determination of sulfonamides in animal tissues, and is currently being used to automate the multiresidue analysis of β-lactam drug residues in animal tissues, milk, etc.

Why hasn't the extraction or isolation stage been included in the automated procedure? Because there is not yet a suitable commercially available system that will both homogenize tissue samples and efficiently dislodge all tissue samples from the probe before the next sample is homogenized. In addition, there is no inexpensive setup available that can combine the homogenization step with the solvent extraction procedures, including centrifugation, to separate the residual tissue plugs from the supernatant as rapidly and efficiently as an analyst can. Therefore, until an inexpensive commercial system designed to accomplish these two critical steps is available, tissue analyses will continue to be done as it is currently being done at HAL. For fluid samples such as milk, plasma or serum, and urine, which require no homogenization or centrifugation steps, research is being done at HAL on automating the whole sample preparation and analysis procedure.

Moats (*90*) was the first to describe the application of an automated on-line trace enrichment procedure into the sample cleanup stage when analyzing β-lactams. In his semiautomated technique, an automated fraction collector collects predetermined fractions of the chromatographic effluent from 2 mL of crude extract. Each fraction collected is then taken off and concentrated in a rotary evaporator and loaded into an LC system provided with an automated sampler. This method has been used to determine ampicillin (*90*), cloxacillin and penicillin V (*112*), and ampicillin and amoxicillin (*132*) in milk.

Chromatographic Resolution of β-Lactam Drugs

Because an extract obtained at this point will rarely contain only the analyte(s) of interest, it is recommended that the components of the extract be resolved from one another by ion exchange, GC, LC, or CZE technique prior to detection and quantification. Most of the methods developed after the 1990 meeting for determining β-lactam residues in biological matrices have relied on LC techniques for separation. The rest of the discussion will, therefore, be limited to LC methods that include normal-phase, reversed-phase, ion-exchange, and capillary zone electrophoresis applications. In a recently published survey on the use of various LC modes of separation, Majors (*136*) reported that out of 1,000 LC–GC users polled, with a 16.3% response rate, use of reversed-phase and normal-bonded phase was 47.2% and 13.5%, respectively, while the use of adsorption (i.e., liquid–solid) and ion exchange were 9.8% and 16.6%, respectively. Size exclusion, chiral columns, and hydrophobic interaction modes showed smaller usage percentages (*136*). For an extensive treatment of the mechanisms that govern the separation of β-lactam

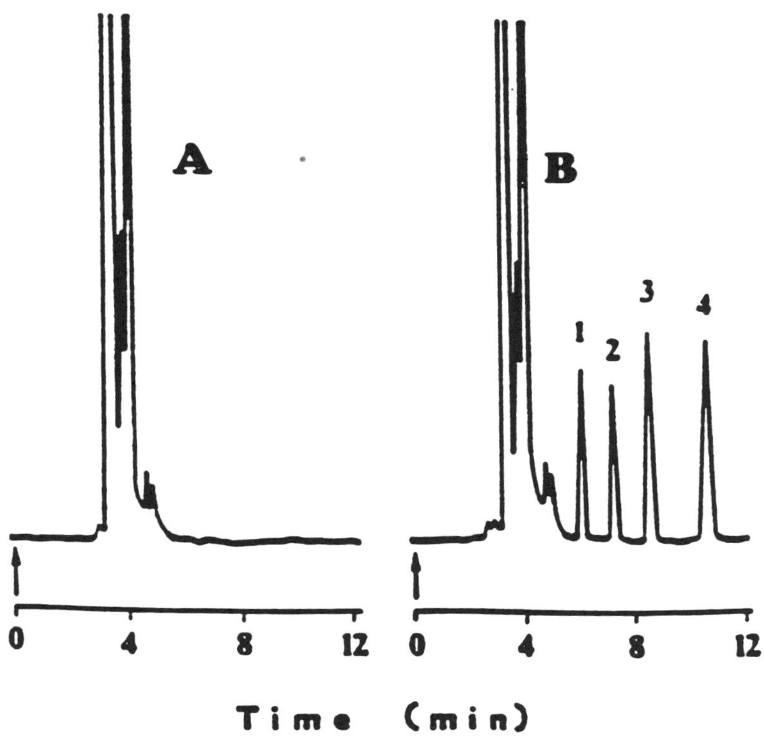

Figure 4 Ion-pair liquid chromatographic separation of penicillins in a blank milk sample (A) and a sample (B) spiked with 30 ppb of penicillin G (1), penicillin V (2), oxacillin (3), and cloxacillin (4).
LC conditions: mobile phase, 5 mM TBA-HSO_4 in acetonitrile-0.02 M phosphate buffer (38:62 v/v) pH 6; Nucleosil C_{18} column (5 μm); column temp, 40°C; flow rate 1 mL/min wavelength, 210 nm; detector sensitivity, 0.020 AUFS; chart speed, 0.5 cm/min; injection volume, 100 μL. See reference (*113*).

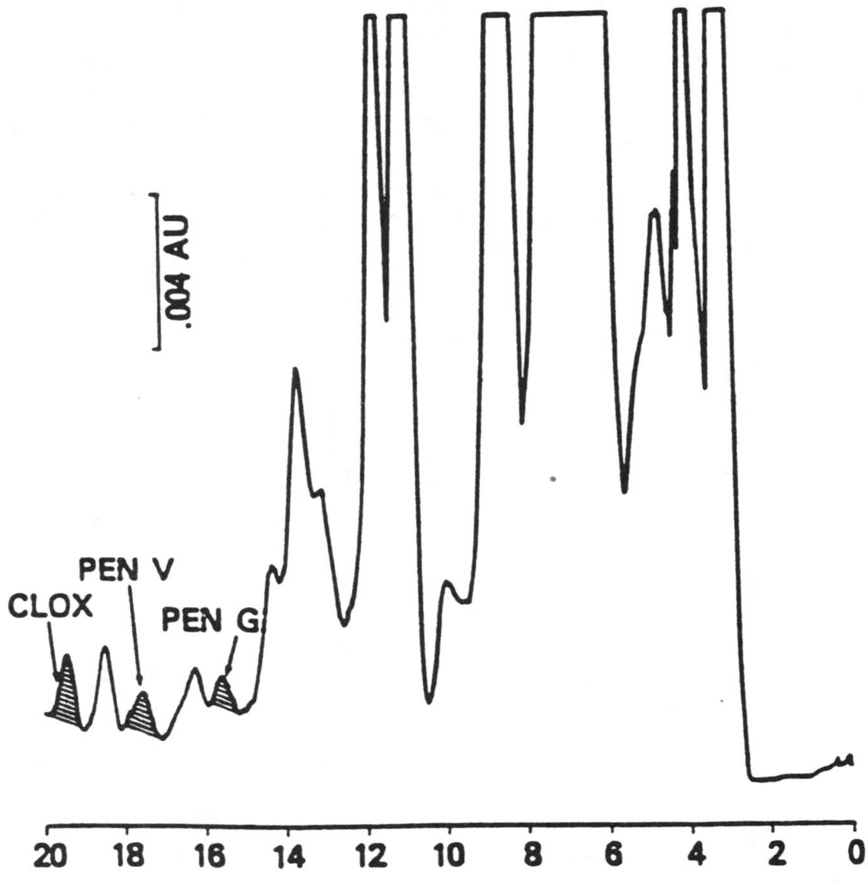

Figure 5 Reversed-phase chromatography of 2 mL equivalent milk fortified with 0.01 ppm penicillin G, penicillin V, and cloxacillin. Mobile phase 0.01 M phosphate–acetonitrile (80:20–40:60); flow rate 1 mL/min, 0.02 AUFS at 220 nm; C_{18} 10 μm reversed-phase column. See reference (*58*).

antibiotics by normal-bonded phase, reversed-phase, and ion-pair reversed-phase, ion exchange chromatography, and capillary zone electrophoresis, the reader is referred to excellent reviews (*117–119*, *137–141*) and fundamental papers (*142–145*) that have already been published on the subject. Figures 4–6 depict typical separations of β-lactams by ion-pair reversed-phase, reversed-phase, and capillary zone electrophoresis, respectively.

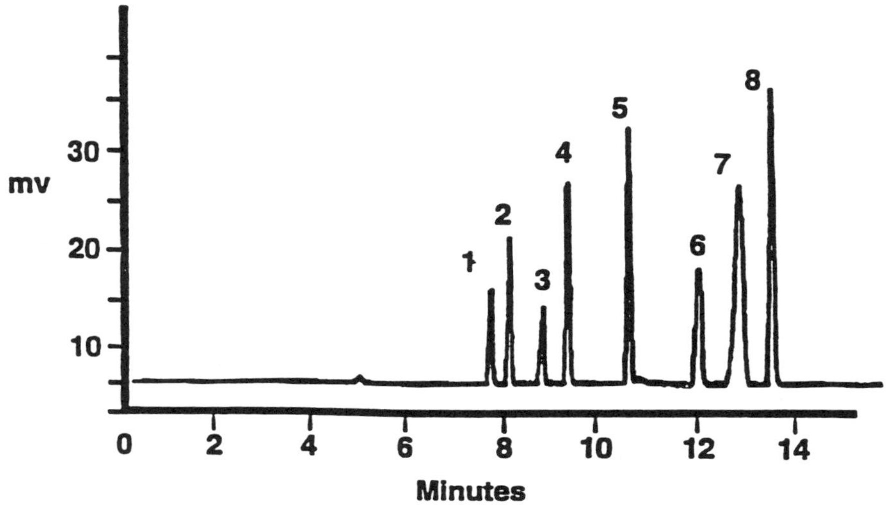

Figure 6 Micellar electrokinetic capillary chromatography of eight penicillins at +18 kV d.c. voltage with UV detection at 214 nm in an electrolytic buffer consisting of a 0.02 M NaH_2PO_4, 0.02 M $Na_2B_4O_7$ (pH 9) with 50 mM sodium dodecyl sulfate as modifier. Capillary: 75 m i.d. × 65 cm; Injection: 5 s hydrostatic. Peaks: amoxicillin (peak 1), ampicillin (peak 2), 6-aminopenicillanic acid (peak 3), oxacillin (peak 4), cloxacillin (peak 5), ticarcillin (peak 6), nafcillin (peak 7), dicloxacillin (peak 8). See reference (*93*).

Once β-lactams have been efficiently extracted from a biological matrix with recoveries greater than 70% and a chromatographic system has been developed to resolve the composition of the extract into individual components, then the detection techniques that will be used to quantify these analytes need to be examined. Because the discussion is being limited to LC techniques, only those LC detection techniques that minimize matrix interferences to improve and enhance detection sensitivities will be considered. These enhancement techniques include derivatization reactions and the selection of suitable detectors that may be either concentration or mass-flow types.

Detection Techniques for β-Lactam Drug Residues

The most popular mode of detecting β-lactams after LC separation is by concentration-type ultraviolet-visible (UV-vis) and fluorescence detectors. The mass spectrometer, a mass-flow rate type detector, is currently being extensively used as a specific detector for residue analysis now that significant improvements in interface technology allow mass spectrometers to be directly coupled to LC instruments. A mass-flow rate detector measures a signal that is proportional to the mass flow rate (i.e., the mass per unit time) through the detector cell. Even though the signal produced in a detector of this type is independent of the detector cell volume, it is imperative that the connecting tubing be kept short in order to prevent extracolumn peak broadening. Electrochemical detectors have also been used to analyze cloxacillin (*146*), cephalosporins (*147*, *148*), and penicillins (*149*). A concentration-type detector, on the other hand, measures a signal that is proportional to the concentration (i.e., the mass per unit volume) of analyte in the detector cell. Because this is the analyte concentration in the detector cell, the dimensions of the cell volume are very important. In the design of such detectors then, consideration is given to the dimensions of the cell volume and cell volumes are kept as small as practicable. Because mass-flow rate detectors produce peak areas that are flow-independent but peak heights that are flow dependent, using peak areas for quantitation is advantageous. However, because concentration-type detectors produce peak areas that are flow-dependent but peak heights that are flow-independent it is more practical to use peak heights instead of peak areas for quantitation.

Concentration-Type Detectors in Liquid Chromatographic Analyses of β-Lactams

UV-vis and fluorescent detectors are the most commonly used concentration type detectors for the LC analyses of β-lactams. Both precolumn and postcolumn reaction techniques have been used to determine β-lactam drug residues in food. Precolumn reactions have been used to improve UV detection sensitivities of methods developed to analyze β-lactam antibiotic drug residues. For the derivatization reaction to be useful, the derivative must be formed readily and quantitatively, be stable, and exhibit increased UV absorbance properties.

UV-vis Detection After Precolumn Chemical Modification of β-Lactams

Imidazole or 1,2,4-triazole catalyzed formation of mercaptide complexes of penicillenic acids. The most extensively used precolumn derivatization reaction to detect β-lactam antibiotic residues in animal-derived foods, is based on the imidazole-catalyzed isomerization of benzylpenicillin into benzylpenicillinic acid (*25*, *30*, *32*, *43–47*, *57*, *60*, *66*, *67*, *84*, *95*, *105*, *108*, *109*, *123*, *124*, *128–132*) described by Hans Bundgaard (*43*). In the presence of mercuric chloride, a self-catalyzed nucleophilic attack by the imidazole base occurs on the β-lactam bond and forms *N*-penicilloylimidazole (Scheme 5). This intermediate product then forms a stable mercuric mercaptide complex. In neutral solutions, several

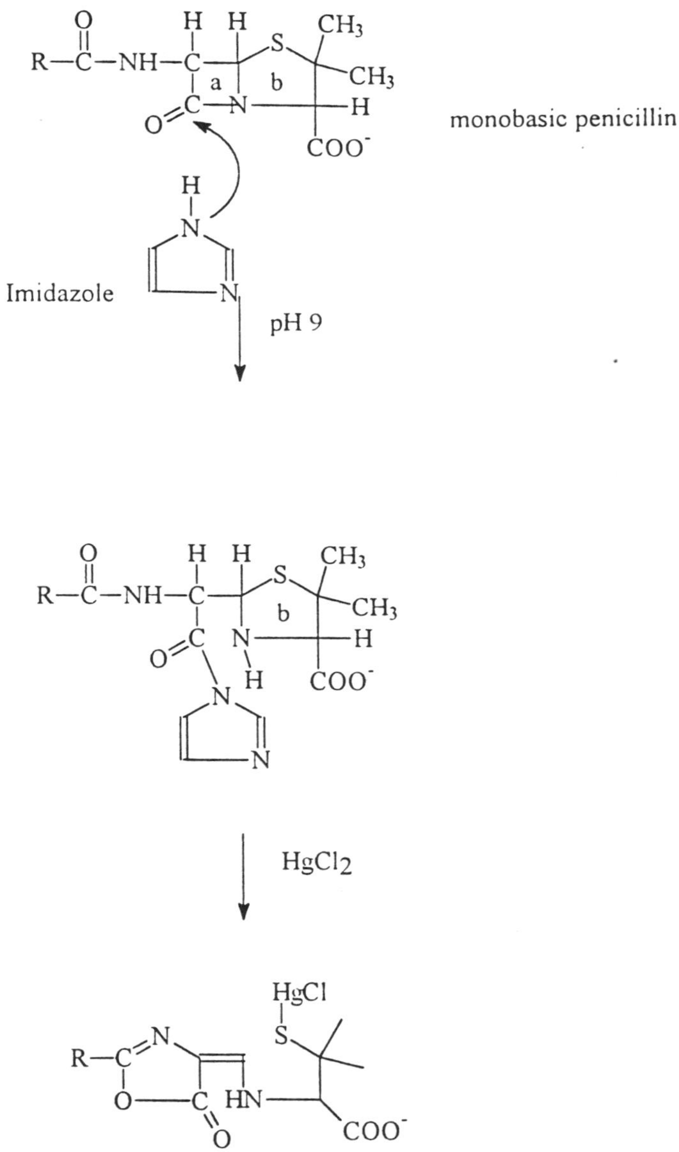

Scheme 5 Mechanism for the imidazole base catalyzed formation of penicillenic acid mercury mercaptide. See reference (*68*).

Table 11 UV Absorption Properties of Penicillenic Acid Mercuric Mercaptide Derivatives

Compound	Molar absorptivity at λ_{max}; λ_{max} in the presence of 1,2,4-Triazole[a] ($L\ mol^{-1}\ cm^{-1}$; nm)	Molar absorptivity at λ_{max}; λ_{max} in the presence of imidazole[b] ($L\ mol^{-1}\ cm^{-1}$; nm)
Penicillin G sodium	22,000; 327	26,000; 325
Penicillin V potassium	— ; 325	26,500; 325
Phenethicillin potassium	— ; 325	26,800; 325
Propicillin potassium	— ; 325	26,600; 325
Oxacillin sodium	— ; 343	27,600; 343
Dicloxacillin sodtum	21,000; 346	28,000; 343
Methicillin sodium	— ; 335	26,500; 335
Carbenicillin sodium	21,000; 328	26,700; 325
Ampicillin sodium	21,000; 323	—[c] ; 311
Ampicillin sodium	21,000; 325	— ; 325
Amoxicillin	21,000; 324	— ; 325
Sulbenicillin	21,000; 331	— ; —[d]
Ticarcillin	21,000; 327	— ; —[d]

Derivatization reactions with 1,2,4-Triazole were complete in 10 min at 60°C; corresponding reactions in imidazole at 60°C took about 30 min.
[a] See reference (*64*).
[b] See reference (*43*).
[c] These are the reaction parameters when ampicillin is reacted without initial acetylation of the free amino group in the side chain.
[d] This parameter was not measured for this compound.

penicillins form stable mercuric mercaptide complexes of the corresponding penicillenic acids in quantitative yield (*44*) that exhibit intense molar absorbencies with absorbance maxima at wavelengths >300 nm.

Native β-lactams do not have chromophores with significant UV absorption properties. For example, the molar absorbance of amoxicillin in 0.1 M HCl solution is 9,500 $L\ mol^{-1}\ cm^{-1}$ at its maximum absorbance 229 nm (*70*). Table 11 shows the molar absorptivities and the wavelength of maximum absorption for several penicillenic acid mercuric mercaptides formed by reacting penicillins with mercuric chloride in the presence of either imidazole (*43*) or 1,2,4-triazole (*68*) at 60°C (Scheme 6). When 1,2,4-triazole is the base, stable mercuric mercaptide

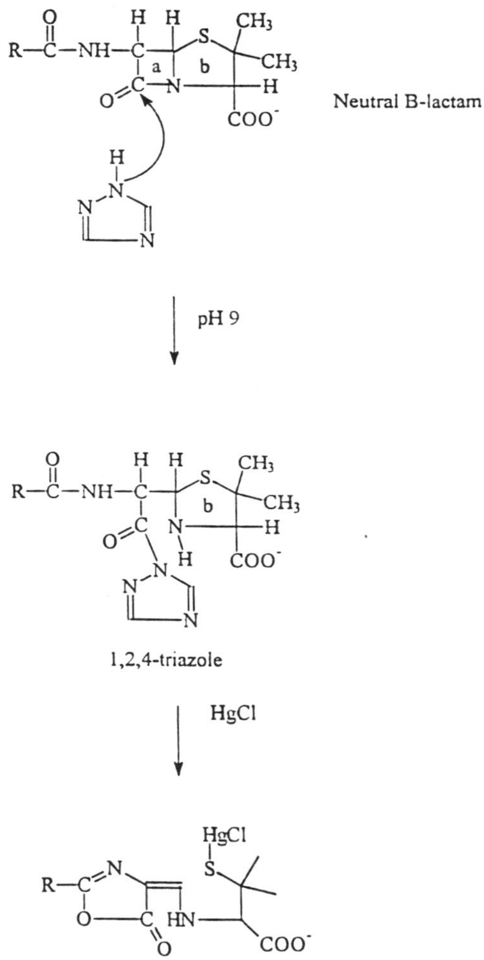

Penicillenic acid mercury(II) mercaptide

Scheme 6 Mechanism for the formation of mercury mercaptide derivatives of penicillenic acid in the presence of 1,2,4-triazole. See reference (*68*).

derivatives are formed more rapidly than when imidazole is present (*68*). When mercuric chloride is replaced with nickel, zinc, or cadmium salts, however, no product appears to be formed, and spectrophotometric studies reveal that no absorption at 325 nm or any other wavelength takes place. An intense absorption is observed at 325 nm when silver, gold, or mercury (II) salts are used. It is believed that the inability of these other metal ions to produce stable penicillenic acid derivatives may be a reflection of their relative "hardness" and, consequently, their poor tendency to complex with "soft" ligands, such as penicillenic acids, which are thought to coordinate metal ions through the sulfa atom. Irrespective of the metal ion used in the derivatization reagent, both chromatographic efficiency and detection sensitivity are enhanced when the reversed-phase LC mobile phase solutions contain in increasing order ethylenediaminetetraacetic acid (EDTA), sodium thiosulfate, and sodium cyanide as ion-pair reagents. However, because sodium cyanide is extremely toxic, sodium thiosulfate is satisfactory in most cases for the analysis of penicillins.

Most penicillins exhibit maximum UV detection absorbance at 210–230 nm, which is also the wavelength range that many endogenous compounds in extracts from biological matrices exhibit intense absorbencies. At a pH of 3.4, native Pen G exhibits a molar absorbance of 38,000 L mol^{-1} cm^{-1} at 193 nm compared with that of its mercury mercaptide derivative of 20,000 L mol^{-1} cm^{-1} at a maximum absorbance of 325 nm. Despite the high molar absorbance recorded for native penicillin at 190 nm, this wavelength is not useful for determining native penicillin in complex matrices such as biological fluids and animal tissues, because it is also the wavelength at which most proteins and endogenous materials in these matrices show intense absorption. When penicillins are dissolved in solution, they ionize into penicillinic acid and a counter ion. When an aqueous solution of sodium penicillin G is injected into an LC system, the LC detector should find only a single peak due to the penicillinic acid end of the molecule because the sodium counter ion will be transparent to the UV detector. When procaine penicillin G, on the other hand, is injected into the LC system, an additional peak is expected due to the absorption by procaine, the counter ion for this compound. The more UV-absorbing compounds injected into the LC system, the more complex and less specific the chromatogram can become. Because the main objective in these analyses is to determine the activity or concentration of the active drug and not that of the counter or spectator ions in the matrix, it is advantageous if the detector or precolumn derivatization reaction was designed to respond exclusively to the active form of the drug.

There are many advantages provided by a precolumn derivatization reaction over one that directly examines the UV absorbance of the native penicillins without derivatization. The bathochromic shift produced in the UV spectra of the penicillins permits these analytes to be detected at a wavelength that is transparent to many of the typical endogenous compounds in biological matrices and provides a high degree of specificity in detecting individual penicillins. Additionally, the derivatization reactions are quantitative, simple, rapid to perform, and cause the reaction products to exhibit relatively higher absorbance properties when compared to the low UV-absorbing characteristics of the native penicillins. Because the reaction is specific for only those compounds that have an intact penicillin nucleus, the method is highly selective and counter ions used in preparing the original formulations become transparent to the derivatization reaction. As a result, these ions do not complicate the LC chromatograms as their native forms would.

This derivatization technique is also applicable to analyzing amphoteric penicillins, such as ampicillin and amoxicillin. When an aqueous solution of ampicillin {D(-)-[α-aminobenzylpenicillin]} is treated with imidazole or 1,2,4-triazole, it rearranges itself, at least partly, to form the corresponding penicillenic acid (λ_{max} 322 nm). In the presence of mercuric chloride, an unstable product with an absorption maximum at 311 nm is formed. This product is believed to be α-aminobenzylpenicillinic acid in which both the thiol and the side chain amino group have competed for the available mercuric chloride. If, however, the side chain amino group in ampicillin is initially acetylated to form α-acetamidobenzylpenicillin, and the resulting product is reacted with mercuric chloride in the presence of imidazole or 1,2,4-triazole, a stable product (α-acetamidobenzylpenicillinic acid mercuric mercaptide) is formed (Scheme 7) that exhibits an absorption maximum at 325 nm and a molar absorbance comparable to those of other monobasic penicillins (*47*). This means that this method of derivatization, with a slight modification, can quantitatively determine all the penicillins including the amphoteric analogs–an obvious advantage over other methods that are suitable for only the neutral penicillins. The concept has been used at HAL to develop a multiresidue LC method (Figure 7) for determining amoxicillin, ampicillin, penicillin G, and cloxacillin residues in animal tissues and milk using penicillin V as an internal standard (Boison and Keng, unpublished results). The detection limits for applying this concept to determining β-lactam drug residues have been in the low μg/kg or μg/L range.

Determining penicillins using precolumn derivatization with 1-hydroxybenzotriazole. A slight modification of the precolumn derivatization reaction in which benzoic anhydride is used instead of acetic anhydride, 1-hydroxybenzotriazole is used instead of imidazole or 1,2,4-triazole has been used by Shah et al. (*83*, *84*) for determining penicillins G, V, X, and K, penicillin G penicilloic acid, and 6-aminopenicillinic acid (6-APA) in fermentation broth has been discussed.

Determining penicillin G using copper (II) acetate as a complexing agent. Saha (*72*) describes a method in which copper (II) acetate is used as a chromogenic reagent for determining benzylpenicillin spectrophotometrically in pharmaceutical preparations. In this method, copper forms a 2:1 green complex with benzylpenicillin at pH 6–6.8 with a molar absorptivity of 0.24×10^3 L mol^{-1} cm^{-1} at its maximum absorbance (λ_{max} 650 nm). The green complex formed is claimed to be stable for nearly 25 min after its formation, therefore, such a complex would not be suitable for determining benzylpenicillins in a large number of samples.

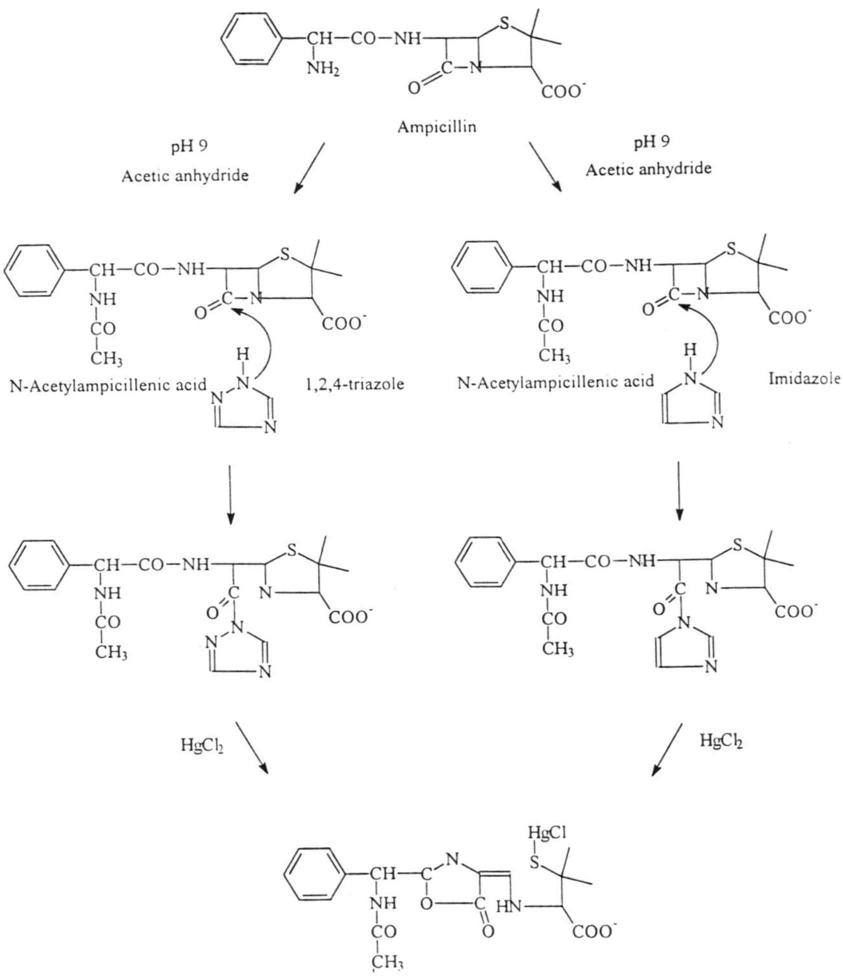

Scheme 7 Mechanism for the imidazole or 1,2,4-triazole catalyzed reaction for the formation of mercury mercaptide derivatives of amphoteric penicillins after initial *N*-acetylation. See reference (*47*).

This method has not yet been applied in a chromatographic method for determining benzylpenicillin or any other penicillins.

Utilizing π-acceptors for the spectrophotometric determination of some penicillins. Askal et al. (*103*) describe a method in which 2,3-dichloro-5,6-dicyano-*p*-benzoquinone (DDQ) or 7,7,8,8-tetracyanoquinodimethane (TCNQ) acting as

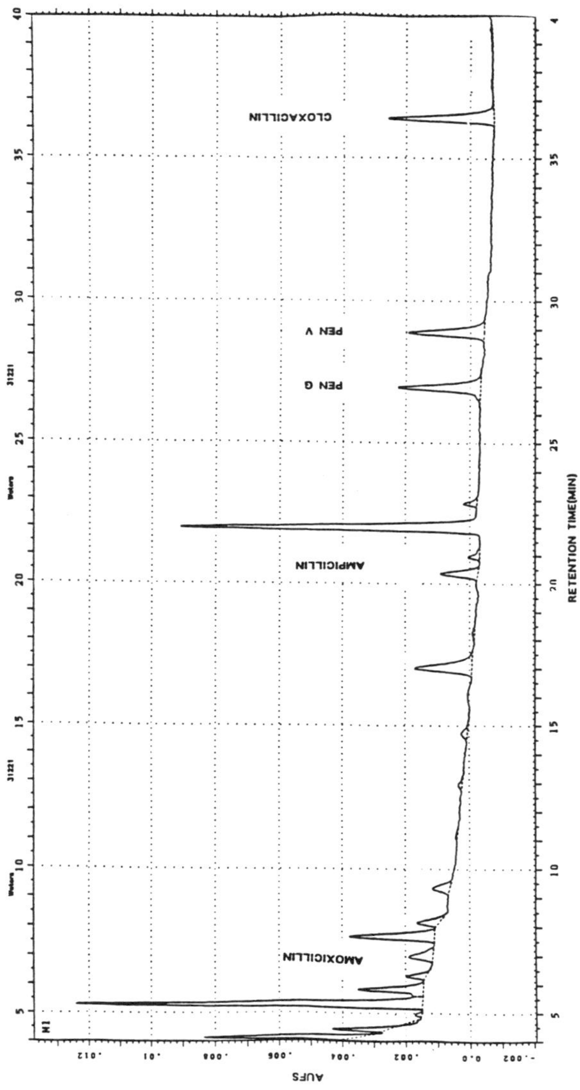

Figure 7 Gradient LC analysis of the β-lactams, amoxicillin, ampicillin, penicillin G, penicillin V (internal standard), and cloxacillin from a fortified animal tissue sample (20 ppb of each drug except the internal standard) as their mercaptide derivatives. Mobile phase: ACN/0.1 M; phosphate buffer, pH 6.5; gradient conditions: initial time, 18% ACN; 8 min, 25% ACN; 25 min, 35% ACN; 30 min, 100% ACN; 33 min, 18% ACN; flow rate, 1 mL/min; 250 × 4.6 mm Spherisorb 5 ODS (5 μm) column; λ = 325 nm at a sensitivity of 0.005 AUFS. (Boison and Keng, unpublished data).

π-acceptors react with penicillins, which behave as *n*-electron donors, to produce highly colored radical anions. The reaction of DDQ with different penicillins forms an intense orange-red complex that exhibits an absorption maximum at 460 nm with molar absorptivities ranging 2,960–4,600 L mol^{-1} cm^{-1}, while TCNQ forms complexes with molar absorptivities ranging from 18,600 to 30,400 L mol^{-1} cm^{-1}, with an absorption maximum at 842 nm. This method has not yet been applied in a chromatographic method for determining any penicillins.

Fluorescence Detection after Precolumn Derivatization Reactions

Fluorescence detection methods are usually favored over UV methods in most LC analyses because of their inherently greater sensitivity. The native β-lactams, however, do not have any fluorophores, and so derivatization is necessary if fluorophores are to be introduced into the compounds. For compounds that possess primary and secondary amino functionalities, this is easily accomplished. For the derivatization step to be useful in determining β-lactams, the derivatizing reagent should be able to produce highly fluorescent derivatives without itself being fluorescent, and be able to react rapidly with primary and secondary amino groups to form stable products. *o*-Pthaldaldehyde-mercaptoethanol has been used to detect and quantify β-lactams (*57*) and even though the reaction is rapid and the reagent is nonfluorescent, it reacts only with primary amines and produces an unstable product thus limiting its usefulness.

Reaction of Penicillins with Dansylhydrazine

In 1978, Tsuji et al. (*150*) described a spectrophotometric procedure for characterizing penicillins that introduces a fluorophore into the side chain of the penicillins. In this procedure, the β-lactam ring is enzymatically hydrolyzed to form the respective penicilloate product. Mercuric chloride is then added, which reacts with the penicilloate to liberate the side chain with a terminal aldehyde. The penilloaldehyde product is then extracted into methylene chloride and reacts with dansyl hydrazine to form the fluorolabeled side chains. Munns et al. (*63*) describe the LC conditions used to separate and determine eight neutral penicillins in milk previously characterized by Tsuji et al. (*150*). Because the success of the method hinges on the ability to completely hydrolyze the β-lactam ring, the method could achieve lower recoveries for compounds such as the isoxazolyl penicillins (oxacillin, cloxacillin, and dicloxacillin) that hydrolyze rather poorly under enzymatic conditions. It is no wonder then that Munns et al. reported recoveries as high as 79.4% for benzylpenicillin and as low as 26.4% for dicloxacillin. Unless a more reliable and efficient technique can be found for hydrolyzing the β-lactam ring, it is not likely that this method, with such a wide range of recoveries for the drugs of interest, will be useful for routine analysis.

Precolumn Derivatization Reactions with Dansylaziridine

Unlike Tsuji et al. (*150*) and Munns et al. (*63*), Orford et al. (*106*) have used alkaline hydrolysis as a more effective method of hydrolyzing the β-lactam ring to form the thiolate anion. This is then allowed to react with dansylaziridine at pH 12 to form a stable, fluorescent-labeled product.

Precolumn Derivatization Reactions with 9-Fluoroenylmethyl-chloroformate

Shah and Adlard (*83*) have described a novel and sensitive precolumn fluorometric derivatization reaction for determining β-lactams and their biosynthetic intermediates in fermentation media. In this procedure, cephalosporins and penicillins containing a primary or secondary amino side chain were reacted with excess 9-fluoroenylmethylchloroformate (FMOC-Cl) in borate buffer (Scheme 8). The reaction, which occurs rapidly at room temperature, is suitable for the LC

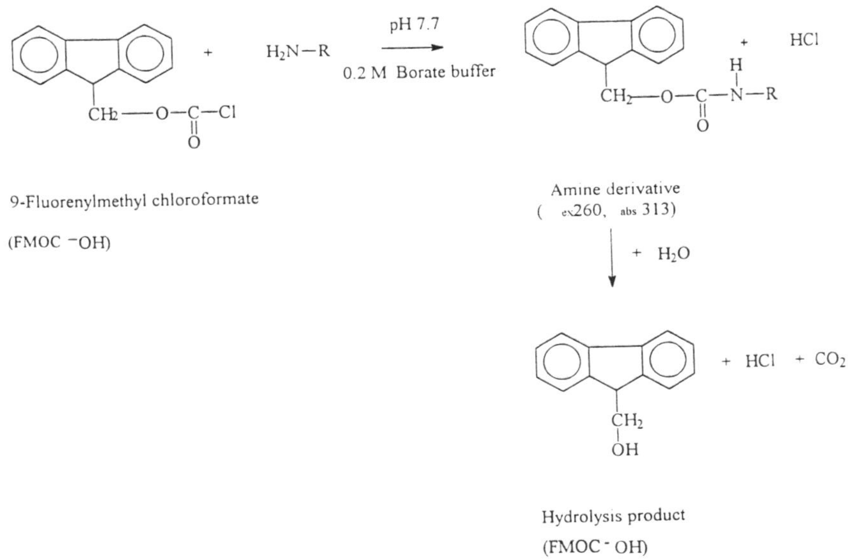

Scheme 8 Mechanism for the reaction of primary and secondary amines with FMOC-Cl in the presence of 0.2 M borate buffer (pH 7.7). See reference (83).

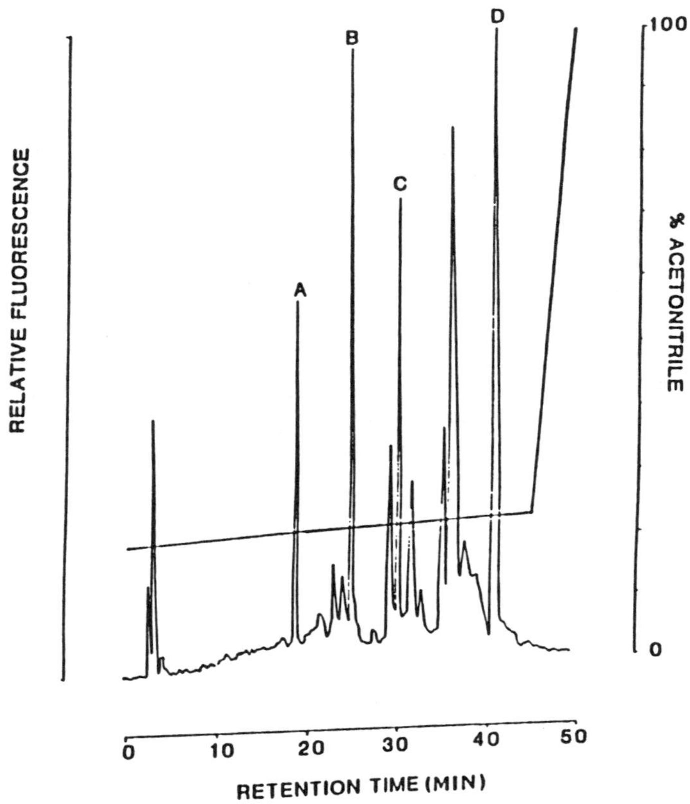

Figure 8 Reversed-phase gradient LC analysis of FMOC-C1 derivatives of cephalosporins and ampicillin at 5 μg/mL in filtered fermentation broth. Injection volume, 10 μL; detector sensitivity, 16; λ_{ex}, 260 nm; λ_{em}, 311 nm; 150 × 4 mm i.d. Hypersil ODS, 3 μm; column operated at a flow rate of 1 mL/min. Mobile phase: ACN/50 mM; acetate buffer, pH 4.1. Gradient conditions: initial time, 18% ACN, 20 min; 20% ACN, 45 min; 23% ACN, 50 min; 100 % ACN. Peak identification: (peak 1) deacetylcephalosporin C, (peak 2) deacetoxycephalosporin C, (peak 3) cephalosporin C, (peak 4) ampicillin. See reference (*83*).

determination of penicillins and cephalosporins in a complex mixture such as a fermentation broth (see Figure 8). Detection limits of 10–50 μg/L were determined for the compounds analyzed. It remains to be seen how effective such a method would be for tissue and milk analysis.

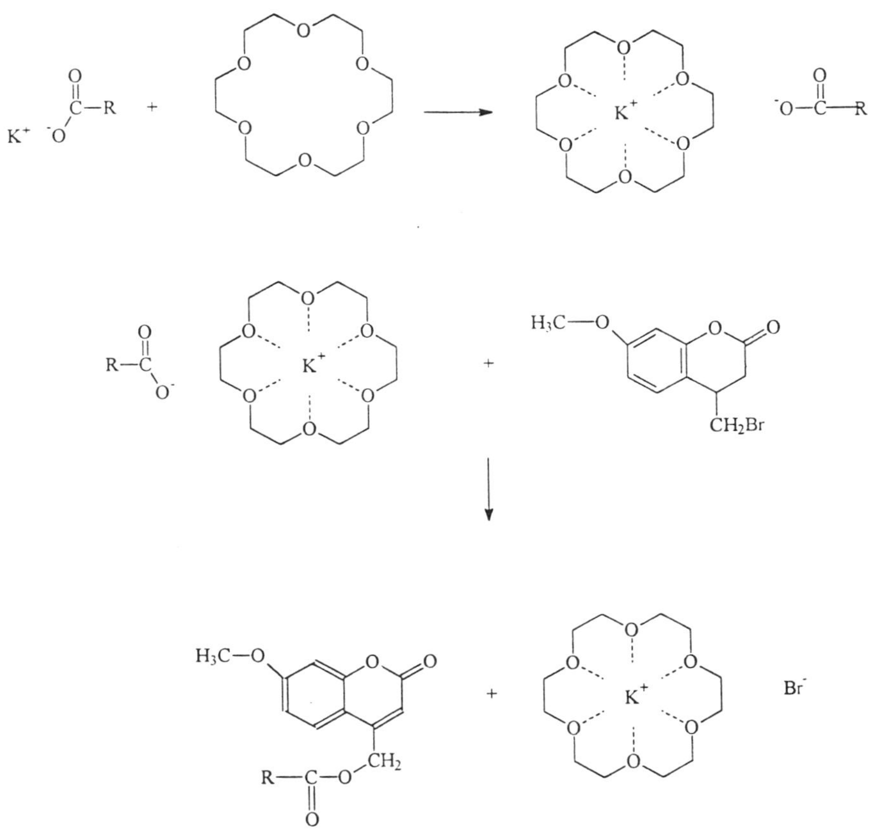

Scheme 9 Reaction mechanism for the crown ether catalyzed esterification of monobasic penicillins with BrMmc. See reference *(91)*

Precolumn Derivatization with 4-Bromomethyl-7-methoxycoumarin

Berger and Petz (*91*) recently reported a method that uses the 18-crown-6 ether catalyzed reaction of benzylpenicillin salts to form stable 4-bromomethyl-7-methoxycoumarin esters in quantitative yield (Scheme 9) in about 30 min at 55°C (*91*). This elegant method has been used for determining methicillin, penicillins G and V, oxacillin, nafcillin, cloxacillin, and dicloxacillin as their potassium salts using diphenylamine as internal standard (Figure 9). Even though this method has not been tested for its applicability toward amphoteric penicillins, as little as 200 pg of derivatized penicillins can be detected in milk extracts indicating that the method shows a lot of promise for the determination of these drugs.

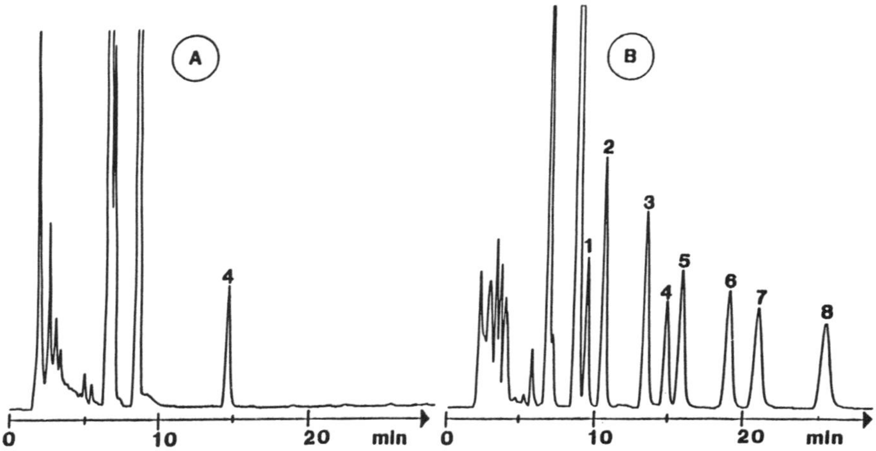

Figure 9 Multiresidue analysis of monobasic penicillins in milk as their 4-bromomethyl-7-methoxycoumarin derivatives.
A: Blank extract; B: fortification with 10 μg/kg each of methicillin (peak 1), benzylpenicillin (peak 2), phenoxymethylpenicillin (peak 3), diphenylamine (internal standard) (peak 4), oxacillin (peak 5), nafcillin (peak 6), cloxacillin (peak 7), and dicloxacillin (peak 8). Injection volume, 10 L; 250 × 4 mm Lichrospher RP 18 (4 μm) column; flow rate, 1 mL/min. Mobile phase: acetonitrile–0.05 M sodium acetate buffer (pH 4.5) (5.3:4.7, v/v); λ_{ex} 320 nm, λ_{em} 400 nm. See reference *(91)*.

Postcolumn Derivatization Reactions

Unlike precolumn derivatization reactions, postcolumn reactions require extra pumps for adding derivatization reagents to the effluent after column separation, reaction coils, mixing tees, debubblers, a thermostated bath, and a peristaltic pump. This array of equipment is aligned to permit the postcolumn reaction products to be fed directly into a detector. For this extra step to be useful in any determinative procedure for analyzing β-lactams, the postcolumn reaction selected must be rapid, quantitative, and reproducible, and the system design must be simple enough to permit other analysts to perform the assay. Additionally, it should present significant advantages over analytical methods that use less expensive LC systems for analyzing these drugs.

UV-vis Detection After Postcolumn Degradation Reactions of Penicillins

Two general techniques have been used for postcolumn reactions involving penicillins and include an initial degradation of the separated penicillins with either sodium hypochlorite (*76*) or sodium hydroxide (*56*, *79–81*). The hydrolyzed penicillins then react with mercuric chloride and EDTA. The degraded penicillins and their metabolites are detected at wavelengths >300 nm and the detection limits were generally greater than 25 µg/L in plasma and 200 µg/L in urine, two matrices to which this technique has been applied. Because these methods still require the same extensive sample cleanup required in other techniques, this is not a favorable method unless there is absolutely no other method available for determining these compounds.

Mass-Flow Rate Type Detectors for Analyzing β-Lactams

Mass spectrometers have been used as specific detectors for analyzing β-lactams either by direct probe analyses, direct liquid introduction (DLI) (*107*), or by coupling GC, LC, or CZE instruments directly to mass spectrometers (GC–MS, LC–MS, or CZE–MS). The role of mass spectrometers in analyzing antibiotics cannot be over-emphasized. They are currently the only detection system capable of providing unequivocal, on-line spectrometric identification of an individual analyte at the very low residue concentrations that meet the regulatory enforcement requirements for confirmatory analysis in animal-derived foods.

Fundamental studies have been conducted to characterize the fragmentation pattern for penicillins and cephalosporins under different MS ionization source conditions. The mass spectra of several penicillins have been investigated using electron impact ionization, chemical ionization, fast atom bombardment (*86*), thermospray ionization (*37*, *77*, *78*, *85*, *95*, *104*, *110*), plasmaspray ionization (*116*), particle beam ionization (*94*), and quite recently, atmospheric pressure ionization (*105*) and electrospray ionization (*125*, *126*) techniques. Additionally, tandem mass spectrometric techniques (MS–MS) have been used to characterize β-lactams (*82*). The results obtained from these fundamental studies have been used to identify and determine these β-lactams in animal-derived foods by GC, LC, or CZE techniques in which the mass spectrometer is coupled to the chromatographic equipment. For example, pyrolysis GC–MS has been used to characterize some penicillins and cephalosporins from pharmaceutical preparations. Generally, mass spectra of penicillins obtained by GC-MS yield no detectable molecular ions. Because GC–MS techniques fail to successfully analyze these polar nonvolatile, thermally-labile β-lactams, these compounds are now being analyzed by LC–MS methods. Boison et al. (*110*), who have detected 5 µg/L Pen G in milk, have also used Thermospray (TSP) LC–MS to provide confirmatory analyses for penicillin residues in plasma and animal tissues (Figure 10). Voyksner et al. (*94*) and Tyczkowska et al. (*36*, *37*, *85*, *88*, *95*, *127*) have also done extensive work developing MS methods for determining β-lactams

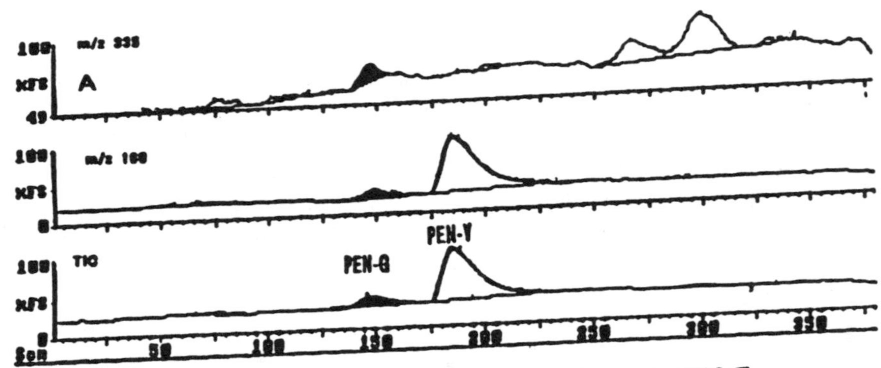

BOVINE MUSCLE SPIKED WITH 10 ppb PEN-G

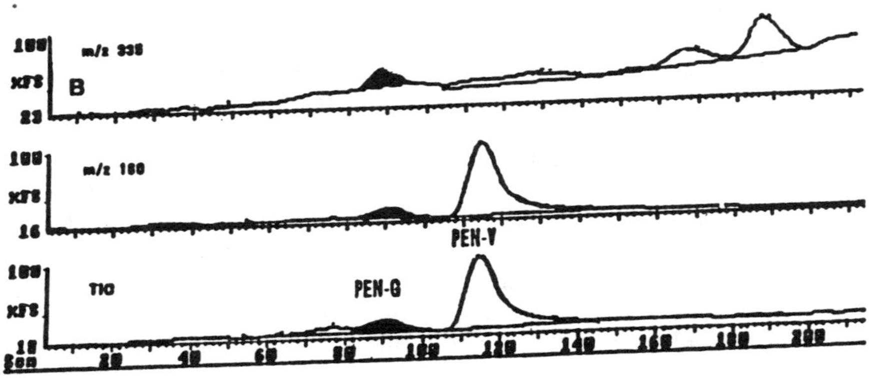

BOVINE PLASMA SPIKED WITH 10 ppb PEN-G

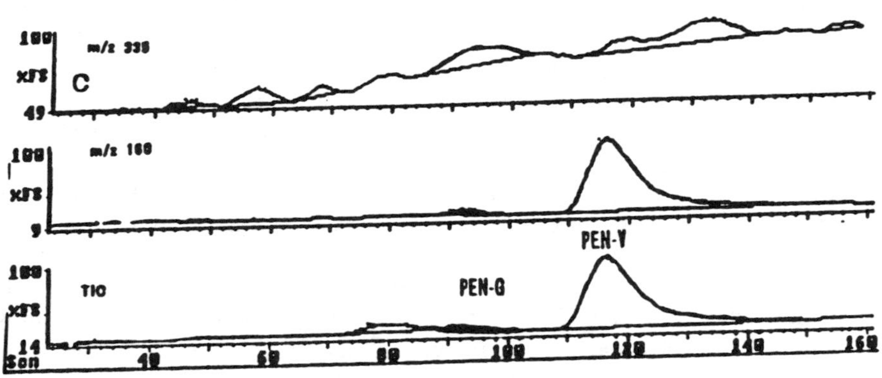

BULK TANK MILK SPIKED WITH 5 ppb PEN-G

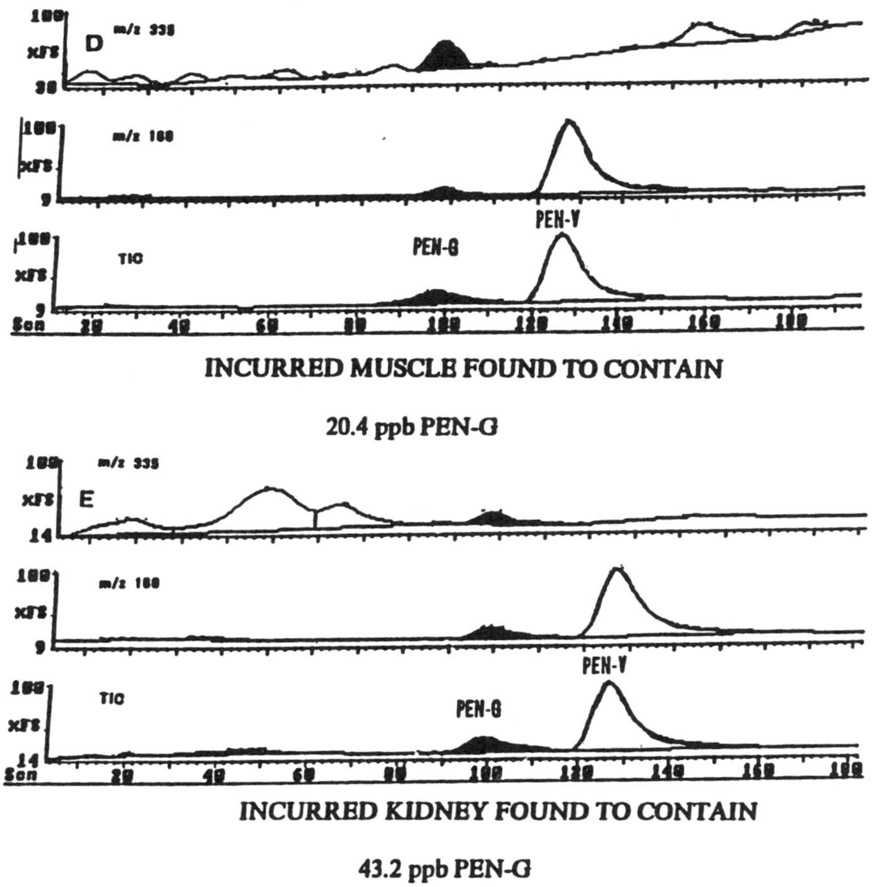

Figure 10 Total and selected ion current chromatograms showing the presence of penicillin G residues at trace levels in different food matrices.
LC conditions: Waters Nova-Pak C_{18}, 150 × 3.9 mm (4 μm) column; flow rate, 0.9 mL/min; mobile phase, 0.15 M NH_4OAc-ACN; gradient conditions, initial (0% ACN), 10–20 min (30% ACN), 20–30 min (100% ACN).
MS conditions: thermospray capillary probe tip, 255°C; ion source, 180°C. VG (Fisons) Trio 2 mass spectrometer was scanned from m/z 100–500 while selectively monitoring the characteristic ions at m/z 160 and 335. See reference (*110*).

in biological fluids and animal tissues. Most of their methods demonstrate that the mass spectrometer can be used to reliably determine and confirm β-lactam residues in these matrices down to 10–100 μg/L or μg/kg concentration levels. Figure 11 shows thermospray LC–MS spectra of commonly used penicillins obtained in aqueous mobile phase (acetonitrile/0.1 M ammonium acetate buffer with a pH of 6.5) on a VG Trio 2 single quadrupole LC–MS instrument with a thermospray LC–MS interface. These drugs generally show intense $[M + 1]^+$ ions with little or no fragmentation under thermospray LC–MS conditions. When fragmentation does occur, only one or two other characteristic ions of significant intensity are

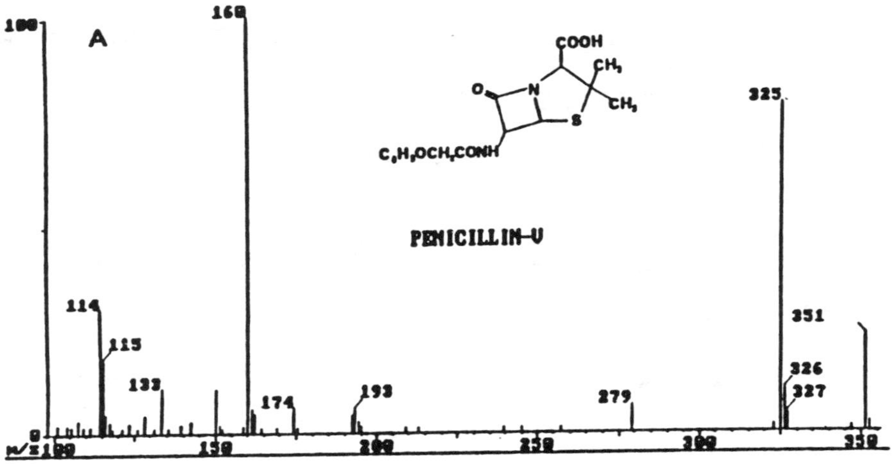

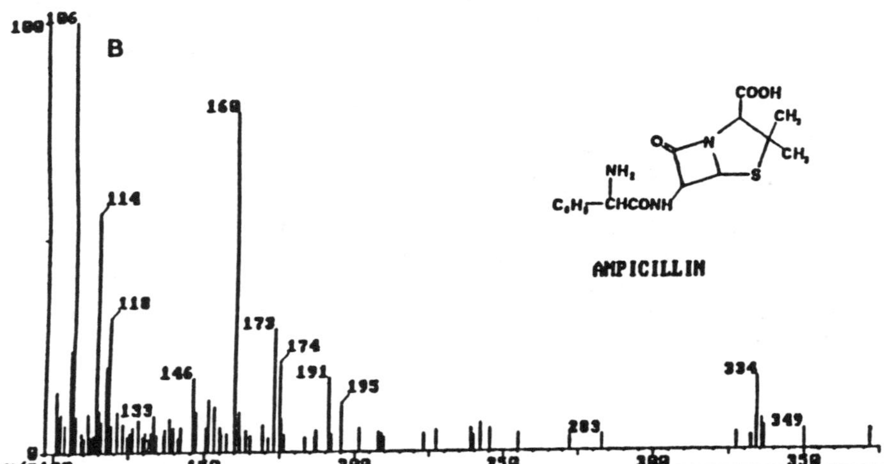

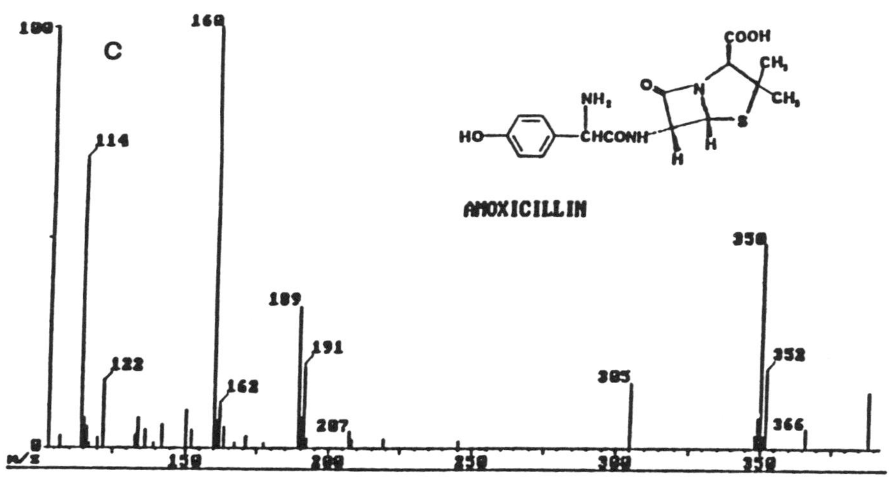

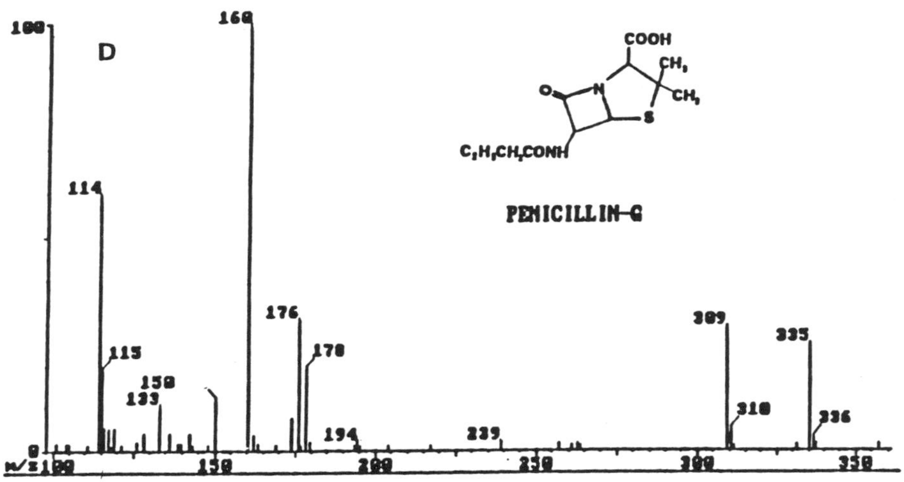

Figure 11 Thermospray LC—mass spectra of common β-lactam antibiotics obtained by loop injections on a VG Trio 2 mass spectrometer. Conditions: mobile phase, 0.15 M ammonium acetate–acetonitrile; flow rate, 1 mL/min; mass spec conditions as indicated in Figure 10.

observed. While the $[M+1]^+$ ion provides sufficient molecular weight data suitable for determining which β-lactam drug is present in an injected sample, the absence of significant characteristic fragment ions in the TSP mass spectra of these compounds makes the technique less suitable than Electrospray Ionization (ESI–MS), where cone voltages can be manipulated to induce more fragmentation, for the confirmation of these drugs in regulatory analysis.

Applications of Developed Analytical Methodology

At the 1990 JECFA meeting, the absence of suitable residue data for benzylpenicillin residues in edible animal tissues was brought to attention. The following sections will review the new residue data that have been determined for the deposition of the Pen G in biological fluids and animal tissues in response to the need identified at the JECFA meeting.

When drugs are administered to animals at the label dose, their dispositions in plasma and other biological fluids are usually very well characterized. Pen G is, in practice, administered at dosages (extra-label usage) far in excess of what is usually prescribed on the label *(151–154)* (see Table 12). In Canada the approved label dose for procaine Pen G usage is 7,500 IU/kg body weight given IM once daily with a preslaughter withdrawal period of 5 days. For a 1:1 commercial formulation of procaine Pen G and benzathine Pen G, the approved dose is 8,600 IU/kg body weight IM with a 14 day withdrawal period. In the United States, the

Table 12 Pharmacokinetic Parameters Determined by Microbiological Assays for the Depletion of Penicillin after IM Administration of Procaine Pen G

Dose administered (IU/kg)	C_{max} (μg/mL)	T_{max} (h)	$t_{1/2}$ (h)	Ref.
3,585	0.354	6	ND	*(151)*
5,470	0.538	6	ND	*(151)*
6,600	2.49	1	3.7	*(152)*
23,000	2.695	6	ND	*(151)*
30,000	2.09	4	6	*(153)*
30,000	1.55	2	4.84	*(154)*
33,000	7.2	1.3	4.7	*(152)*

C_{max}: the maximum plasma concentration; T_{max}: the time at which maximum plasma concentration was achieved; $t_{1/2}$: the disappearance half-life; ND: not determined approved dose for procaine Pen G is 6,600 IU/kg body weight IM with a 10 day preslaughter withdrawal period. When benzathine-procaine Pen G is administered, the label dose is 8,800 IU/kg body weight subcutaneously (SC) with a withdrawal period of 30 days.

It has been reported by Mechor et al. (*155*) that procaine Pen G administered at a dose of 65,000 IU/kg IM once daily was effective for treating acute undifferentiated bovine respiratory disease (BRD). Bateman et al. (*156*) also reported that a dose of 45,000 IU/kg procaine Pen G administered SC once daily was effective in treating bovine pneumonia. The practice of using higher than recommended doses appears to be endorsed even in academic circles where procaine Pen G doses ranging from 20,000–30,000 IU/kg body weight IM or SC are recommended for treating bovine pneumonic pasteurelosis (*157*). Susceptibility studies also indicate that serum concentrations of penicillins ranging from 0.5–2 μg/mL are required (*158*) to treat some penicillin-sensitive bacteria in cattle.

With the exception of one depletion study conducted by Moats et al. (*73*), which used a quantitative LC method, all the available data on the pharmacokinetic parameters for the disposition of Pen G in cattle (see Table 12) have been determined by microbiological assays (*151–154*). Although procaine Pen G has been administered at much higher doses than approved, there was very little pharmacokinetic information to guide therapy when used in an extra-label fashion. Therefore, pharmacokinetic parameters for Pen G administered to cattle at extra-label doses were determined using the LC methods developed.

Disposition of Benzylpenicillin in Animal Plasma

Table 12 shows that the highest plasma or serum concentrations, C_{max}, measured in previous studies for the administration of Pen G to cattle ranged from 0.3–7.2 μg/mL at dosages from 0.5–10 times the label dose. To determine whether extra-label use of Pen G can provide the plasma concentration required to counteract the effect of susceptible bacteria, Papich et al. (*124*) measured the disposition of Pen G administered to feedlot steers. The results of the study, summarized in Tables 13 and 14, show that when beef steers were injected with 66,000 IU/kg of procaine Pen G, the highest C_{max} was attained when injections were made IM into the neck (see Table 13). The plasma concentration was lowest when the same dose of the drug was administered SC into the neck. Additionally, Papich et al. found that the half-life, $t_{1/2}$, was shortest for the IM neck injections when compared with other injection sites. It was also noted that if serum concentrations 0.5–2 g/mL are required to treat some penicillin-sensitive bacteria in cattle (*158*), then it might be more effective to administer procaine Pen G at a dose much higher, perhaps as high as 66,000 IU/kg, than has been usually recommended on the label. Table 14 shows that when a 1:1 combination of benzathine procaine Pen G formulation was administered IM to beef steers at 24,000 IU/kg, the C_{max} was not significantly different when only procaine Pen G was administered IM to beef steers at the same

Table 13 Pharmacokinetic Parameters for the Depletion of Procaine Pen G in Feedlot Steers[a]

Animal group	Dose (IU/kg)	Route and injection site	C_{max} (µg/mL)	T_{max} (h)	$t_{1/2}$(h)
Group 1	24,000	IM, gluteal	0.99 ± 0.04	5.33 ± 0.67	15.74
Group 2	66,000	IM, gluteal	2.63 ± 0.27	6.00 ± 0.00	15.96
Group 3	66,000	IM, neck	4.24 ± 1.08	6.00 ± 0.00	8.85
Group 4	66,000	SC, neck	1.85 ± 0.27	5.33 ± 0.67	18.08

IM: intramuscular; SC: subcutaneous; C_{max}: the maxium plasma concentration; T_{max}: the time at which maximum plasma concentration was achieved; $t_{1/2}$: the disappearance half-life. There were three steers in each group ($n = 3$).

[a] See reference (*124*).

Table 14 Pharmacokinetic Parameters for the Depletion of Benzathine and Procaine Pen G Combination or Benzathine Pen G Alone in Plasma of Yearling Beef Steers[a]

Pen G drug formulation	Route of administration	Dose (IU/kg)	C_{max} (µg/mL)	T_{max} (h)	$t_{1/2}$ (h)
Benz-procaine	IM	8,600 {Cdn label dose, (n = 5)}	0.58 ± 0.15	1.8 ± 0.37	40.6 ± 2.1
Benz-procaine	IM	24,000 (n = 5)	1.06 ± 0.14	2.2 ± 0.5	57.7 ± 6.9
Benz-procaine	SC	8,800 {US label dose, (n = 5)}	0.44 ± 0.02	2.2 ± 0.2	29.3 ± 3.8
Benzathine alone	IM	12,000 (n = 5)	0.23 ± 0.01	17.6 ± 6.1	46.1 ± 4.2

IU: International Units; C_{max}: the maximum plasma concentration; T_{max}: the time at which maximum plasma concentration was attained; $t_{1/2}$: the disappearance half-life; Benz-procaine: Benzathine–procaine combination formulation; IM: Intramuscular; n : the number of animals; SC: subcutaneous.

[a] See reference (*156*).

dose as in Table 13. In comparing plasma concentrations obtained when benzathine procaine Pen G combination drug was administered with those when benzathine Pen G alone was administered, Papich et al. observed that the Pen G plasma concentrations, especially within the first 24 h after drug administration, were primarily due to absorption from procaine Pen G alone. The C_{max} obtained from an injection of benzathine Pen G alone was only 0.23 ± 0.01 μg/mL, which falls well below the levels believed to be inhibitory for most disease-causing bacteria in cattle.

Tissue Residue Studies

The use of Pen G at extra-label doses poses significant challenges to regulatory agencies. Recommended withdrawal periods are useful only when drugs are administered at their prescribed label dosages. Because there are no algorithms for predicting or estimating withdrawal periods when Pen G is used in an extra-label fashion, Papich and co-workers (*124, 159*) and Korsrud et al. (*123*, *128*, *131*) conducted tissue depletion studies to determine appropriate withdrawal periods for extra-label drug use. Experiments were also conducted to determine if the administration of procaine Pen G, alone or in combination with benzathine Pen G, in this fashion results in the persistence of significant residues in edible tissues, as well as the injection sites, at slaughter to cause public health concerns. At the extra-label doses examined, IM injections of 24,000 and 66,000 IU/kg, we determined from tissue depletion studies that withdrawal periods of 10 and 21 days, respectively (compared with 5 days at the label dose), were required for penicillin residues to fall below the established MRL for penicillin.

Tissue Residues at Injection Sites

In a study to determine the persistence of drug residues at injection sites, Korsrud et al. (*128*) treated a set of steers with a single IM injection of a 1:1 commercial formulation of benzathine procaine Pen G at a dose of 24,000 IU/kg (extra-label), the Canadian label dose, or the U.S. label dose and then sacrificed them according to the withdrawal periods indicated in Table 15. Another group of steers was treated with a single IM injection of 12,000 IU/kg benzathine Pen G alone and slaughtered 14 days later. The results of the tissue analyses for penicillin residues at the injection sites from the treated animals shown in Table 15 clearly indicate that, regardless of the dose of the mixed formulation administered either at the Canadian or the U.S. label dose, penicillin residues at levels far in excess of established MRLs were found in injection-site tissues. Pen G residues at concentrations 24 times the MRL were still present in the injection-site tissues 50 days following IM drug administration of the mixed formulation. Injection sites from the group of animals treated IM with benzathine Pen G alone and slaughtered 14 days thereafter contained penicillin residues that were 156 times above the MRL. From the previous work at the same dosage with procaine Pen G alone, we

Table 15 Penicillin G Levels Detected in Injection-Site Muscle Tissues of Steers Administered Either Procaine Pen G Alone or in Combination with Benzathine Pen G[a] Using Liquid Chromatography[b]

Drug formulation and route of administration	Dose (IU/kg)	Days to slaughter after last injection (days)	Pen G concn ± S.E.M. (μg/kg)
Benz-procaine Pen G, IM	8,600 (Cdn label dose)	14	3,200 ± 720
Benz-procaine Pen G, IM	8,600 (Cdn label dose)	30	2,100 ± 800
Benz-procaine Pen G, IM	24,000	14	104,000 ± 55,600
Benz-procaine Pen G, IM	24,000	30	1,200 ± 380
Benz-procaine Pen G, SC	8,800 (US label dose)	30	1,600 ± 1,000
Benzathine Pen G alone	12,000	14	7,800 ± 1,200

Benz-procaine Pen G: Benzathine–Procaine–Pen G combination formulation; IM: Intramuscular; SC: Subcutaneous; IU: International Units; S.E.M.: Standard estimate of the mean (n = 5–7).
[a] See reference (*128*).
[b] See reference (*108*).

concluded that the persistence of residues at the injection sites can be attributed primarily to the benzathine Pen G, the long-acting form of the drug, in the formulation.

Conclusion

It is clear from the methods developed since the 1990 JECFA meeting that there are quite a few suitable single-analyte residue methods now available for the routine determination and even confirmation of benzylpenicillin and other β-lactam veterinary drug residues in animal-derived foods. These methods have also enabled some very significant residue data the depletion of benzylpenicillin in animal tissues to be determined. There are also multiresidue methods that appear to be suitable for analyzing β-lactams in milk and animal tissues. There is, however, a clear absence of any attempt to subject some of these methods to exhaustive collaborative studies so these methods can be accorded official method testing status. Regulatory scientists need to initiate interlaboratory studies to evaluate some of the suitable methods. As governments move toward the harmonization of protocols for evaluating veterinary drugs, proficiency testing of chemical analytical laboratories, etc., it will only seem fit for us to work toward harmonizing third-party recognized chemical procedures for analyzing β-lactam drug residues in animal-derived foods. In order to be able to do this, it must be remembered that, unlike some of the other antibiotics that have demonstrated stability in biological samples, β-lactam are inherently unstable. Therefore, more research is required to determine the ways and means by which samples containing these veterinary drugs can be shipped from one laboratory to another without significant degradation if interlaboratory collaborative studies or proficiency tests are to be undertaken.

Acknowledgments

I would like to acknowledge the excellent contribution of the Research Team comprised of Drs. Korsrud, Yates, and MacNeil (Health of Animals Laboratory, Saskatoon), Drs. Papich (University of North Carolina Veterinary College), Janzen (Western College Of Veterinary Medicine, University Of Saskatchewan, Saskatoon), Dr. Cohen (Department of Animal Science, University Of Saskatchewan, Saskatoon), Drs. Landry, Lambert, Man Sen Yong, and Messier (Bureau Of Veterinary Drugs, Health Canada, Ottawa). I also acknowledge the excellent contributions made by my associates, Ms. Lily Keng and Wayne Chan, throughout the development of some of the methods reported in this chapter. I acknowledge the tremendous technical support and enthusiastic participation in our experimental studies by the following: Ms. Gobeil, Messrs Covey, Pinsonneault, Nykiforuk, Epp, Marshall, Stonechild (Health of Animals Lab, Saskatoon), Drs. Klemmer, and Graham and other meat inspection staff at the Intercontinental Meat Packing plant, Saskatoon. I thank Mr. Gerhardt and Mr. Asea for their excellent skills in drawing the chemical structures shown in the figures and diagrams in this manuscript. I thank and acknowledge Health Canada for financially supporting the depletion studies. Finally, I thank my wife, Betty, and my two children, Lloyd and Beth, for their patience, endurance, and encouragement during the preparation of this manuscript.

References

1. G. Brotzu, "Richerche su di un nuovo Antibiotico" in *Lavori Dell'Instituto D'Igiene du Cagliari*, **1**: 1, 1948.
2. E. P. Abraham, *"Discovery, Chemistry, and Antibacterial Properties of Cephalosporins"* in β-*Lactam Antibiotics*, ed. M. Salton and G. D. Shockman (New York, NY: Academic Press, 1981), 311.
3. I. M. Miller, E. O. Stapley, and L. Chaiet, Bacteriol. Proc., **8**: 32, 1962.

4. R. Nagarajan, L. D. Boeck, M. Gorman, R. L. Hamill, C. E. Higgins, M. M. Hoehn, W. M. Stark, and J. G. Whitney, *J. Am. Chem. Soc.*, **93**: 2308, 1971.
5. E. O. Stapley, M. Jackson, S. Hernandez, S. B. Zimmerman, S. A. Currie, S. Mochales, J. M. Mata, H. B. Woodruff, and D. Hendlin, *Antimicrob. Agents Chemother.*, **2**: 122, 1972.
6. A. K. Miller, E. Celozzi, B. A. Pelak, E. O. Stapley, and D. Hendlin, *Antimicrob. Agents Chemother.*, **2**: 281, 1972.
7. H. Imanka, J. Hasoda, K. Jomon, I. Veda, D. Morin, and H. Sakai, *Japan Patent 49-26488,* March 8, 1974.
8. M. Kojima, T. Shomura, H. Watanabe, Y. Kando, K. Oobayashi, S. Inoue, and T. Niide, *Japan Kokai*, **75**: 155, 1975.
9. H. Fukase, T. Hasegawa, K. Hatano, H. Iwasaki, and M. Yoneda, *J. Antibiotics*, **29**: 113, 1976.
10. Meiji Seika Kaisha Ltd., *U. S. Patent 3,974,035*, August 10, 1976.
11. J. L. Strominger, *Harvey Lect.*, **64**: 179, 1970.
12. R. B. Sykes and N. H. Georgopapadakou, β-Lactam Antibiotics, ed. M. Salton and G. D. Shockman (New York, NY: Academic Press, 1981), 199.
13. D. Greenwood, *Antibiotics of the β-Lactam Group*, ed. R. N. Grüneberg (New York, NY: Research Studies Press, 1982), 37.
14. G. C. Brander and P. R. Ellis, *Animal and Human Health: The Control of Disease,* (London: Bailler Tindall, 1976).
15. J. R. Walton, *Zbl. Vet. Med. A*, **30**: 81, 1983.
16. W. R. Van Dresser and J. R. Wilcke, *J. Am. Vet. Med. Assoc.*, **194**: 1700, 1989.
17. C. D. C. Salisbury, C. E. Rigby, and W. Chan, *J. Agric. Food Chem.*, **37**: 105, 1989.
18. J. K. Shearer and S. F. Sundlof, "Drag Residues: Important Concern in Firestock Production," paper presented at International Conference on Livestock in the Tropics, Gainesville, FL, 1989.
19. I. Tscheuschner, *Z. Haut- Geschlechtskr.*, **47**: 591, 1972.
20. H. J. Schwartz and T. H. Sher, *Ann. Allergy*, **52**: 342, 1984.
21. K. Wicker and R. E. Reisman, *J. Allergy Clin. Immunol.*, **66**: 155, 1980.
22. C. Howes, "Meat Victim Seeks Warnings," *Calgary Herald*, 15 April 1989, sec. B1.
23. World Health Organization, *36th Report of the Joint FAO/WHO Expert Committee On Food Additives*, Rome February 5–14, 1990 (Geneva: World Health Organization, 1990), 37.
24. J. M. Dewdney and R. G. Edwards, *J. Royal Soc. Med.*, **77**: 866, 1984.
25. J. M. Dewdney, L. Maes, J. P. Raynaud, F. Blanc, J. P. Scheid, T. Jackson, S. Lens, and C. Verschueren, review of *Food Chem. Toxicol.*, **29**: 477, 1991.
26. J. P. Hou and J. W. Poole, *J. Pharm. Sci.*, **60**: 503, 1971.

27. J. P. Hou and J. W. Poole, *J. Pharm. Sci.*, **58**: 1510, 1969.
28. H. D. C. Rapson and A. E. Bird, *J. Pharm. Pharmacol.*, **15**: 222T, 1963.
29. B. Wiese and K. Martin, *J. Pharm. Biomed. Anal.*, **7**: 67, 1989.
30. B. Wiese and K. Martin, *J. Pharm. Biomed. Anal.*, **7**: 107, 1989.
31. J. O. Boison, G. O. Korsrud, M. G. Papich, J. D. MacNeil, and W. D. G. Yates, *J. AOAC Int.*, **75**: 974, 1992.
32. B. Wiese and K. Martin, *J. Pharm. Biomed. Anal.*, **7**: 95, 1989.
33. J. O. K. Boison, L. Keng, and J. D. MacNeil, *J. AOAC Int.*, **77**: 565, 1994.
34. P. K. Bhattacharyya and W. M. Cort, *Analytical Profiles of Drug Substances*, vol. 7, ed. K. Florey (New York, NY: Academic Press Inc., 1978), 35.
35. F. Kreuzig, *Analytical Profiles of Drug Substances*, vol. 11, ed. K. Florey (New York, NY: Academic Press, Inc., 1982), 466.
36. K. L. Tyczkowska and A. L. Aronson, *Proceedings 39th ASMS Conference on Mass Spectrometry & Allied Topics*, May 19–24 (Nashville, TN: American Society for Mass Spectrometry, 1991), 589.
37. K. L. Tyczkowska, R. D. Voyksner, and A. L. Aronson, *J. Chromatogr.*, **594**: 195, 1992.
38. G. D. Chisholm, *Scand. J. Infect. Dis.* Suppl., **14**: 118, 1978.
39. R. R. Tight, R. B. Prior, R. L. Perkins, and C. A. Rotilie, *Antimocrob. Agents Chemother.*, **6**: 133, 1975.
40. S. Kawai and S. Hashiba, *Bunseki Kagaku (Jap. Anal.)*, **13**: 1223, 1964.
41. P. B. English, *The Vet. Record*, **77**: 810, 1965.
42. C. Hishta, D. L. Mays, and M. Garafalo, *Anal. Chem.*, **43**: 1530, 1971.
43. H. Bundgaard, *Tetrahedron Lett.*, **48**: 4613, 1971.
44. H. Bundgaard, *Dan. Tidsskr. Farm.*, **46**: 29, 1972.
45. H. Bundgaard, *Dan. Tidsskr. Farm.*, **46**: 85, 1972.
46. H. Bundgaard and K. Ilver, *J. Pharm. Pharmacol.*, **24**: 790, 1972.
47. H. Bundgaard, *J. Pharm. Pharmacol.*, **26**: 385, 1974.
48. Y. Fujimoto, T. Suzuki, and Y. Hoshino, *J. Chromatogr.*, **105**: 99, 1975.
49. T. Suzuki, Y. Fujimoto, Y. Hoshino, and A. Tanaka, *J. Chromatogr.*, **105**: 95, 1975.
50. A. M. DePaolis, S. E. Katz, and J. D. Rosen, *J. Agric. Food Chem.*, **25**: 1112, 1977.
51. H. Wu, M. Masada, and T. Uno, *J. Chromatogr.*, **137**: 127, 1977.
52. B. Meesschaert, P. Adriaens, and H. Eyssen, *J. Chromatogr.*, **136**: 162, 1977.
53. T. Nakagawa, J. Haginaka, M. Masada, and T. Uno, *J. Chromatogr.*, **154**: 264, 1978.
54. D. G. Groothius, ASJPAM Van Miert, G. Ziv, and J. F. M. Nouws, *J. Vet. Pharmacol. Ther.*, **1**: 81, 1978.
55. J. F. M. Nouws and G. Ziv, *J. Vet. Pharmacol. Ther.*, **1**: 47, 1978.

56. J. Carlqvist and D. Westerlund, *J. Chromatogr., Biomed. Appl.*, **164**: 373, 1979.
57. M. E. Rogers, M. W. Adlard, G. Saunders, and G. Holt, *J. Liq. Chromatogr.*, **6**: 2019, 1983.
58. W. A. Moats, *J. Agric. Food Chem.*, **31**: 880, 1983.
59. W. A. Moats, *J. Chromatogr.*, **317**: 311, 1984.
60. M. E. Rogers, M. W. Adlard, G. Saunders, and G. Holt, *J. Chromatogr.*, **297**: 385, 1984.
61. W. J. Irwin, J. M. Hempenstall, A. Li Wan Po, and A. H. Andrews, *J. Chromatogr.*, **287**: 85, 1984.
62. A. Aboul Khier, G. Blaschke, and M. El Sadek, *Anal. Lett.*, **17**: 1659, 1984.
63. R. K. Munns, W. Shimoda, J. E. Roybal, and C. Vieira, *J. Assoc. Off. Anal. Chem.*, **68**: 968, 1985.
64. R. H. Rumble and M. S. Roberts, *J. Chromatogr.*, **342**: 436, 1985.
65. J. Haginaka and J. Wakai, *Analyst*, **110**: 1185, 1985.
66. H. Terada and Y. Sakabe, *J. Chromatogr.*, **348**: 379, 1985.
67. H. Terada, M. Asanoma, and Y. Sakabe, *J. Chromatogr.*, **318**: 299, 1985.
68. J. Haginaka, J. Wakai, H. Yasuda, and T. Uno, *Anal. Sciences*, **1**: 73, 1985.
69. J. O. Miners, *J. Liq. Chromatogr.*, **8**: 2827, 1985.
70. J. H. G. Jonkman, R. Schoenmaker, and J. Hempenius, *J. Pharm. Biomed. Anal.*, **3**: 359, 1985.
71. J. E. Hutchins, K. Tyczkowska, and A. Aronson, *J. Assoc. Off. Anal. Chem.*, **69**: 757, 1986.
72. U. Saha, *Analyst*, **111**: 1179, 1986.
73. W. A. Moats, E. W. Harris, and N. C. Steele, *J. Agric. Food Chem.*, **34**: 452, 1986.
74. T. Nagata and M. Saeki, *J. Assoc. Off. Anal. Chem.*, **69**: 448, 1986.
75. W. F. Smyth, C. Ayling, and J. G. Smyth, *Anal. Proc.*, **23**: 84, 1986.
76. J. Haginaka and J. Wakai, *Anal. Chem.*, **58**: 1896, 1986.
77. M. M. Seigel, R. K. Isensee, and D. J. Beck, *Anal. Chem.*, **59**: 989, 1987.
78. K. Tyczkowska, D. P. Aucion, D. C. Richardson, and A. L. Aronson, *J. Liq. Chromatogr.*, **10**: 2613, 1987.
79. J. Haginaka and J. Wakai, *J. Pharm. Pharmacol.*, **39**: 5, 1987.
80. A. M. Lipczynski, *Analyst*, **112**: 411, 1987.
81. J. Haginaka, J. Wakai, H. Yasuda, T. Uno, K. Takahashi, and T. Katagi, *J. Chromatogr.*, **400**: 101, 1987.
82. S. Suwanrumpha, D. A. Flory, R. B. Freas, and M. L. Vestal, *Biomed. Environ. Mass Spectrom.*, **16**: 381, 1988.
83. A. J. Shah and M. W. Adlard, *J. Chromatogr.*, **424**: 325, 1988.
84. A. J. Shah, M. W. Adlard, and G. Holt, *Analyst*, **113**: 1197, 1988.

85. K. Tyczkowska and A. L. Aronson, *J. Chromatogr. Biomed. Appl.*, **427**: 103, 1988.
86. J. Martin, R. Mendez, and A. Negro, *J. Liq. Chromatogr.*, **11**: 1707, 1988.
87. A. F. Casy, C. Cryer, and E. M. A. Ominde, *J. Pharm. Biomed. Anal.*, **7**: 1121, 1989.
88. K. Tyczkowska, R. D. Voyksner, and A. L. Aronson, *J. Chromatogr. Biomed. Applic.*, **490**: 101, 1989.
89. U. Meetschen and M. Petz, *J. Assoc. Off. Anal. Chem.*, **73**: 373, 1990.
90. W. A. Moats, *Proceedings of the EuroResidue Conference on Residues of Veterinary Drugs in Food*, Noordwijkerhout, Netherlands, May 21–23, ed. N. Haagsma, A. Ruiter and P. B. Czedik-Eysenberg (ADDIX, Wijk bij Duurstede, 1990), 280.
91. K. Berger and M. Petz, *Proceedings of the EuroResidue Conference on Veterinary Drugs in Food*, Noordwijkerhout, Netherlands, May 21—23, ed. N. Haagsma, A. Ruiter and P. B. Czedik-Eysenberg (ADDIX, Wijk bij Duurstede, 1990), 118.
92. P. S. Jaglan, B. L. Cox, T. S. Arnold, M. F. Kubicek, D. J. Stuart and T. J. Gilbertson, *J. Assoc. Off. Anal. Chem.*, **73**: 26, 1990.
93. P. Rahn, *Application Notes*, Waters Chromatography (Millford, MA: Nations Chromatography, 1990).
94. R. D. Voyksner, C. S. Smith and P. C. Knox, *Biomed. Environ. Mass Spectrom.*, **19**: 523, 1990.
95. K. L. Tyczkowska, A. L. Aronson, D. J. A. Grindstaff and R. D. Voyksner, *Proceedings of the 38th ASMS Conference on Mass Spectrometry & Allied Topics*, June 3-8, Tucson, Arizona (East Lansing, MI: American Society for Mass Spectrometry, 1990), 1049.
96. P. S. Jaglan, M. F. Kubicek, T. S. Arnold, and V. L. Hubbard, "High Pressure Liquid Chromatographic Analysis of Ceftiofur Residues in Bovine Kidney following Intramuscular Treatment with Five Doses of 2.2 mg/kg/day of ^{14}C-Ceftiofur Hydrochloride" in *Technical Report 788-9760-88-014 Upjohn Company* (Kalamazoo, MI: Upjohn Company, 1988).
97. Commission of the European Communities, *The Rules Governing Medicinal Products In the European Community*, **6**: 1991.
98. *Canada Gazette, Food and Drug Regulations, Amendment, Part II*, No. 9, **125**: 1991, 1478.
99. S. D. Brynes and M. S. Yong, *Proceedings EuroResidue II Conference of Residues in Food*, May 3-5, ed. N. Haagsma, A. Ruiter and P. B. Czedik-Eysenberg (ADDIX Wijk bij Duurstede, 1993), 226.
100. P. Guillot, "Etude Pharmacocinetique de la Penicillin G chez le Bovin" (Ph. D. Thesis, University of Rennes, Rennes, 1983).
101. A. McCracken and J. J. O'Brien, *Res. Vet. Sci.*, **21**: 240, 1976.
102. R. J. Heitzman, *Anal. Chim. Acta*, **275**: 17, 1993.

103. H. F. Askal, G. A. Saleh and N. M. Omar, *Analyst*, **116**: 387, 1991.

104. S. Pleasance, R. Bailey, P. Blay, D. North, and M. A. Quilliam, *Proceedings 39th ASMS Conference on Mass Spectrometry & Allied Topics*, Nashville, TN, May 19–24 (East Lansing, MI: American Society for Mass Spectrometry, 1991), 1362.

105. J. D. MacNeil, G. O. Korsrud, J. O. Boison, M. G. Papich, and W. D. G. Yates, *J. Food Prot.*, **54**: 37, 1991.

106. C. D. Orford, D. Perry, and M. W. Adlard, *J. Liq. Chromatogr.*, **14**: 2665, 1991.

107. C. M. Moore, K. Sato, and Y. Katsumata, *J. Chromatogr.*, **539**: 215, 1991.

108. J. O. Boison, C. D. C. Salisbury, W. Chan, and J. D. MacNeil, *J. Assoc. Off. Anal. Chem.*, **74**: 497, 1991.

109. J. O. Boison, G. O. Korsrud, J. D. MacNeil, L. Keng, and M. G. Papich, *J. Chromatogr. Biomed. Appl.*, **576**: 315, 1992.

110. J. O. Boison, J. Patterson, J. D. MacNeil, R. Gedir, and L. Keng, *Proceedings 40th ASMS Conference on Mass Spectrometry & Allied Topics*, May 31–June 5 (Washington D. C.: American Society for Mass Spectrometry, 1992), 920.

111. R. G. Aoyama, D. D. Kitts, H. M. Burt, and K. M. McErlane, *J. Chromatogr.*, **626**: 271, 1992.

112. W. A. Moats and R. Malisch, *J. Assoc. Off. Anal. Chem.*, **75**: 257, 1992.

113. D. J. Fletouris, J. E. Psomas, and A. J. Mantis, *J. Agric. Food Chem.*, **40**: 617, 1992.

114. H. Mineo, S. Kaneko, I. Koizumi, K. Asida, and F. Akahori, *Vet. Hum. Toxicol.*, **34**: 393, 1992.

115. M. J. Dunn and D. Hahn, *J. Chromatogr.*, **595**: 185, 1992.

116. Y. Ohki, T. Nakamura, H. Nagaki, and T. Kinmoshita, *Biol. Mass Spectrom.*, **21**: 133, 1992.

117. J. O. Boison, *J. Chromatogr.*, **624**: 171, 1992.

118. W. A. Moats, *Analysis of Antibiotic Residues in Food Products of Animal Origin*, ed. V. K. Agarwal (New York, NY: Plenum Press, 1992), 133.

119. M. Petz, *Analysis of Antibiotic Residues in Food Products of Animal Origin*, ed. V. K. Agarwal (New York, NY: Plenum Press, 1992), 147.

120. S. Chulavatnatol and B. G. Charles, *J. Chromatogr. Biomed. Appl.*, **615**: 91, 1993.

121. C. Hendrix, Y. Zhu, C. Wijsen, E. Roets, and J. Hoogmartens, *J. Chromatogr.*, **634**: 257, 1993.

122. K. Borner, E. Borner, and H. Lode, *J. Chromatogr. Biomed. Appl.*, **615**: 174, 1993.

123. G. O. Korsrud, J. O. Boison, M. G. Papich, W. D. G. Yates, J. D. MacNeil, E. D. Janzen, R. D. H. Cohen, D. A. Landry, G. Lambert, M. S. Yong, and J. R. Messier, *Can. J. Vet. Res.*, **57**: 223, 1993.
124. M. G. Papich, G. O. Korsrud, J. O. Boison, W. D. G. Yates, J. D. MacNeil, E. D. Janzen, R. D. H. Cohen, and D. A. Landry, *J. Vet. Pharmacol. Ther.*, **16**: 317, 1993.
125. C. E. Parker, J. R. Perkins, K. B. Tomer, Y. Shida, and K. O'Hara, *J. Chromatogr. Biomed. Appl.*, **616**: 45, 1993.
126. R. F. Straub and R. D. Voyksner, *J. Chromatogr.*, **647**: 167, 1993.
127. K. L. Tyczkowska, R. D. Voyksner, K. L. Anderson, and A. L. Aronson, *J. Chromatogr. Biomed. Appl.*, **614**: 123, 1993.
128. G. O. Korsrud, J. O. Boison, M. G. Papich, W. D. G. Yates, J. D. MacNeil, E. D. Janzen, R. D. H. Cohen, J. J. McKinnon, D. A. Landry, G. Lambert, M. S. Yong, J. R. Messier, and L. Ritter, *Proceedings EuroResidue II Conference on Residues of Veterinary Drugs in Food*, May 3–5, ed. N. Haagsma, A. Ruiter, and P. B. Czedik-Eysenberg (ADDIX, Wijk bij Duurstede, 1993), 419.
129. W. Chan and C. D. C. Salisbury, *Proceedings EuroResidue II Conference on Residues of Veterinary Drugs in Food*, May 3–5, ed. N. Haagsma, A. Ruiter, and P. B. Czedik-Eysenberg (ADDIX, Wijk bij Duurstede, 1993), 236.
130. J. D. MacNeil, J. R. Patterson, C. D. C. Salisbury, G. O. Korsrud, and J. O. Boison, *Proceedings EuroResidue II Conference on Residues of Veterinary Drugs in Food*, May 3–5, ed. N. Haagsma, A. Ruiter, and P. B. Czedik-Eysenberg (ADDIX Wijk bij Duurstede, 1993) 469.
131. G. O. Korsrud, J. O. Boison, M. G. Papich, W. D. G. Yates, J. D. MacNeil, E. D. Janzen, J. J. McKinnon, D. A. Landry, G. Lambert, M. S. Yong, and L. Ritter, *Food Addit. Contam.*, **11**: 1, 1994.
132. W. A. Moats, *J. AOAC Int.*, **77**: 41, 1994.
133. S. A. Barker, A. R. Long, and C. R. Short, *Proceedings of the 6th symposium on Vet. Pharmacol. and Therapeutics*, ed. W. Huber (Blacksburg, VA: American Academy of Veterinary Pharmacology and Therapeutics, 1988), 55.
134. S. A. Barker, A. R. Long, and C. R. Short, *J. Chromatogr.*, **475**: 353, 1989.
135. *Anal. Proc.*, **30**: 321, 1993.
136. R. E. Majors, *Liq. Chrom. Gas Chrom.*, **9**: 686, 1991.
137. J. G. Dorsey, W. T. Cooper, J. F. Wheeler, H. G. Barth, and J. P. Foley, *Anal. Chem.*, **66**: 500R, 1994.
138. S. M. Staroverov and A. Y. Fadeev, *J. Chromatogr.*, **544**: 77, 1991.
139. S. H. Hansen, P. Helboe, and M. Thomsen, *J. Chromatogr.*, **544**: 53, 1991.
140. T. C. Pinkerton, *J. Chromatogr.*, **544**: 13, 1991.
141. B. Feibush and C. T. Santasania, *J. Chromatogr.*, **544**: 41, 1991.

142. H. Huang, J. Wu, and M. Chen, *J. Chromatogr. Biomed. Appl.*, **564**: 195, 1991.
143. H. Zou, Y. Zhang, and P. Lu, *J. Chromatogr.*, **545**: 59, 1991.
144. C. P. Desilets, M. A. Rounds, and F. E. Regnier, *J. Chromatogr.*, **544**: 25, 1991.
145. A. A. Petrauskas and V. K. Svedas, *J. Chromatogr.*, **585**: 3, 1991.
146. L. J. Nunez-Vergara, M. M. Silva, and J. A. Squella, *J. Assoc. Off. Anal. Chem.*, **65**: 1403, 1982.
147. B. Ogereve, V. Hudnik, and S. Gomiseek, *Fresenius Z. Anal. Chem.*, **330**: 59, 1988.
148. H. Fabre and W. Th. Kok, *Anal. Chem.*, **60**: 136, 1988.
149. E. Kirchmann and L. E. Welch, *J. Chromatogr.*, **633**: 111, 1993.
150. A. Tsuji, E. Miyamoto, and T. Yamana, *J. Pharm. Pharmacol.*, **30**: 811, 1978.
151. R. A. Huebner, *Cornell Vet.*, **42**: 457, 1952.
152. R. H. Teske, L. D. Rollins, and G. G. Carter, *J. Am. Vet. Med. Assoc.*, **160**: 873, 1972.
153. J. Luthman and S. O. Jacobsson, *Acta. Vet. Scand.*, **27**: 313, 1986.
154. B. Bengtsson, A. Franklin, J. Luthman, and S. O. Jacobsson, *J. Vet. Pharm. Therap..*, **12**: 37, 1989.
155. G. D. Mechor, G. K. Jim, and E. D. Janzen, *Can. Vet. J.*, **29**: 483, 1988.
156. K. G. Bateman, S. W. Martin, P. E. Shewen, and P. I. Menzies, *Can. Vet. J.*, **31**: 689, 1990.
157. D. C. Blood and O. M. Radostits, *Veterinary Medicine*, 7th ed. (Toronto, Canada: Bailliére, Tindall, 1989), 669.
158. C. A. Hjerpe and T. A. Routen, *Bovine Practitioner*, **9**: 97, 1976.
159. M. G. Papich, G. O. Korsrud, J. O. Boison, W. D. G. Yates, J. D. MacNeil, E. D. Janzen, J. J. McKinnon, and D. A. Landry, *Am. J. Vet. Res.*, **55**: 825, 1994.

Chapter 9

Chemical Analysis of Aminoglycoside Antibiotics

Craig D. C. Salisbury
Agriculture Canada, Health of Animals Laboratory,
Saskatoon, Saskatchewan, Canada S7N 5E3

Contents

Introduction

The aminoglycoside antibiotics are produced by members of the genus *Streptomyces* ("mycins") and *Micromonospora* ("micins") (*1*). Widely used in human and veterinary medicine, these broad spectrum antibiotics are effective against *Escherichia coli*, *Staphyloccocus aureus*, *Klebsiella*, *Proteus*, *Serratia*, *Salmonella* and enterobacter infections. Gentamicin and amikacin are also effective against *Pseudomonas* species (*1*, *2*). Beginning with streptomycin in 1944, the aminoglycosides were one of the earliest antibiotic groups studied and developed, and second only to penicillin in attaining widespread use.

Structurally, aminoglycosides consist of two or more sugars, usually aminosugars, attached to an aminocyclitol ring with glycoside linkages (linked through oxygen). The central aminocyclitol, a saturated ring substituted with amine and hydroxyl groups, is 2-deoxystreptamine in most aminoglycosides. For the streptomycins, where the amine functions of the streptamine moiety are part of guanidine groups, the aminocyclitol is streptidine. The structures of the aminocyclitols are shown in Figure 1.

In food-animal production, the most commonly used aminoglycosides are gentamicin, neomycin, dihydrostreptomycin, and streptomycin, although other members of this family are used to a lesser degree. The discovery of other naturally occurring aminoglycosides and the development of semisynthetic derivatives have resulted in drugs with lower toxicity and greater effectiveness. The need to turn to other members of the aminoglycoside family is determined by the growing resistance of bacteria to the established drugs (*3*, *4*). Structures of some selected aminoglycosides are shown in Figure 2.

Aminoglycosides are used as feed and water additives, or they may be administered as injectables. Although they are poorly absorbed orally, parenteral administrations can result in high, persistent residues (*5*, *6*). As a result, when

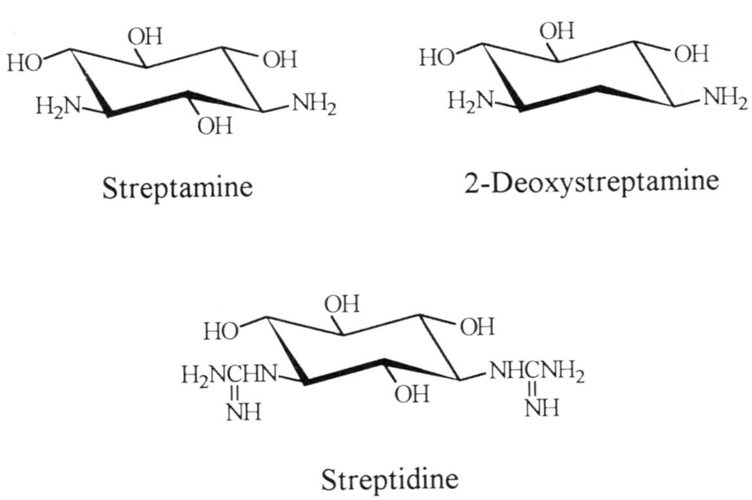

Figure 1 Aminocylitol base groups found in aminoglycoside antibiotics

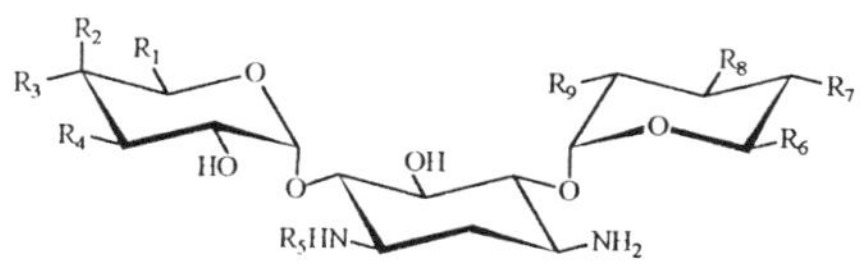

I

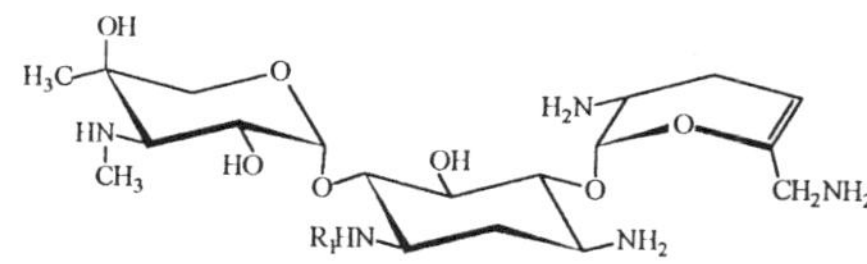

II

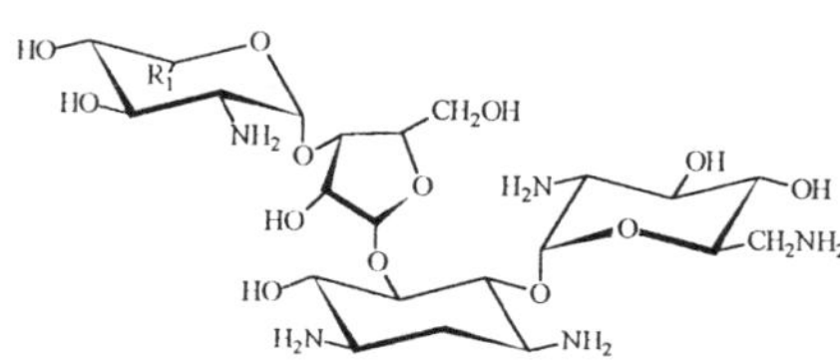

III

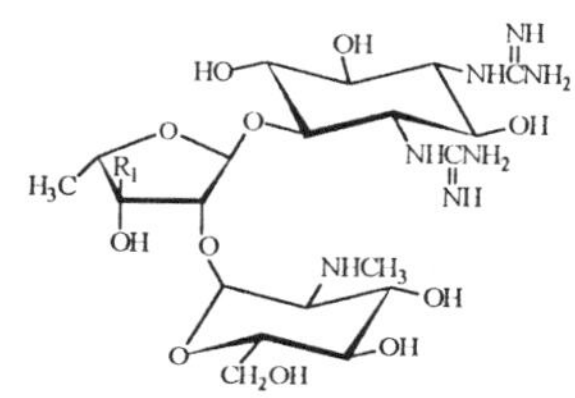

IV

Compound	Structure	R1	R2	R3	R4	R5	R6	R7	R8	R9
Amikacin	I	CH_2OH	H	OH	NH_2	*	CH_2NH_2	OH	OH	OH
Dibekacin	I	CH_2OH	H	OH	NH_2	H	CH_2NH_2	H	H	NH_2
Gentimicin C1	I	H	OH	CH_3	$NHCH_3$	H	CHCH3NHCH3	H	H	NH_2
Gentamicin C1a	I	H	OH	CH_3	$NHCH_3$	H	CH_2NH_2	H	H	NH_2
Gentamincin C2	I	H	OH	CH_3	$NHCH_3$	H	$CHCH_3NH_3$	H	H	NH_2
Kanamycin A	I	CH_2OH	H	OH	NH_2	H	CH_2NH_2	OH	OH	OH
Tobramycin	I	CH_2	H	OH	NH_2	H	CH_2NH_2	OH	H	NH_2
Netilmycin	II	CH_2CH_3								
Sisomycin	II	H								
Neomycin B	III	CH_2NH_2								
Dihydro-streptomycin	IV	CH_2OH								
Streptomycin	IV	CHO								

*$COCHOH(CH_2)_2NH_2$

Figure 2 Structures of some common aminoglycoside antibiotics

aminoglycosides are injected into food animals the withdrawal period can be significant, with some longer than 60 days (*7*, *8*).

When administered, aminoglycosides are eliminated from the body by renal glomerular filtration and excretion in the urine. As a result of this elimination process, residue levels are much higher in urine than in serum (*1*, *4*). Aminoglycosides are not metabolized in the body but rather they are bound and excreted as the parent compound. Approximately 10% of the plasma concentration of aminoglycosides is bound to plasma proteins (5), while 15–25% of the milk concentration is bound to milk proteins (*9*). Bound levels in milk are highest when mammary gland infection is present.

Residues in the body tend to be concentrated in the kidney, particularly in the cortex (*10*), with 40% of the total antibiotic burden found in the kidneys (*4*). This distribution is due to the aminoglycosides, which are polybasic cations, binding strongly to acidic phospholipids. These compounds are found in the kidneys and in cochlear tissues, causing these drugs to be both nephrotoxic and ototoxic (*11–14*). Levels in the remaining tissues rise and fall with increasing and decreasing plasma levels. Binding competition among the various aminoglycosides increases with increasing numbers of available amino groups (*15*, *16*).

The tendency for aminoglycosides to preferentially accumulate in the kidney is quite dramatic. Human tissue analysis for gentamicin residues showed kidney levels to be 161 times greater than muscle levels (*17*). The analysis of bovine and porcine kidney and muscle pairs from suspect Canadian slaughter animals at our laboratory has indicated that there is an average difference in residue levels of 150:1 and 164:1, respectively, for dihydrostreptomycin.

Aminoglycosides are water soluble, polar, and weakly basic compounds. They are stable at both high and low pH levels, and are heat resistant (*18*). They lack useful chromophores and fluorophores, and are not volatile, so detection by modern instrumental techniques usually requires derivatization.

Complicating factors in analyzing these antibiotics are due to formulations of aminoglycosides consisting of more than one component, and because the structures of related aminoglycosides differ only slightly. Commercially available gentamicin, for example, contains three distinct yet structurally similar components: C_1, C_{1a}, and C_2. Sisomicin and gentamicin C_{1a} differ by only one double bond. As a result, chromatographic conditions require significant fine-tuning if complete resolution and separation of all peaks is desired.

Chemical methods for determining aminoglycoside residues are available for serum, plasma, urine, and milk, and enable food animals to be tested prior to slaughter or after milk collection. Significantly fewer methods have been reported for determining residues in tissue, and most of these lack the sensitivity required to detect the low ppb levels likely to occur in muscle.

Most of the current chemical methods are based on high-performance liquid chromatography (LC). A few thin-layer chromatographic (TLC), and gas chromatographic (GC) techniques have been reported, but their use has declined with the advent of modern liquid chromatographic techniques in the late 1970s.

Thin-Layer Chromatography

Langner and Teufel (*19*) used TLC to identify antibiotic residues, including aminoglycosides, in animal tissues. Using an assortment of mobile phases, systems were developed to separate the various antibiotics studied. Spray reagents were also tested to enable visualization and detection of the aminoglycosides, which do not fluoresce naturally. However, the sensitivity obtained in this manner could not match that achieved using a bioassay system for detection.

Combining TLC separation with bioassay detection led to development of TLC/bioautography, a technique whereby antibiotics separated on TLC plates are detected by overlaying the plates with medium seeded with bacteria, or by placing the TLC plate face-down onto the seeded medium. The antibiotics diffuse into the medium, and following incubation, are visualized by zones of bacterial growth inhibition. Dihydrostreptomycin, kanamycin, and neomycin were determined in the tissues and urine of swine and calves using this technique (*20*). Residues of the three drugs were confirmed in kidney and urine samples of treated animals, but not in the corresponding muscle. From 1984 to 1991 we applied the TLC/bioautography technique to the tissues of suspect Canadian slaughter animals in our laboratory (*21*). However, the method's poor sensitivity for the aminoglycosides (30–200 ppm) limited its effective use to kidney only. Testing indicated that low and variable extraction efficiency of the drugs from tissues and poor response of the organism (*B. subtilis*) were responsible for the poor sensitivity.

In the method of Saito et al. (*22*), TLC separation prior to detection with a flame ionization detector was used to determine kanamycin in bovine bile and urine.

Gas Chromatography

Gas chromatography was used to determine gentamicin, tobramycin, netilmicin and amikacin in serum (*23*, *24*). Deproteinized and dehydrated samples were derivatized in a two-stage process with trimethylsilylimidazole and heptafluorobutyrylimidazole and then detected by electron capture.

Kanamycin was determined in bovine muscle following extraction with trichloroacetic acid (TCA) and purification on an ion exchange column (*25*). The extract was derivatized with heptafluorobutyrylimidazole, followed by trimethylsilylation. Detection was by electron capture. Recoveries from tissue ranged from 74–86%, with a detection limit of 100 ppb.

Liquid Chromatography

Tables 1 and 2 summarize liquid chromatography (LC) methods for determining aminoglycoside antibiotics residues in biological fluids and tissues.

Sample Preparation

Biological fluids. Because aminoglycosides are not strongly bound to the proteins of biological fluids, many effective extraction and detection methods have been reported. Most were developed for use in clinical laboratories to monitor human patients for high levels of drugs during treatment, but the analytical techniques are applicable to animal studies as well.

Proteins must be removed from biological fluids because they can interfere with extraction, derivatization, chromatography, and detection steps of analytical methods. Deproteinization is usually accomplished by mixing the fluid with a protein precipitator, such as acetonitrile (*26–37*), methanol (*38–41*), trichloroacetic acid (*42*, *43*), or perchloric acid (*44–48*). Following deproteinization in this manner, the extract, buffered to an alkaline pH, can be further purified by washing with methylene chloride (*26–28*, *31*, *32*, *36*) or chloroform (*35*) to remove extraneous compounds of lower polarity. The elevated pH level ensures the drug is retained in the aqueous phase.

Removing proteins and other matrix constituents can also be accomplished with small solid-phase extraction (SPE) columns packed with weak cation-exchange materials such as Amberlite CG-50 (*49*, *50*) and CM Sephadex C-25 (*51–57*), or with silica (*58–62*) or C18 (*35*, *63*, *64*) sorbents. Ultrafiltration serves as a mechanical means of protein removal (*65*, *66*).

Biological fluids containing aminoglycoside residues have been purified using column enrichment LC techniques (*41*, *48*, *67*). Usually, this involves trapping the analytes onto an LC precolumn, rinsing coextracted materials to waste, and then flushing the concentrated analytes onto the analytical column. In a somewhat different approach, Tamai et al. (*46*) used a weak perchloric acid solution as an initial mobile phase to denature and trap serum proteins on a precolumn packed with polyvinyl resin. Gentamicin was not retained, and it passed through to the analytical column. An LC column that separates proteins from the analyte was used to determine neomycin residues in milk (*50*). The silica-based polymeric packing is shielded by a hydrophilic network permitting the analytes to penetrate to the polymeric surface while excluding proteins, which pass through the column unretained.

Tissues. Extraction of aminoglycosides from tissues, particularly kidney, is considerably more difficult than from fluids, and as a result, there are fewer methods reported (Table 2). The sensitivity of some of these methods is not sufficient to detect residues likely to be found in muscle tissue, and therefore they

Text continues on page 322

Table 1 Summary of Literature Methods for Liquid Chromatographic Determination of Aminoglycoside Residues in Biological Fluids

Fluid	Analytes	Sample treatment	Derivatization	LC column technique	Detection	Detection limit (ppm)	Recovery (%)	Year (Ref.)
Serum	Gentamicin	Silica SPE	Precolumn: OPA	C18	Fluor.	0.5	80–105	1977 (*58*)
Plasma	Gentamicin	CH_3CN deprot., CH_2Cl_2 extn	Precolumn: Dansyl Cl	C18	Fluor.	—[a]	—	1977 (*26*) 1978 (*36*)
Plasma	Netilmicin	CH_3CN deprot., CH_2Cl_2 extn	Precolumn: Dansyl Cl	C18	Fluor.	0.5	—	1977 (*27*)
Serum	Amikacin, Gentamicin, Tobramycin	Ion exchange SPE	Postcolumn: OPA	C18, ion-pairing	Fluor.	—	89–100	1977 (*51*) 1978 (*52*)
Serum	Amikacin	Silica SPE	Precolumn: OPA	C18	Fluor.	—	—	1978 (*59*)
Serum	Gentamicin, Netilmicin, Tobramycin	CH_3CN deprot., CH_2Cl_2 extn	Precolumn: OPA	C18	Fluor.	0.5–1	92–100	1979 (*28*)
Serum	Tobramycin	Silica SPE	Precolumn: OPA	C18	Fluor.	0.5	96–110	1979 (*60*)

Table 1 Summary of Literature Methods for Liquid Chromatographic Determination of Aminoglycoside Residues in Biological Fluids—*continued*

Fluid	Analytes	Sample treatment	Derivatization	LC column technique	Detection	Detection limit (ppm)	Recovery (%)	Year . (Ref.)
Serum, Urine	Tobramycin	CH_3CN deprot., CH_2Cl_2 extn	Precolumn: OPA	CN	Fluor.	0.2	—	1980 (*31*)
Serum	Gentamicin	CH_3CN deprot., CH_2Cl_2 extn	Precolumn: BSC	C18	Spectro.	0.2	—	1980 (*29*)
Serum	Gentamicin	CH_3CN deprot.	Precolumn: FDNB	C18	Spectro.	0.3	83-84	1980 (*30*)
Plasma, Urine	Gentamicin	CH_3CN deprot., CH_2Cl_2 extn	Precolumn: FLCM	SCX	Fluor.	1	—	1981 (*32*)
Serum	Gentamicin, Sisomicin	CH_3CN deprot.	Precolumn: FDNB	C18	Spectro.	—	83–84	1981 (*33*)
Serum	Tobramycin	CH_3CN deprot.	Precolumn: FDNB	C18	Spectro.	0.5	—	1981 (*34*)
Serum	Gentamicin, Netilmicin, Tobramycin	Ion exchange, SPE	Precolumn: OPA	C18	Fluor.	0.5	80-90	1982 (*49*)
Plasma, Serum	Amikacin	CH_3OH deprot., Silica SPE	Precolumn: FDNB	C18	Spectro.	—	—	1982 (*38*)

Table 1 Summary of Literature Methods for Liquid Chromatographic Determination of Aminoglycoside Residues in Biological Fluids—*continued*

Fluid	Analytes	Sample treatment	Derivatization	LC column technique	Detection	Detection limit (ppm)	Recovery (%)	Year . (Ref.)
Plasma, Urine	Gentamicin	CH_3CN deprot., $CHCl_3$ extn (plasma), C18 SPE (urine)	Precolumn: OPA	C8	Fluor.	0.5	85	1982 *(35)*
Serum	Gentamicin	Silica SPE	Precolumn: OPA	C18	Fluor.	0.5	92–100	1982 *(61)*
Serum	Gentamicin	Ultrafiltration	Precolumn: OPA	C18, ion-pairing	Fluor.	0.2	101	1982 *(65)*
Serum	Gentamicin	CH_3OH deprot.	Postcolumn: OPA	C18, ion-pairing	Fluor.	0.5	97–103	1982 *(39)*
Serum	Gentamicin	Centrifugation	Precolumn: TNBS	C8	Spectro.	0.5	97–99	1983 *(100)*
Plasma, Urine	Gentamicin	Ion exchange SPE	Postcolumn: OPA	C18, ion-pairing	Fluor.	—	67–69	1983 *(53)*
Serum	Amikacin	C18 SPE	Precolumn: TNBS	C8	Spectro.	—	93–98	1984 *(63)*

Table 1 Summary of Literature Methods for Liquid Chromatographic Determination of Aminoglycoside Residues in Biological Fluids—*continued*

Fluid	Analytes	Sample treatment	Derivatization	LC column technique	Detection	Detection limit (ppm)	Recovery (%)	Year . (Ref.)
Serum	Tobramycin	CH_3OH deprot.	Postcolumn: OPA	C18, ion-pairing	Fluor.	0.3	98-103	1984 (*40*)
Serum	Astromicin, Micronomicin, Netilmicin, Sisomicin	Ion exhchange SPE	Postcolumn: OPA	C8, C18, ion-pairing	Fluor.	—	—	1984 (*54*)
Serum	Amikacin, Dibekacin, Gentamicin, Netilmicin, Sisomycin, Tobramycin	Centrifugation	Oncolumn: OPA	C18, column enrichment	Fluor.	—	—	1984 (*67*)
Serum	Streptomycin	C18 SPE	None	C18, ion-pairing	Spectro.	2	—	1985 (*64*)
Peri-lymph	Dihydro-streptomycin, Kanamycin, Neomycin, Streptomycin	Lyophilization	Precolumn: Benzoyl Chloride	Normal phase	Spectro.	—	—	1985 (*101*)

Table 1 Summary of Literature Methods for Liquid Chromatographic Determination of Aminoglycoside Residues in Biological Fluids—*continued*

Fluid	Analytes	Sample treatment	Derivatization	LC column technique	Detection	Detection limit (ppm)	Recovery (%)	Year . (Ref.)
Plasma	Sisomicin	Ion exchange, SPE	Precolumn: OPA	C18	Fluor.	—	99	1985 *(55)*
Serum	Dibekacin, Kanamycin .	PCA deprot	Postcolumn: OPA	C18, ion-pairing	Fluor.	—	99	1985 *(44)*
Serum	Streptomycin	PCA deprot.	Postcolumn: NQS	C18, ion-pairing	Fluor.	—	100	1986 *(45)*
Plasma	Sisomicin	Ion exchange SPE	Precolumn: OPA–TGA	C18, column enrichment	Spectro.	—	—	1986 *(56)*
Serum	Gentamicin	None	Postcolumn: OPA	C18, column enrichment, ion-pairing	Fluor.	0.5	100–104	1986 *(46)*
Serum	Streptomycin	PCA deprot.	Postcolumn: NHN	C18, ion-pairing	Fluor.	1.0	100	1987 *(47)*
Serum	Gentamicin	Silica SPE	Precolumn: OPA	C18	Fluor.	0.02–0.08	—	1987 *(62)*
Bile, Urine	Kanamycin, Neomycin	Ion exchange SPE	Precolumn: FDNB	C18	Spectro.	0.3	83-97	1987 *(57)*

Table 1 Summary of Literature Methods for Liquid Chromatographic Determination of Aminoglycoside Residues in Biological Fluids—*continued*

Fluid	Analytes	Sample treatment	Derivatization	LC column technique	Detection	Detection limit (ppm)	Recovery (%)	Year . (Ref.)
Serum	Netilmicin	CH_3CN deprot.	Precolumn: FDNB	C18	Spectro.	0.5	78-81	1988 *(37)*
Serum	Sisomycin	CH_3OH deprot.	Precolumn: OPA	C18, column enrichment, ion-paring	Fluor.	0.06	98	1988 *(41)*
Milk	Neomycin	Centrifugation, TCA deprot.	Postcolumn: OPA	C8, ion-pairing	Fluor.	0.15	83–108	1989 *(42)*
Milk	Neomycin	Ion-exchange SPE	Precolumn: OPA	Protein exclusion column, gradient	Fluor.	0.05	94–102	1990 *(50)*
Plasma, Urine	Neomycin	TCA deprot.	Postcolumn: OPA	C8, ion-pairing	Fluor.	—	91–105, 75–104	1991 *(43)*
Plasma, Urine	Amikacin	Ultrafiltration	Precolumn: FDNB	C18	Spectro.	0.5, 10.0	97	1992 *(66)*

Table 1 Summary of Literature Methods for Liquid Chromatographic Determination of Aminoglycoside Residues in Biological Fluids—*continued*

Fluid	Analytes	Sample treatment	Derivatization	LC column technique	Detection	Detection limit (ppm)	Recovery (%)	Year . (Ref.)
Milk	Dihydro-streptomycin, Streptomycin	PCA deprot.	Postcolumn: NQS	C8, column enrichment, ion-pairing	Fluor.	0.02, 0.01	32–65	1994 (*48*)

Fluor: flourescence; Spectro: spectrometry; Deprot: deprotinization; Extn: extraction; SPE: solid-phase extraction; OPA: o-phthalaldehyde; TCA: trichloroacetic acid; PCA: perchloric acid; NHN:ninhydrin; MSPD: matrix solid-phase dispersion; CN: cyanopropylsilyl; NQS: β-naphthoquinone-4-sulfonate; Fluor: flourescence; Spectro: spectrometry; Deprot: deprotinization; Extn: extraction

[a] Not reported.

Table 2 Summary of literature methods for determination of aminoglycoside residues in tissues using liquid chromatography

Tissue	Analytes	Sample treatment	Derivatization	LC Column/ Technique	Detection	Detection Limit (ppm)	Recovery (%)	Year (Ref.)
Kidney	Gentamicin	Buffer extn, cell lysis, ion exchange SPE	Postcolumn: OPA	C18, ion-pairing	Fluor.	—[a]	—	1983 (53)
Muscle	Destomycin A, Neomycin	TCA deprot., ion exchange SPE	Postcolumn: OPA	C18, ion-pairing	Fluor.	0.1, 0.2	82, 72	1986 (73)
Muscle, Kidney	Kanamycin	TCA deprot., ion exchange SPE	Precolumn: OPA	C18	Fluor.	0.04	80–95	1986 (72)
Muscle, Kidney	Streptomycin	PCA deprot., C8 SPE	Postcolumn: NHN	C18	Fluor.	0.5	67	1988 (75)
Muscle	Gentamicin	Buffer extn, Heat deprot., ion exchange and Silica SPE	Precolumn: OPA	C18, ion-pairing, gradient	Fluor.	0.2	69–105	1989 (76)
Kidney	Neomycin	MSPD: CN	Postcolumn: OPA	C18, ion-pairing	Fluor.	—	84–94	1991 (80)

Table 2 Summary of Literature Methods for Determination of Aminoglycoside Residues in Tissues Using Liquid Chromatography—*continued*

Tissue	Analytes	Sample treatment	Derivatization	LC Column/ Technique	Detection	Detection Limit (ppm)	Recovery (%)	Year (Ref.)
Kidney	Dihydro-streptomycin, Hygromycin B, Spectinomycin, Streptomycin	MSPD: CN	None	C18, ion-pairing	PAD	—	—	1992 (81)
Kidney	Neomycin	Buffer extn, Heat deprot.	Post-column: OPA	C18, ion-pairing	Fluor.	0.5	80–115	1985 (77), 1993 (78)
Muscle, Kidney	Dihydro-streptomycin, Streptomycin	PCA deprot., ion exchange SPE	Post-column: NQS	C8, ion-pairing, column enrich-ment	Fluor.	0.01, 0.02	61, 55	1994 (74)

SPE: solid-phase extraction; OPA: o-phthalaldehyde; TCA: trichloroacetic acid; PCA: perchloric acid; NHN:ninhydrin; MSPD: matrix solid-phase dispersion; CN: cyanopropylsilyl; NQS: β-naphthoquinone-4-sulfonate; Fluor: flourescence; Spectro: spectrometry; Deprot: deprotinization; Extn: extraction
[a] Not reported.

are only suitable for determining the higher residue levels found in kidney. In many cases the limiting factor for detection is the presence of coextracted materials that produce interfering peaks in the chromatogram.

A successful tissue extraction method should render the tissue-bound aminoglycosides soluble, remove most or all of the proteins, eliminate other matrix interferences, and provide a satisfactory detection limit and reproducible recovery.

Tissue extraction techniques for determining aminoglycoside residues have been investigated for use with a commercially-available fluorescence immunoassay (TDx; Abbott Laboratories, Diagnostics Division, Irving, TX) (*18, 68*). Simple homogenization of the tissue with a buffer and TCA precipitation of proteins in homogenized tissue were compared with sodium hydroxide digestion of the tissue (*69*). In renal tissue, homogenization in buffer recovered 59% of the added gentamicin, TCA precipitation recovered 84%, sodium hydroxide digestion of homogenized tissue recovered 86%, and sodium hydroxide digestion of unhomogenized tissue recovered 90% (*18*). The values were similar for skeletal muscle, with mean recoveries of 46%, 70%, 77%, and 80%, respectively.

Reports of tissue extraction techniques used for microbiological assay methods indicate that TCA precipitation of proteins from tissue homogenates increased recovery over simple homogenization (*70*), and that including the TCA during the homogenization step increased recovery even more (*71*).

Using TCA or perchloric acid (PCA) during homogenization can deproteinize and solubilize the aminoglycosides. Following centrifugation, the relatively clear supernatant can be further purified using ion-exchange chromatography with Amberlite CG-50 (*72, 73*) or aromatic sulfonic acid (*74*) sorbent, or reversed-phase chromatography, with C8 sorbent (*75*). Lacharte et al. (*53*) reported that because deproteinizing kidney samples was unsatisfactory using acid or organic solvent treatments, an alkaline buffer extract instead was applied directly to a CM Sephadex C-25 solid-phase extraction column.

The above results suggest that high or low pH conditions are required in order to free the bound aminoglycosides from the tissue and to keep them in solution. At a moderate pH level, almost complete recovery of gentamicin from muscle and kidney tissue was obtained only by soaking homogenized tissue in buffer four to five times over a period of 14 to 30 days (*17*), which is impractical for routine analytical use.

In order to apply extraction techniques designed for bioassay and immunoassay to chemical analysis, additional cleanup steps must be taken to remove coextracted materials. Depending on the nature of these additional steps, whether they are ion exchange chromatography, solid-phase extraction chromatography, solvent partitioning, or some other means of purification, it is probable that analyte will be lost at each step. This is particularly true for the alkaline hydrolysis technique, due to the large amounts of tissue constituents liberated, and the need for their removal. Studies in our laboratory indicate that while the initial extraction of the analytes is nearly complete by hydrolysis, subsequent cleanup steps cause overall recoveries to be similar to those obtained by simple homogenization or acid precipitation techniques.

Heat deproteinization following homogenization in alkaline buffer has also been used to purify extracts from tissue for chemical analysis. The extract is then acidified and passed through CM Sephadex C-25 and silica gel SPE columns (*76*), or is analyzed directly by LC (*77, 78*).

The matrix solid-phase dispersion (MSPD) technique (*79*) has been used for determining aminoglycosides in tissue (*80, 81*), using cyanopropylsilyl (CN) sorbent as the solid phase.

Employing the various extraction and cleanup techniques to purify tissue samples can still permit coextracted materials to pass through the system, causing problems with the chromatography. Peaks in the chromatogram from these materials can significantly decrease the sensitivity of the method (*73, 75, 76, 80, 81*), or even limit the number of analytes that can be detected (*75, 76*).

While developing a method for determining streptomycin and dihydrostreptomycin residues in tissue at our laboratory (*74*), we wanted to avoid complex and laborious sample preparation. Homogenizing tissue in a perchloric acid solution produced a clear final extract, but we found that sensitivity and specificity were compromised by the presence of coextracted materials in the chromatograms. Ion-exchange solid-phase extraction with sulfonic acid sorbent removed these interferences, but about 30% of the analytes were lost. Elution of the analytes from the SPE column with aqueous eluent also resulted in unacceptable dilution of the analytes. To counteract this, column enrichment of the sample extract was used to reconcentrate the analytes and further purify the extract. With the column enrichment technique, a relatively large volume of sample is injected onto a short LC column. Selecting the appropriate mobile phase allows coextracted materials to pass through the column to waste, while the analytes are concentrated at the head of the column. The mobile phase is then switched to one with a stronger composition, and the analytes are flushed onto the analytical column to be separated prior to postcolumn derivatization and detection. Using a simple solvent switching valve and an autosampler capable of injecting a 2 mL sample volume, the system can be easily automated.

By combining SPE cleanup and column enrichment, the resulting chromatograms (see Figure 3) were clean enough to enable detector sensitivity to be increased, and provided detection limits of 10–20 ppb, levels which have not been attained by other methods. Recoveries were quite reproducible (mean coefficient of variation <7%), lessening the importance of obtaining a high recovery value.

Investigators have reported problems with losing significant amounts of aminoglycoside in solution stored in glass vessels (*52, 83*) and during extracting and purifying the sample (*71*). Storage losses can be reduced by using polypropylene vessels and by preparing standards in acidic or basic buffers (*52, 71*). During our investigations into the recovery of neomycin and tobramycin from tissues, problems were encountered with analyte losses not only in glass storage vessels, but also to glass fiber and paper filters used to remove proteins during sample preparation. Silylation of the filters did not alleviate the problem, and filtration as a means of tissue matrix removal had to be abandoned.

Derivatization Procedures

The presence of primary amine groups in most of the aminoglycoside antibiotics enables a number of derivatives to be readily formed. The most common derivatizing reagent used in aminoglycoside detection is *o*-phthalaldehyde (OPA) (*84*, *85*). OPA can be used either as a precolumn or a postcolumn reagent, with the derivative monitored by fluorescence detection. For precolumn derivatization, the reaction can be conducted in solution, or oncolumn if the analytes are bound to ion-exchange (*49*, *50*) or silica (*58–60*, *62*) sorbents during sample cleanup. Postcolumn derivatization with OPA requires a postcolumn reaction unit or a separate LC reagent pump, mixing tee, and reaction coil.

Other reagents used for producing a fluorescent derivative of aminoglycosides include dansyl chloride (*26*, *27*, *36*) and fluorescamine (*32*). OPA has also been used to produce a UV-absorbing derivative by including thioglycollic acid in the reaction mixture (*56*, *86*).

Several UV-absorbing derivatives can also be formed with aminoglycosides containing primary amino groups. Applied mainly to biological fluids, they are useful because the reactions are quick, they can be used with small sample volumes, and the derivatives are stable, which enables precolumn derivatization. The reagents include 1-fluoro-2,4-dinitrobenzene (FDNB) (*87*), and 2,4,6-trinitrobenzene-1-sulfonic acid (TNBS).

For the streptidine-based aminoglycosides, streptomycin and dihydrostreptomycin, derivatization occurs at the guanidino functions, so standard primary amine derivatizing reagents are not suitable. β-naphthoquinone-4-sulfonate (NQS) (*88*) and ninhydrin (NHN) have been used to produce fluorescent derivatives postcolumn in fluid and tissue samples.

There are also several methods that detect aminoglycoside antibiotics without derivatization. Streptomycin was determined in serum following sample purification on a C18 SPE column by UV absorbance at 195 nm (*64*). Applying this method to tissue samples would be impractical, if not impossible, because of the coextracted materials that mask detection at low wavelengths. Pulsed amperometric detection of underivatized dihydrostreptomycin, hygromycin B, spectinomycin, and streptomycin was reported for tissue (*81*). The postcolumn addition of sodium hydroxide to the column effluent was required for detection. The method was not suitable for monitoring low levels of residues.

Chromatography

Liquid chromatographic separation of aminoglycosides from the sample matrix and from each other is almost exclusively done on reversed-phase LC systems. The separation mechanism is dependent on the derivatization technique, precolumn or postcolumn.

With postcolumn methods, the polar aminoglycosides are poorly retained on reversed-phase columns, but retention can be increased by adding ion-pairing

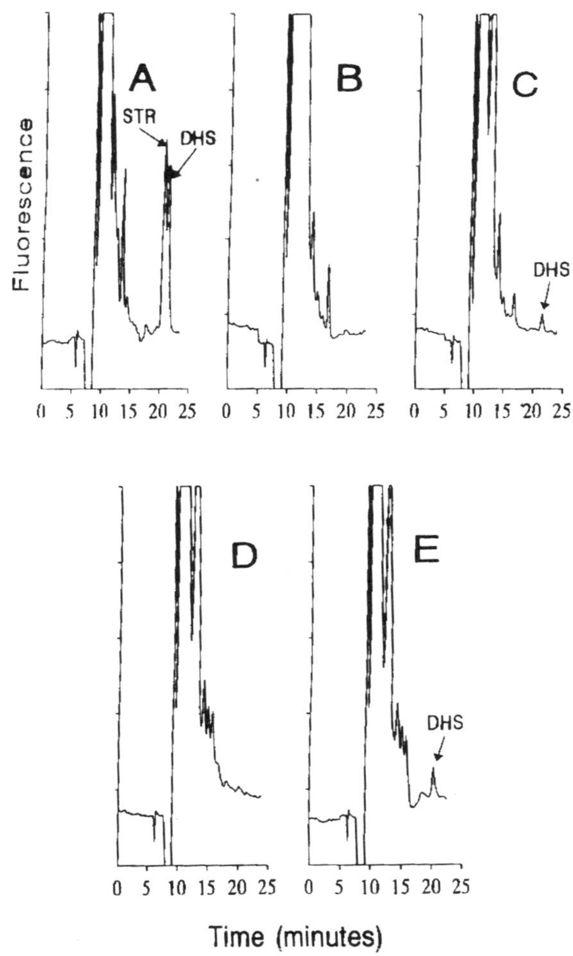

Figure 3 Chromatograms of streptomycin and dihydrostreptomycin in tissue

reagents. Selectivity and separation can be altered by adding alkyl sulfonates (such as pentane, hexane, heptane, octane, etc.), to the mobile phase. The ion-pairing reagent must be chosen so that the concentration of organic modifier in the mobile phase (usually methanol, acetonitrile, or both) produces good separation of the analyte peaks, but is not great enough to cause precipitation of the ion-pairing reagents.

Investigators have discovered that the presence of sodium sulfate in the mobile phase is required for good peak shape and good chromatography. With sodium sulfate present, decreasing the concentration of ion-pair reagent decreases retention time of aminoglycosides, as predicted by ion-pair theory (*53*). However, without the presence of sodium sulfate, decreasing the concentration of ion-pair reagent led to an increase in retention time on standard C18 packing materials (*81*). With ultra-end-capped silica-based sorbent or with polystyrene-divinylbenzene material, this anomaly was not seen. Sodium sulfate appears to compete with the aminoglycosides for free silanol groups present in most C18 column packing materials (*51*, *53*).

Precolumn derivatization of aminoglycosides produces lower polarity compounds that can be separated on a reversed-phase system without ion-pairing reagents. Instead, buffers can be used to change the pH of the mobile phase in an ion-suppression mode (*89*), altering retention and improving peak shape.

Confirmation Procedures

There are very few confirmatory methods reported for aminoglycoside residues in fluids or tissues. Shaikh et al. (*77*) used peak heights of a neomycin derivative at two different emission wavelengths to establish a ratio for standards and then used it to test the identity of suspect neomycin peaks in samples.

Mass spectrometric analysis is preferred to confirm chemically analyzed positive samples. Not surprisingly, with only a few methods reported for aminoglycoside residue analysis in tissues and fluids by gas chromatography, there have been no GC–MS confirmative methods reported for animal-based samples. Any such method would require derivatization of the aminoglycoside, because of their nonvolatile nature. LC–MS, on the other hand, is particularly suited to analyzing nonvolatile analytes.

Requiring ion exchange or ion-pair reagents to provide the separation and the resolution of aminoglycosides in reversed-phase systems can be problematic for LC–MS because these nonvolatile reagents are not compatible with current interfaces. However, volatile perfluorinated carboxylic acids (*90–93*) and ammonium salts (*94*) have been used to separate aminoglycosides and other ionic compounds prior to LC–MS. McLaughlin and Henion (*81*) reported an ion-spray LC–MS method for confirming dihydrostreptomycin, hygromycin B, spectinomycin, and streptomycin residues in bovine kidney using pentafluorobutyric and heptafluorobutyric acids as ion-pair reagents. Confirmation was possible, but not at the levels required for regulatory work because of sample matrix interferences and extract dilution.

Recently, McLaughlin et al. (*82*) reported a multiresidue confirmation method for aminoglycoside antibiotics in bovine kidney that uses ion spray LC–MS–MS, with tobramycin as an internal standard. Samples were extracted by matrix solid-phase dispersion using an end-capped 40 µm cyanopropyl packing material. The eluates were concentrated and analyzed using a gradient LC mobile phase containing pentafluoropropionic acid. The method is applicable to dihydrostreptomycin, neomycin B, the four major components of the gentamicin C complex (as tested on fortified and incurred tissues), and can also detect streptomycin and hygromycin while monitoring for three target ions in the selected ion monitoring mode. Spectinomycin could not be detected using the established criteria. Detection limits for dihydrostreptomycin, neomycin B, and the gentamicin C components were 0.1 ppm or lower, with recoveries ranging from 46% for neomycin to 75% for dihydrostreptomycin in bovine kidneys. Although the method requires equipment that is not routinely available in many regulatory laboratories, the complexity of the task facing the analyst and laboratory charged with providing confirmation of aminoglycoside residues is clearly demonstrated.

Conclusion

Reviews of the methods for determining aminoglycoside antibiotics in biological fluids (*95–97*) and fluids and tissues (*98, 99*) have been previously published.

Methods for determining residues of aminoglycosides in biological fluids are numerous and are applicable to a wide range of analytes and matrices. Sensitivities of some of these methods permit quantitative monitoring at low levels and would be suitable for monitoring residues in live animals. Chemical methods for residues in tissue are not as common or as universal, and very few are capable of detecting residues in muscle at the levels likely to be seen in treated animals (i.e., <100 ppb). Developing suitable methods is hampered by the ionic nature of the aminoglycosides and by their tendency to bind strongly to tissues and labware. Analysts may have to accept analyte recoveries that are lower than the standard levels considered acceptable for residue methods, due to losses during sample cleanup steps. Cleaner extracts, however, should enable lower sensitivities at the detection stage, and allow for LC–MS confirmation.

References

1. J. E. Riviere, A. L. Craigmill, and S. F. Sundlof, *Handbook of Comparative Pharmacokinetics and Residues of Veterinary Antimicrobials* (Boca Raton, FL: CRC Press, Inc., 1991), 263.
2. I. R. Hooper, *Aminoglycoside Antibiotics*, ed. H. Umezawa and I .R. Hooper (New York, NY: Springer-Verlag, 1982), 1.
3. H. Umezawa and S. Kondo, *Aminoglycoside Antibiotics*, ed. H. Umezawa and I. R. Hooper (New York, NY: Springer-Verlag, 1982), 267.
4. A. M. Ristuccia, *Antimicrobial Therapy*, ed. A. M. Ristuccia and B. A. Cunha (New York, NY: Raven Press, 1984), 305.
5. J. M. Bowen, *CRC Handbook Series in Zoonoses. Section D. "Antibiotics, Sulfonamides, and Public Health,"* vol. 1, ed. J. H. Jukes, H. L. Dupont, and L. M. Crawford, 1979, 289.
6. M. Barza and R. T. Scheife, *J. Maine Med. Assoc.*, **68**: 194, 1977.
7. Canadian Animal Health Institute, *Compendium of Veterinary Products* (Hensall, ON: North American Compendiums Ltd., 1993).
8. S. F. Sundlof, J. E. Riviere, and A. L. Craigmill, *Food Animal Residue Avoidance Databank Trade Name File,* (Gainesville, FL: Institute of Food and Agricultural Services, University of Florida, 1988).
9. G. Ziv and F. G. Sulman, *Res. Vet. Sci.*, **17**: 68, 1974.
10. J. R. Lockyer, A. Bucknall, and I. C. Shaw, *Euroresidue Conference On Residues of Veterinary Drugs in Food,* ed. N. Haagsma, A. Ruiter, and P. B. Czednik-Eysenberg (The Netherlands: Noordwijkerhout, 1990), 254.
11. T. Koeda, K. Umemura, and M. Yokota, *Aminoglycoside Antibiotics*, ed. H. Umezawa and I. R. Hooper (New York, NY: Springer-Verlag, 1982), 293.
12. H. D. Humes, J. M. Weinberg, and T. C. Knauss, *Am. J. Kidney Dis.*, **11**: 5, 1982.
13. R. E. Brummett and K. E. Fox, *The Aminoglycosides: Microbiology, Clinical Use, and Toxicology*, ed. A. Whelton and H. C. Neu (New York, NY: Marcel Dekker, Inc., 1982), 419.
14. G. R. Hodges, *The Aminoglycoside Antibiotics: A Guide to Therapy*, ed. W. G. Barnes and G. R. Hodges (Boca Raton, FL: CRC Press, 1984), 153.
15. C. M. Kunin, *J. Infect. Dis.*, **121**: 55, 1970.
16. M. Sastrasinh, T. C. Knauss, J. M. Weinberg, and H. D. Humes, *J. Pharmacol. Exp. Ther.*, **222**: 350, 1982.
17. J. J. Schentag, W. J. Jusko, J. W. Vance, T. J. Cumbo, E. Abrutyn, M. DeLattre, and L. M. Gerbracht, *J. Pharmacokinet. Biopharm.*, **5**: 559, 1977.
18. S. A. Brown, D. R. Newkirk, R. P. Hunter, G. G. Smith, and K. Sugimoto, *J. Assoc. Off. Anal. Chem.*, **73**: 479, 1990.
19. H. J. Langner and U. Teufel, *Die Fleischwirtschaft*, **52**: 1610, 1972.
20. H. Yoshimura, O. Itoh, and S. Yonezawa, *Jpn. J. Vet. Sci.*, **44**: 233, 1982.

21. C. D. C. Salisbury, C. E. Rigby, and W. Chan, *J. Agric. Food Chem.*, **37**: 105, 1989.
22. T. Saito, H. Kaneshima, and M. Nakano, *J. Hokkaido Hyg. Res.*, **35**: 111, 1985.
23. J. W. Mayhew and S. L. Gorbach, *J. Chromatogr.*, **151**: 133, 1978.
24. J. W. Mayhew and S. L. Gorbach, *Antimicrob. Agents Chemother.*, **14**: 851, 1978.
25. K. Nakaya, A. Sugitani, and F. Yamada, *Shokuhin Eiseigaku Zasshi*, **26**: 443, 1985.
26. G. W. Peng, M. A. F. Gadalla, A. Peng, V. Smith, and W. L. Chiou, *Clin. Chem.*, **23**: 1838, 1977.
27. G. W. Peng, G. G. Jackson, and W. L. Chiou, *Antimicrob. Agents Chemother.*, **12**: 707, 1977.
28. S. Bäck, I. Nilsson-Ehle, and P. Nilsson-Ehle, *Clin. Chem.*, **25**: 1222, 1979.
29. N. Larsen and K. Marinelli, *J. Chromatogr.*, **221**: 182, 1980.
30. D. M. Barends, J. S. F. Van der Sandt, and A. Hulshoff, *J. Chromatogr.*, **182**: 201, 1980.
31. D. B. Haughey, D. M. Janicke, M. Adelman, and J. J. Schentag, *Antimicrob. Agents Chemother.*, **17**: 649, 1980.
32. S. E. Walker and P. E. Coates, *J. Chromatogr.*, **223**: 131, 1981.
33. D. M. Barends, C. L. Zwaan, and A. Hulshoff, *J. Chromatogr.*, **222**: 316, 1981.
34. D. M. Barends, C. L. Zwaan, and A. Hulshoff, *J. Chromatogr.*, **225**: 417, 1981.
35. J. D'Souza and R. I. Ogilvie, *J. Chromatogr.*, **232**; 212, 1982.
36. W. L. Chiou, R. L. Nation, G. W. Peng, and S. M. Huang, *Clin. Chem.*, **24**: 1846, 1978.
37. S. Dionisotti, F. Bamonte, M. Gamba, and E. Ongini, *J. Chromatogr.*, **434**: 169, 1988.
38. L. T. Wong, A. R. Beaubien, and A. P. Pakuts, *J. Chromatogr.*, **231**: 145, 1982.
39. H. Kubo and T. Kinoshita, *J. Chromatogr.*, **227**: 244, 1982.
40. H. Kubo, T. Kinoshita, Y. Kobayashi, and K. Tokunaga, *J. Liq. Chromatogr.*, **7**: 2219, 1984.
41. H. Matsunaga, T. Fujimoto, R. Tawa, and S. Hirose, *Chem. & Pharm. Bull. (Tokyo)*, **36**: 1565, 1988.
42. B. Shaikh and J. Jackson, *J. Liq. Chromatogr.*, **12**: 1497, 1989.
43. B. Shaikh, J. Jackson, G. Guyer, and W. R. Ravis, *J. Chromatogr.*, **571**: 189, 1991.
44. H. Kubo, Y. Kobayashi, and T. Nishikawa, *Antimicrob. Agents Chemother.*, **28**: 521, 1985.
45. H. Kubo, Y. Kobayashi, and T. Kinoshita, *Anal. Chem.*, **58**: 2653, 1986.

46. G. Tamai, H. Imai, and H. Yoshida, *Chromatographia*, **21**: 519, 1986.
47. H. Kubo, H. Li, Y. Kobayashi, and T. Kinoshita, *Anal. Biochem.*, **162**: 219, 1987.
48. G. C. Gerhardt, C. D. C. Salisbury, and J. D. MacNeil, *J. AOAC Int.*, **77**: 765, 1994.
49. J. Marples and M. D. G. Oates, *J. Antimicrob. Chemother.*, **10**: 311, 1982.
50. V. K. Agarwal, *J. Liq. Chromatogr.*, **13**: 2475, 1990.
51. J. P. Anhalt, *Antimicrob. Agents Chemother.*, **11**: 651, 1977.
52. J. P. Anhalt and S. D. Brown, *Clin. Chem.*, **24**: 1940, 1978.
53. G. Lacharte, G. Nicot, C. Gonnet, J. Tronlhet, L. Merle, J. P. Valette, Y. Nouaille, and R. Desroches, *Analysis*, **11**: 168, 1983.
54. T. Kawamoto, I. Mashimo, S. Yamauchi, and M. Watanabe, *J. Chromatogr.*, **305**: 373, 1984.
55. K. Koshide, R. Tawa, S. Hirose, and T. Fujimoto, *Clin. Chem.*, **31**: 1921, 1985.
56. K. Koshide, T. Fujimoto, H. Matsunaga, R. Tawa, and S. Hirose, *Yakugaku Zasshi*, **106**: 574, 1986.
57. T. Saito, H. Kaneshima, M. Nakano, and T. Takahashi, *J. Food Hyg. Soc. Jpn.*, **28**: 317, 1987.
58. S. K. Maitra, T. T. Yoshikawa, J. L. Hansen, I. Nilsson-Ehle, W. J. Palin, M. C. Schotz, and L. B. Guze, *Clin. Chem.*, **23**: 2275, 1977.
59. S. K. Maitra, T. T. Yoshikawa, C. M. Steyn, L. B. Guze, and M. C. Schotz, *Antimicrob. Agents Chemother.*, **14**: 880, 1978.
60. S. K. Maitra, T. T. Yoshikawa, J. L. Hansen, M. C. Schotz, and L. B. Guze, *Am. J. Clin. Pathol.*, **71**: 428, 1979.
61. C. J. Delaney, E. O. Kent, A. L. Smith, and J. J. Plorde, *Antimicrob. Agents Chemother.*, **21**: 19, 1982.
62. R. H. Rumble and M. S. Roberts, *J. Chromatogr.*, **419**: 408, 1987.
63. P. M. Kabra, P. K. Bhatnagar, and M. A. Nelson, *J. Chromatogr.*, **307**; 224, 1984.
64. N. Kurosawa, S. Kuribayashi, E. Owada, K. Ito, M. Nioka, M. Arakawa, and R. Fukuda, *J. Chromatogr.*, **343**: 379, 1985.
65. L. Essers, *Eur. J. Clin. Microbiol.*, **1**; 367, 1982.
66. E. A. Papp, C. A. Knupp, and R. H. Barbhaiya, *J. Chromatogr.*, **574**: 93, 1992.
67. L. Essers, *J. Chromatogr.*, **305**: 345, 1984.
68. S. A. Brown, K. Sugimoto, G. G. Smith, and F. B. Garry, *Antimicrob. Agents Chemother.*, **32**: 595, 1988.
69. D. N. Gilbert and S. J. Kohlhepp, *Antimicrob. Agents Chemother.*, **30**: 361, 1986.
70. M. L. Kornguth and C. M. Kunin, *Antimicrob. Agents Chemother.*, **11**: 974, 1977.

71. R. A. Giuliano, G. A. Verpooten, D. E. Pollet, L. Verbist, S. L. Scharpe, and M. E. De Broe, *Antimicrob. Agents Chemother.*, **25**; 783, 1984.
72. K. Nakaya, A. Sugitani, and F. Yamada, *J. Food Hyg. Soc. Jpn.*, **27**: 258, 1986.
73. Y. Yoneda, M. Okada, S. Mizoguchi, Y. Tokonabe, M. Shiroishi, and Z. Yoshimoto, *Shokuhin Eiseigaku Zasshi*, **27**: 369, 1986.
74. G. C. Gerhardt, C. D. C. Salisbury, and J. D. MacNeil, *J. AOAC Int.*, **77**: 334, 1994.
75. A. Okayama, Y. Kitada, Y. Aoki, S. Umesako, H. Ono, Y. Nishi, and H. Kubo, *Bunseki Kagaku*, **37**: 221, 1988.
76. V. K. Agarwal, *J. Liq. Chromatogr.*, **12**: 613, 1989.
77. B. Shaikh, E. H. Allen, and J. C. Gridley, *J. Assoc. Off. Anal. Chem.*, **68**: 29, 1985.
78. B. Shaikh and J. Jackson, *J. AOAC Int.*, **76**: 543, 1993.
79. S. A. Barker, A. R. Long, and C. R. Short, *J. Chromatogr.*, **475**: 353, 1989.
80. F. J. Schenk, *U.S. FDA Laboratory Information Bulletin*, vol. 7, issue no. 4, No. 3559, (Rockville, MD: U.S. Food and Drug Administration, 1991).
81. L. G. McLaughlin and J. D. Henion, *J. Chromtogr.*, **591**: 195, 1992.
82. L. G. McLaughlin, J. D. Henion, and P. J. Kijak, *Biol. Mass Spectrom.*. **23**: 417, 1994.
83. L. Josephson, P. Houle, and M. Haggerty, *Clin. Chem.*, **25**: 298, 1979.
84. M. Roth, *Anal. Chem.*, **43**: 880, 1971.
85. M. Roth and A. Hampai, *J. Chromatogr.*, **83**: 353, 1973.
86. M. Freeman, P. A. Hawkins, J. S. Loran, and J. A. Stead, *J. Liq. Chromatogr.*, **2**: 1305, 1979.
87. D. M. Barends, J. C. A. M. Brouwers, and A. Hulshoff, *J. Pharm. Biomed. Anal.*, **5**; 613, 1987.
88. F. Faure and P. Blanquet, *Clin. Chim. Acta*, **9**: 292, 1964.
89. B. A. Bidlingmeyer, *J. Chromatogr. Sci.*, **18**: 525, 1980.
90. G. Inchaupsé and D. Samain, *J. Chromatogr.*, **303**: 277, 1984.
91. G. Inchaupsé, C. Deshayes, and D. Samain, *J. Antibiot. (Tokyo)*, **38**: 1526, 1985.
92. G. Inchaupsé, P. Delrieu, P. Dupin, M. Laurent, and D. Samain, *J. Chromatogr.*, **404**: 53, 1987.
93. T. A. Getek, M. L. Vestal, and T. G. Alexander, *J. Chromatogr.*, **554**: 191, 1991.
94. A. L. L. Duchateau, B. H. M. Munsters, G. T. C. Kwakkenbos, and R. G. J. Van Leuken, *J. Chromatogr.*, **552**: 605, 1991.
95. S. K. Maitra, T. T. Yoshikawa, L. B. Guze, and M. C. Schotz, *Clin. Chem.*, **25**: 1361, 1979.
96. I. Nilsson-Ehle, *J. Liq. Chromatogr.*, **6(S-2)**: 251, 1983.
97. M. C. Rouan, *J. Chromatogr.*, **340**: 361, 1985.

98. B. Shaikh and E. H. Allen, *J. Assoc. Off. Anal. Chem.*, **68**: 1007, 1985.
99. B. Shaikh and W. A. Moats, *J. Chromatogr.*, **643**: 369, 1993.
100. P. M. Kabra, P. K. Bhatnagar, and M. A. Nelson, *J. Anal. Toxicol.*, **7**: 283, 1983.
101. T. Harada, M. Iwamori, Y. Nagai, and Y. Nomura, *J. Chromtogr.*, **337**: 187, 1985.

Chapter 10

Chemical Analysis of Tetracycline Antibiotics

Hisao Oka

Aichi Prefectural Institute of Public Health, Tsuji-machi, Kita-ku
Nagoya 462, Japan

AND

John Patterson

Health of Animals Laboratory, Agriculture & Agri-FoodCanada, 116
Veterinary Road, Saskatoon, Saskatchewan, Canada S7N 5E3

Contents

Introduction

Tetracycline antibiotics (TCs) (Figure 1), which are derived from *Streptomyces* spp., have broad spectrum activity against many Gram-positive and -negative bacteria, and are especially effective against *Staphylococcus, Streptococcus, Pneumococcus, Gonococcus, Cholera, Dysentery bacillus, Pertussis, Rickettsia, Chlamydia,* and *Mycoplasma*. TCs are actively transported into the cells of susceptible bacteria where they exert a bacteriostatic effect by inhibiting protein biosynthesis after binding to the 30S ribosomal subparticle. Since the first member of the tetracycline family, chlortetracycline (CTC) was discovered in 1948 (*1*), eight TCs have become commercially available. Of these, oxytetracycline (OTC), tetracycline (TC), CTC, and doxycycline (DC) are commonly applied to food-producing animals (including honeybees) as veterinary drugs and feed additives because of their broad spectrum activity and cost effectiveness. However, other TCs may also be administered to animals. For example, minocycline (MINO) has the highest biological activity among TCs and is active against tetracycline-resistant pathogens, methacycline (MTC) and demeclocycline (DMCTC) are more stable than tetracycline and give higher, more sustained blood levels with smaller, less frequent dosing, and rolitetracycline (PRMTC) is readily soluble in water, making it more suitable for injection than other TCs. Their wide spread utilization has led to increasing resistance, so acute monitoring by public health agency is required. As with most antibiotics, microbiological assays have commonly been used for measuring TCs in food, but they are time consuming, they cannot identify or differentiate certain TCs, and their precision is variable. Therefore, precise chemical analysis methods are clearly needed.

Pharmaceutical preparations of TC contain small amounts of impurities, namely 4-epitetracycline (ETC), anhydrotetracycline (ATC), 4-epianhydrotetracycline (EATC), and CTC. The permitted concentrations of impurities are listed in various pharmacopoeias (*2–4*). Because CTC is more active than TC,

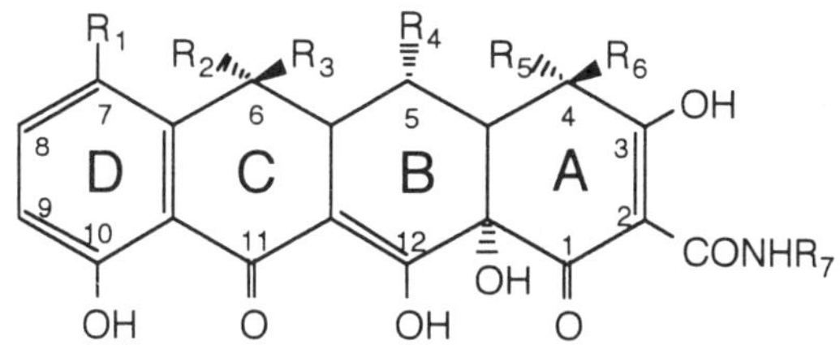

I

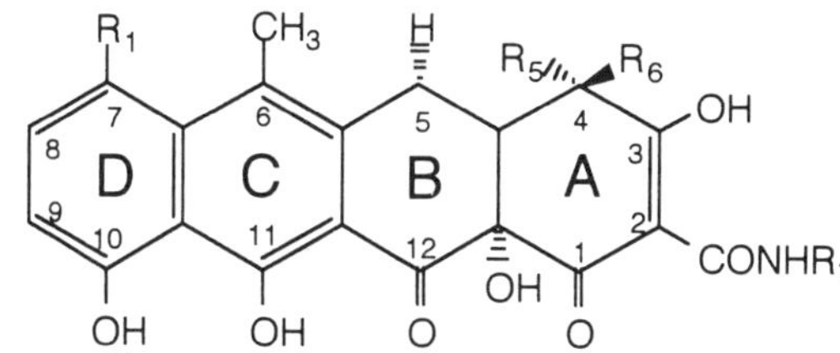

II

Compound	Structure	R_1	R_2	R_3	R_4	R_5	R_6	R_7
Tetracycline (TC)	I	H	CH_3	OH	H	$N(CH_3)_2$	H	H
Oxytetracycline (OTC)	I	H	CH_3	OH	OH	$N(CH_3)_2$	H	H
Chlortetracycline (CTC)	I	Cl	CH_3	OH	H	$N(CH_3)_2$	H	H
Doxycycline (DC)	I	H	CH_3	H	OH	$N(CH_3)_2$	H	H
Minocycline (MINO)	I	$N(CH_3)_2$	H	H	H	$N(CH_3)_2$	H	H
Methacycline (MTC)	I	H	$=CH_2$		OH	$N(CH_3)_2$	H	H
Demeclocycline (DMCTC)	I	Cl	H	OH	H	$N(CH_3)_2$	H	H
Rolitetracycline (PRMTC)	I	H	CH_3	OH	H	$N(CH_3)_2$	H	*
4-Epitetracycline (ETC)	I	H	CH_3	OH	H	H	$N(CH_3)_2$	H
Anhydrotetracycline (ATC)	II	H	-	-	-	$N(CH_3)_2$	H	H
4-Epianhydrotetracycline (EATC)	II	H	-	-	-	H	$N(CH_3)_2$	H

* 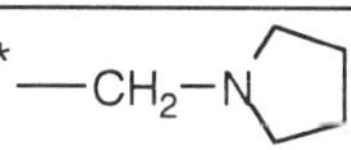

Figure 1 Structures of tetracycline antibiotics

relying only on a microbiological potency test for TC may render misleading results if significant amounts of CTC are present. EATC has been identified as a cause of Fanconi-syndrome (*5–8*). ETC and ATC are inactive as antibiotics, but they may undergo dehydrogenation or epimerization respectively, under storage conditions, and form EATC. Therefore, a precise chemical analysis method for these impurities is also required.

TCs have similar chemical and physicochemical properties. They are amphoteric compounds with characteristic pK_a values and form crystalline hydrates and salts with acids and bases (*9*). Their UV spectra show strong absorption around 270 and 360 nm in neutral and acidic solutions (*10*). TCs are soluble in acids, bases, alcohols, and polar organic solvents and they can be extracted with organic solvents such as *n*-butanol and ethyl acetate. TCs produce also fluoresce in the presence of metal ions or under basic conditions (*11–13*). The stability of these antibiotics is poor under strong acidic and alkaline conditions with reversible formation of epimers, 4-epi-TCs in weak acid (pH 3), and anhydro-TCs under strong acidic (below pH 2) conditions (*14*).

Considering the physicochemical properties of this family of antibiotics, many chemical analysis methods for determining TC residues in food have been reported over the last 20 years or more. However, most of these methods require complicated procedures because of TCs propensity to form chelate complexes with metal ions at β-diketones (C10–12) and carboxyamide (C2) (*15–17*), and to bind with proteins and silanol groups in the stationary phase (*18, 19*). These properties inhibited the development of a simple chemical method for the analysis of TCs. In 1985, it became possible to control these undesirable properties by using ethylenediaminetetraacetic acid (EDTA) and oxalic acid during extraction, cleanup, and separation (*20–24*). Their use has remarkably improved the available analytical methods for TCs in food.

In this chapter, we describe recent developments in the chemical analysis of TCs, including the control of their undesirable properties, and we will make recommendations on applying chemical methods for determining and confirming TCs in food to routine regulatory use.

Thin-Layer Chromatography

Normal-Phase Thin-Layer Chromatography

Table 1 shows that a number of scientists have reported thin-layer chromatographic (TLC) methods to separate TCs using adsorbent layers of kieselguhr (*25–32*), silica gel (*33–40*), and cellulose (*41–44*). In general, TLC is simple and requires no special equipment, however most of the published methods need excessive time for preparing the TLC plate to avoid the binding of TCs with trace metals in the adsorbent. In order to control this, EDTA has been added to the adsorbent when the plate is prepared and to the solvent systems (*25, 27–30, 32–34, 44*). A spraying technique, where precoated TLC plates are sprayed with EDTA

Table 1 Summary of Normal Phase TLC Conditions for the Analysis of TCs

Tetracyclines	Sample	Plate	Solvent system	Extraction and cleanup	Detection	Detection Limit	Ref.
OTC, TC, and CTC	—	Silica gel containing EDTA	*n*-BuOH sat. with H_2O	Dissolution in MeOH	Spray ($FeCl_3$)	—	(*33*)
TC, ATC, and EATC	Tablet	Cellulose	0.1M EDTA-0.1% NH_4Cl $CHCl_3$ sat. with EDTA-NH_4Cl	—	FL 428 nm	—	(*41*)
TC, CTC, ETC, ATC, and EATC	—	Diatomaceous earth containing EDTA, glycerol, and PEG 400	EtOAc sat. with EDTA	—	FL 370 nm	—	(*25*)
OTC, TC, and their impurities	—	Silica gel containing EDTA	*n*-BuOH-EtOH-H_2O (4:7:4)	—	FL 370 nm	—	(*34*)
TC, CTC, ETC, ATC, and EATC	—	Kieselguhr containing EDTA	Me_2CO-H_2O (10:1) Me_2CO-EtOAc-H_2O (20:10:3)	— —	FL 370 nm	—	(*26*)
OTC, TC, CTC, ETC, ATC, and EATC	—	Kieselguhr containing EDTA, glycerol, and PEG 400	MEK sat. MacIlvaine buffer	—	Spray (Fast Blue B)	—	(*27*)
TC, CTC, ETC, ATC, and EATC	—	Kieselguhr containing EDTA and glycerol	CH_2Cl_2-EtOH (9:1)	—	FL 370 nm	—	(*28*)

Table 1 Summary of Normal Phase TLC Conditions for the Analysis of TCs–*contiuned*

Tetracyclines	Sample	Plate	Solvent system	Extraction and cleanup	Detection	Detection Limit	Ref.
TC, CTC, ETC, ATC, and EATC	Capsule and tablet	Kieselguhr containing EDTA	Me_2CO-EtOAc-H_2O (80:40:12)	Dissolution in MeOH	Spray Fast Blue B)	1μ g/spot	*(29)*
OTC, TC, and CTC	Animal tissues	Kieselguhr containing EDTA, glycerol, and PEG 400	EtOAc-*n*-BuOH (9:1) containing EDTA	Extraction with 0.1 *N* HCl and XAD-2 cleanup	FL 370 nm	0.025–0.1 ppm	*(30)*
		Cellulose containing EDTA and NH_4Cl	*n*-BuOH sat. with H_2O				
OTC, TC, and CTC	Milk	Precoated kieselguhr impregnated with phosphate/citric buffer containing glycerol	$CHCl_3$-Me_2CO-phosphate/citric buffer (5:5:2)	Extraction with Me_2CO	Bioautography	0.3 ppm	*(31)*
OTC, TC, CTC, DC, DMCTC, MTC, MINO, and ATC	—	Cellulose	Aq. $MgCl_2$, aq. $CaCl_2$, aq. $BaCl_2$, and aq. $ZnCl_2$	—	Fl 370 nm	0.25 μg/spot	*(42)*

Table 1 Summary of Normal Phase TLC Conditions for the Analysis of TCs–*continued*

Tetracyclines	Sample	Plate	Solvent system	Extraction and cleanup	Detection	Detection Limit	Ref.
OTC, MTC, and DC	—	Cellulose containing MacIlvaine buffer	EtOAc sat. with H_2O	—	Spray ($MgCl_2$) FL 370 nm	—	*(43)*
TC, CTC, ETC, ATC, and EATC	Capsule and powder	Cellulose containing EDTA	$CHCl_3$ sat. with EDTA	Dissolution in MeOH	FL 370 nm	—	*(44)*
TC	—	Precoated silica gel sprayed with EDTA	Me_2CO-EtOAc-H_2O (80:35:15)	—	FL 370 nm	—	*(35)*
TC	Capsule	Kieselguhr containing EDTA	Me_2CO-EtOAc-H_2O (20:10:3)	Dissolution in MeOH	Densitometry (366 nm)	—	*(32)*
OTC, TC, CTC, DC, ETC, ATC, and EATC	—	Precoated silica gel TLC predeveloped with EDTA	$CHCl_3$-MeOH-EDTA (65:20:5) i-PrOH-EtOAc-EDTA(3:4:7) Me_2CO-EDTA (10:1)	—	Densitometry (360 and 450 nm)	0.1 μ g/spot	*(20)*
OTC, TC, and CTC	Fish tissues	Precoated silica gel TLC predeveloped with EDTA	$CHCl_3$-MeOH-EDTA (65:20:5)	Extraction with EDTA-MacIlvaine buffer C18 cartridge pretreated with EDTA cleanup	Densitometry (360 nm)	0.1 ppm	*(21)*

Table 1 Summary of Normal Phase TLC Conditions for the Analysis of TCs–*continued*

Tetracyclines	Sample	Plate	Solvent system	Extraction and cleanup	Detection	Detection Limit	Ref.
TC, CTC, ETC, ATC, and EATC	Capsule, syrup, powder, and tablet	Precoated silica gel TLC predeveloped with EDTA	$CHCl_3$-MeOH-EDTA (65:20:5)	Dissolution in MeOH	Densitometry (360 and 450 nm)	0.1 μg/spot	*(56)*
					Spray (Fast Violet B)	0.1 μg/spot	
OTC, TC, CTC, DC, MINO, MTC, DMCTC, PRMTC, ETC, ATC, and EATC	—	Precoated silica gel TLC predeveloped with EDTA	$CHCl_3$-MeOH-EDTA (65:20:5)		Densitometry (360 and 450 nm)	0.1 μg/spot	*(36)*
					Spray Fast Violet B)	0.1 μg/spot	
OTC, TC, CTC, DC, MINO, MTC, and DMCTC	Honey	Precoated silica gel TLC predeveloped with EDTA	$CHCl_3$-MeOH-EDTA (65:20:5)	C18 certridge pretreated with EDTA and COOH cartridge cleanup	Spray ($MgCl_2$) FL 370 nm	0.1 ppm	*(37)*

Table 1 Summary of Normal Phase TLC Conditions for the Analysis of TCs–*continued*

Tetracyclines	Sample	Plate	Solvent system	Extraction and cleanup	Detection	Detection Limit	Ref.
OTC, TC, CTC, and DC	Animal tissues	Precoated silica gel TLC predeveloped with EDTA	$CHCl_3$-MeOH-EDTA (65:20:5)	Extraction with EDTA-McIlvaine buffer	Densitometry (360 nm)	0.1 ppm	*(38)*
				C18 cartridge pretreated with EDTA cleanup	Spray Fast (Blue BB and pyridine)		
OTC, DC, and MINO	—	Precoated silica gel TLC sprayed with EDTA	CH_2Cl_2-MeOH-H_2O (59:35:6)	—	Densitometry (280 nm)	—	*(39)*
CTC and DMCTC	—	Precoated silica gel TLC sprayed with EDTA	CH_2Cl_2-MeOH-H_2O (59:35:6)	—	Densitometry (280 nm)	—	*(40)*

OTC: oxytetracycline; TC: tetracycline; CTC: chlortetracycline; EDTA: disodium ethylenediaminetetraacetate; ATC: anhydrotetracycline; EATC: 4-epianhydrotetracycline; ETC: 4-epitetracycline; DC: doxycycline; DMCTC: demeclocycline; MTC: methacycline; MINO: minocycline; DMCTC: demeclocycline; PRMTC: rolitetracycline; FL: fluorescence; MEK: methyl ethyl ketone and activated (*35*, *39*, *40*), has been applied to precoated silica gel thin-layer chromatographic plates, however the resolution among the TCs is poor, especially between the impurities (ETC, ATC, EATC, and CTC). An improved method was therefore designed (*20*, *36–38*, *56*) in which the high-performance thin-layer chromatographic (TLC) plate is predeveloped with a saturated Na_2EDTA aqueous solution, and is then activated before applying the sample. Although the predevelopment is time consuming, satisfactory results were obtained with this method. Various solvent systems were investigated and the three solvent systems shown in Figure 2 gave good results. The development time of the chromatograms is about 30 min, which is shorter than with other systems. The solvent system A, chloroform–methanol–5% aqueous Na_2EDTA solution (65:20:5, lower layer) yields a better resolution for seven TCs, especially among TC, OTC, CTC, and DC than that of solvent systems B, isopropanol–ethyl acetate–5% aqueous Na_2EDTA solution (3:4:7, upper layer) and C, acetone–5% aqueous Na_2EDTA solution (10:1). In the case of solvent system B, although the spots of CTC and DC overlap, there is a good separation between CTC and EATC which weakly interact in solvent system A. In solvent system C, a good separation of TC, ETC, ATC, EATC, and CTC is given, but DC and TC overlap. However, depending on the application, it is usually not necessary to fully resolve all seven TCs. For residue analysis in a biological matrix it is necessary to completely separate the

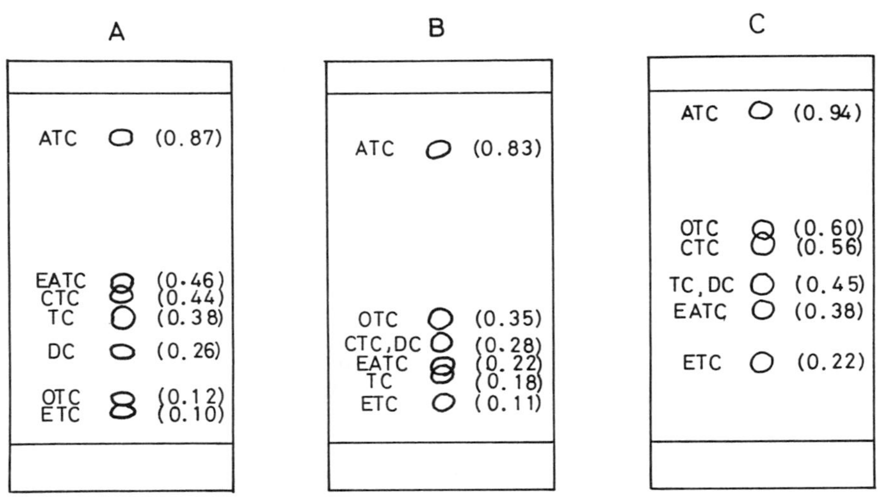

Figure 2 Separation of TCs on silica gel TLC. Plate: Silica gel TLC predeveloped with saturated Na_2EDTA and then activated for 2 hrs at 13°C. Solvent system: (A) Chloroform-methanol-5% Na_2EDTA (65:20:5) (lower layer) (B) Isopropanol-ethyl-acetate-5% Na_2EDTA (3:4:7) (upper layer) (C) Acetone-5% Na_2EDTA (10:1)
SOURCE: Reproduced by permission of Elseriver Science. See reference *(20)*.

parent TCs, whereas for the analysis of impurities in TC pharmaceutical preparations a good separation of TC, ETC, ATC, EATC, and CTC is needed. Therefore, the following solvent systems are recommended for use with silica gel TLC plates: Solvent system A for the analysis of the parent TCs and solvent system C for the analysis of the impurities of TC. In both cases, solvent system B can be used as an alternative. The complementary use of these three solvent systems enables an identification for most purposes.

Reversed-Phase Thin-Layer Chromatography

As summarized in Table 2, all studies have used solvent systems containing oxalic acid in reversed-phase thin-layer chromatography (RP–TLC) to control the undesirable properties of TCs. The spots of TCs show extreme tailing on RP–TLC plates when aqueous solutions of methanol, acetonitrile, and ethanol are used as solvent systems. In several reversed-phase liquid chromatographic (RP–LC) systems, using a mobile phase containing phosphoric (*45, 46*), citric (*47, 48*), tartaric (*48*), or malonic (*48*) acids, or EDTA (*49, 50*) gives good separation of TCs. Although these acids have been applied in RP–TLC, all TCs showed tailing spots. When TCs are purified from culture broth, oxalate is added to prevent the formation of chelate complexes of TCs with metal ions (*51*). Therefore, a mixture of methanol–0.5 M aqueous oxalic acid solution (pH 2.0) was tried as a solvent system on a C18 TLC plate. Separation among TCs was unsatisfactory, but the spots showed no tailing. Therefore, the aqueous oxalic acid solution is essential in the solvent system to eliminate tailing (*22*).

Using methanol–aqueous oxalic acid solution (pH 2.0) (1:1) as a solvent system, the influence of the acid concentration was examined. The shape of the spots was improved with increasing acid concentration, and a minimum of 0.5 M aqueous oxalic acid solution was needed in the solvent system to eliminate tailing.

In order to compare the suitability of C8 and C18 TLC plates for separating TCs using methanol–0.5 M aqueous oxalic acid solution (pH 2.0) (1:1) as a solvent system, TCs were separated on both plates. Although poor resolution was obtained for the separation of impurities of TC using C8 TLC plates, good resolution was obtained for the parent TCs. On the other hand, good resolution was given for the impurities using C18 TLC plates, but it was unsatisfactory for the parent TCs. Therefore, C18 and C8 plates are suitable for the separation of the impurities and the parent TCs, respectively.

The influence of the pH of the aqueous oxalic acid solution (adjusted with 28% aqueous ammonia) in the solvent systems on the R_f values was investigated. When methanol–acetonitrile–0.5 M aqueous oxalic acid solution (1:1:2) was used, all TCs showed tailing spots at pH 1.2. Above pH 3.0, EATC and ATC showed tailing, and the separations between CTC and DC and between ETC and TC were unsatisfactory. CTC and DC overlapped at pH 5.0. A precise separation among the TCs without tailing spots was obtained at pH 2.0.

Table 2 Summary of Reversed-Phase TLC Conditions for the Analysis of TCs

Tetracyclines	Sample	Plate	Solvent system	Extraction and cleanup	Detection	Detection limit	Ref.
TC, CTC, ETC, ATC, and EATC	Capsule, syrup, powder, and tablet	Precoated C8-TLC	MeOH-MeCN-0.5M oxalic acid (pH 2.0, 1:1:2)	Dissolution in MeOH	Densitometry (360 and 450 nm)	0.1 μg/spot	*(56)*
					Spray (Fast Violet B)	0.1 μg/spot	
OTC, TC, CTC, DC, MINO, MTC, DMCTC, PRMTC, ETC, ATC, and EATC	—	Precoated C8-TLC	MeOH-MeCN-0.5 M oxalic acid (pH 2.0, 1:1:4)	—	Densitometry (360 and 450 nm)	0.1 μg/spot	*(36)*
			MeOH-MeCN-0.5 M oxalic acid (pH 2.0, 1:1:2)		Spray (Fast Violet B)	0.1 μg/spot	
OTC, TC, CTC, DC, MINO, MTC, and DMCTC	Honey	Precoated C8-TLC	MeOH-MeCN-0.5 M oxalic acid (pH 3.0, 1:1:4)	C18 cartridge pretreated with EDTA and COOH cartridge cleanup	Spray ($MgCl_2$)	0.1 ppm	*(37)*

Table 2 Summary of Reversed-Phase TLC Conditions for the Analysis of TCs—*continued*

Tetracyclines	Sample	Plate	Solvent system	Extraction and cleanup	Detection	Detection limit	Ref.
OTC, TC, CTC, and DC	Animal tissues	Precoated C8-TLC	MeOH-MeCN 0.5M oxalic acid (pH 2.0, 1:1:4)	Extraction with EDTA-MacIlvaine buffer	Densitometry (360 nm)	0.1 ppm	*(38)*
				C18 cartridge pretreated with EDTA cleanup	Spray (Fast Blue BB and pyridine)		
OTC, TC, CTC, DC, ETC, ATC, and EATC	—	Precoated C8-TLC	MeOH-MeCN-0.5 M oxalic acid (pH 2.0, 1:1:4)	—	Densitometry (360 and 450 nm)	0.1 μg/spot	*(22)*
		Precoated C18-TLC	MeOH-MeCN-0.5 M oxalic acid (pH 2.0, 1:1:2)				
OTC, TC, CTC, and DMCTC	—	Precoated CN-TLC	MeOH-MeCN-0.5M oxalic acid (3:1:6)	—	Densitometry (254 nm)	0.4 μg/spot	*(58)*
OTC, TC, CTC, and DC	Animal tissues	Precoated C8-TLC	MeOH-MeCN-0.5M oxalic acid (pH 2.0, 1:1:4)	Extraction with EDTA-MacIlvaine buffer C18 cartridge pretreated with EDTA cleanup	TLC–FABMS	0.1 ppm	*(59)*

Table 2 Summary of Reversed-phase TLC Conditions for the Analysis of TCs—*continued*

Tetracyclines	Sample	Plate	Solvent system	Extraction and cleanup	Detection	Detection limit	Ref.
OTC, TC, CTC, and DC	Milk	Precoated C8-TLC	MeOH-MeCN-0.5M oxalic acid (pH 2.0, 1:1:4)	Extraction with EDTA-McIlvaine buffer C18 cartridge pretreated with EDTA cleanup	TLC/FABMS	0.05 ppm	*(60)*

TC: tetracycline; CTC: chlortetracycline; ETC: 4-epitetracycline; ATC: anhydrotetracycline; EATC: 4-epianhydrotetracycline; OTC: oxytetracycline; DC: doxycycline; MINO: minocycline; MTC: methacycline; DMCTC: demeclocycline; PRMTC: rolitetracycline; EDTA: disodium ethylenediaminetetraacetate; FL: fluorescence, and FABMS: fast atom bombardment mass spectrometry

Various combinations of methanol, acetonitrile, and 0.5 M aqueous oxalic acid solution (pH 2.0) were tested to identify suitable solvent systems. Unsatisfactory separation between OTC and TC was obtained using methanol–acetonitrile–0.5 M aqueous oxalic acid solution (pH 2.0) (1:1:1). When a 1:1:6 solution was used, the separation between CTC and DC was unsatisfactory. As shown in Figure 3, the parent TCs were precisely separated on C8 TLC plates using a 1:1:4 solution. For separating TC and its impurities, EATC and ATC overlapped when the methanol–acetonitrile–0.5 M aqueous oxalic acid solution at 1:1:4 was used, but separated well on C18 TLC plates using a 1:1:2 solution. Therefore, we recommend the combinations of plate and solvent systems shown in Figure 3, depending on the analytical purpose.

Separation and Determination of Eight Commercially Available Tetracyclines

The above conditions for normal-phase (NP)–TLC and RP–TLC were applied to the analysis of eight parent TCs. As seen in Figure 4, using the above conditions without modification allowed for good separation of the eight TCs on silica gel TLC except for the separation between TC and DMCTC. For RP–TLC, when the above conditions were directly applied to the eight commercially available TCs, the separation between DC and MTC and between TC and DMCTC were not sufficient for their identification. Various combinations of methanol, acetonitrile, and 0.5 M oxalic acid solution (pH 2.0) were investigated, but satisfactory results could not be obtained. Because the separation of TCs in RP–TLC depends on the pH of the solvent system, the effect of pH on the separation of the eight TCs was examined using methanol–acetonitrile–0.5 M oxalic acid solution (1:1:4). The separation between TC and DMCTC improved with an increase in solvent system pH, but no particular change was observed in the separation between DC and MTC. The best resolution among most of the TCs was obtained at pH 3.0. Therefore, the optimal solvent system in RP–TLC on a C8 plate is methanol–acetonitrile–0.5 M aqueous oxalic acid solution at pH 3.0 (1:1:4), and good separation of the eight TCs, except for DC and MTC, is possible in RP–TLC as shown in Figure 4.

Good separations among MINO, CTC, DC, MTC, OTC, and PRMTC were obtained on NP–TLC plates, while good separations among MINO, OTC, PRMTC, TC, DMCTC, and CTC were obtained on RP-TLC plates. Therefore, a combination of NP- and RP-TLC enables us to identify all eight TCs (*36*).

When determining TCs on TLC plates a densitometer was used and the absorbance was measured at 360 nm, which is the absorbance maxima of all TCs. The densitometric profiles of the TCs on RP– and NP–TLC plates that were developed under the conditions described above are shown in Figure 5. Each antibiotic was detected as a sharp peak. The calibration graphs prepared using the peak heights showed linear relationships between 0.1 and 1.0 μg, except for PRMTC. It is well known that PRMTC is rapidly hydrolyzed in aqueous solution

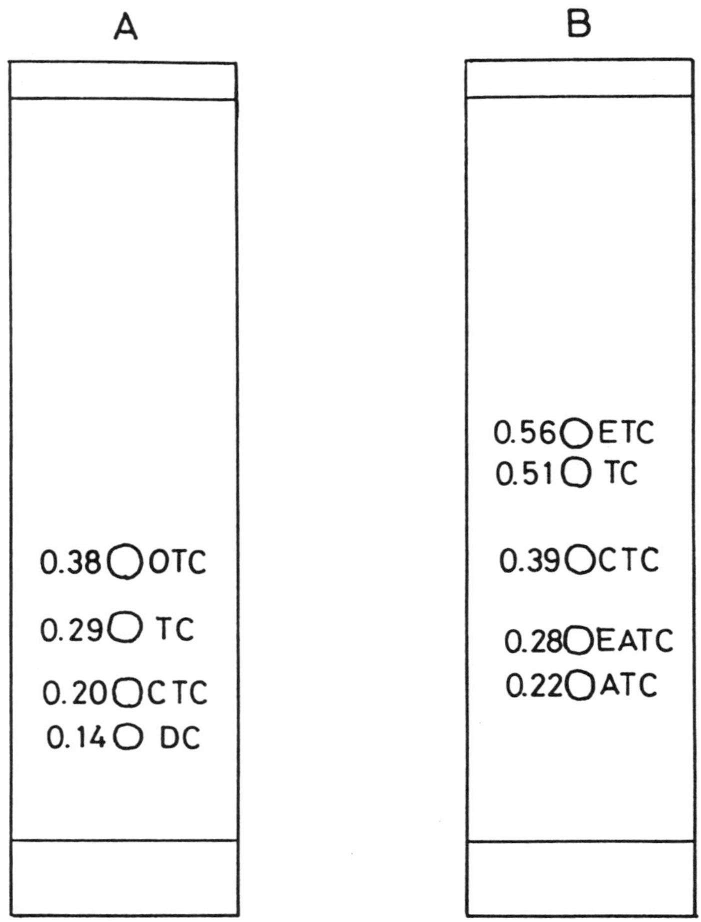

Figure 3 Separation of TCs on reversed-phase TLC.
(A) Plate: C8-TLC. Solvent system: Methanol-acetonitrile-0.5 M oxalic acid (pH 2.0) (1:1:4), (B) Plate: C18-TLC. Solvent system: Methanol-acetonitrile-0.5 M oxalic acid (pH 2.0) (1:1:2)
SOURCE: Reproduced by permission of Elsevier Science. See reference *(22)*.

to give TC (*52–54*) and this hydrolysis occurs even during TLC development. Because PRMTC degraded with the above TLC conditions, it is difficult to determine PRMTC directly by TLC. However, it is possible to determine PRMTC in pharmaceutical preparations as TC after it has completely converted to TC.

For detecting TCs on TLC plates, magnesium chloride (*37, 43*), ferric chloride (*55*), antimony pentachloride (*55*), sulfuric acid (*55*), diazotized *p*-nitroaniline (*27*), modified Sakaguchi reagent (*27*), diphenylpicrylhydrazyl reagent (*27*), and

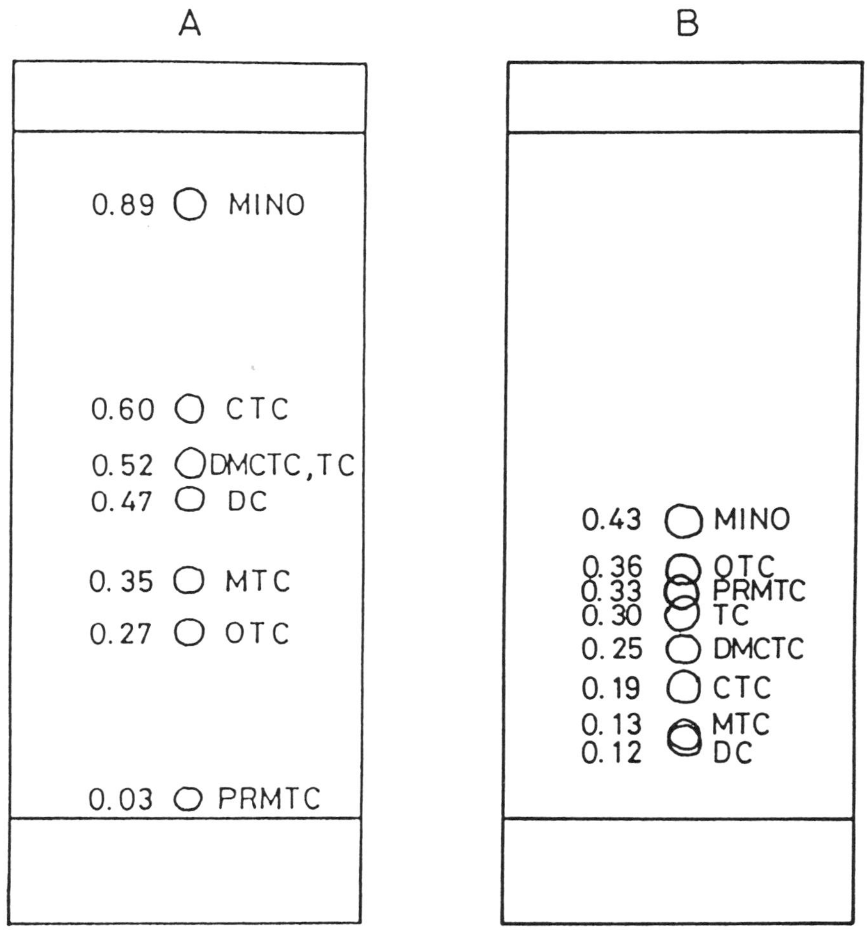

Figure 4 Separation of the eight TCs on normal- and reversed-phase TLC (A) Plate: Silica gel TLC predeveloped with saturated Na_2EDTA and then activated for 2 hr at 130°C. Solvent system: Cholorform-methanol-5% Na_2EDTA (65:20:5) (lower layer) (B) Plate: C8-TLC. Solvent system: Methanol-acetonitrile-0.5 M oxalic acid (pH 3.0) (1:1:4)
SOURCE: **Reproduced by permission of Elsevier Science. See reference *(36)*.**

diazonium salts (*27, 29, 36, 38, 56*) have been used as detection reagents. Highly sensitive detection of TCs on NP– and RP–TLC plates is achieved by observing their fluorescence under a UV lamp (370 nm) after spraying with magnesium chloride (*37*), and observing the color produced after spraying with diazonium salts (*36, 57*) (Tables 3 and 4).

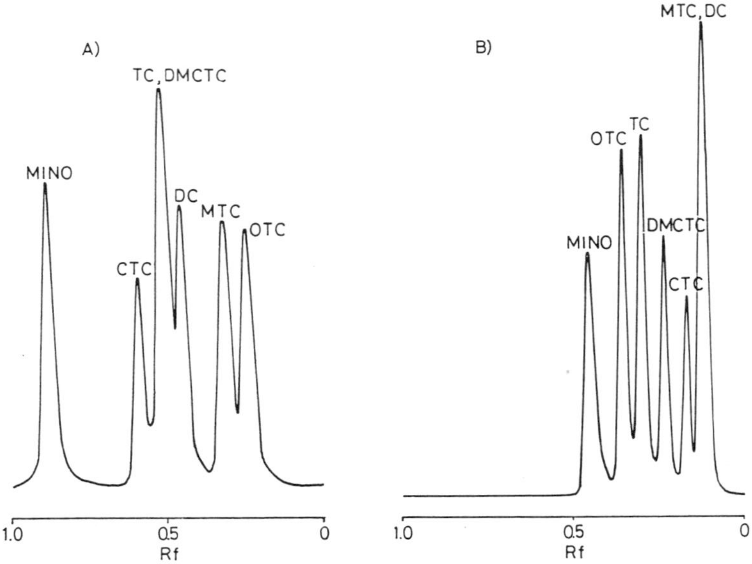

Figure 5 Densitometric profiles of the seven TCs on normal- and reversed-phase TLC.
(A) Silica gel TLC (B) C8 TLC. TLC conditions: See Figure 4. Measurement wavelength: 360 nm
SOURCE: **Reproduced by permission of Elsevier Science. See reference *(36)*.**

As shown in Tables 1 and 2, the TLC methods introduced here have been combined with various cleanup procedures and applied to the analysis of impurities in pharmaceutical preparations as well as to residuals of TCs in milk, honey, and animal tissues.

High-Performance Liquid Chromatography

Optimization of High-Performance Liquid Chromatographic Conditions Using a Mobile Phase Containing Oxalic Acid

Because TCs form chelate complexes with metal ions and adsorb on silanol groups (*22, 23, 61, 62*) in RP columns they are apt to appear as tailing peaks. In order to avoid forming chelate complexes and their adsorption on RP columns, RP column chromatography using mobile phases containing various acids including phosphoric (*45, 46, 63–92*), citric (*47, 48, 50, 62*), tartaric (*48*), and EDTA (*49, 50, 65, 78, 81, 93–100*), and ion-pair chromatography (*62, 78, 82, 88, 90, 101, 102*) have been reported (Table 5). However, even when using these acids in the mobile phase, TCs still show extreme tailing on RP–LC columns and only when a mobile phase containing oxalic acid is used can tailing of TCs be eliminated.

Text continues on page 371

Table 3 Detection Limits of TCs Using Silica Gel TLC with Different Spray Reagents

	Detection limit (μg)									
Spray reagent	OTC	TC	CTC	DC	MINO	DMCTC	MTC	ETC	ATC	EATC
2.0% Fast Red ITR	0.1 (Y)	0.05 (Or)	0.1 (Or)	0.1 (Y)	0.1 (Y)	0.05 (Y)	0.05 (Or)	0.05 (Or)	0.1 (R)	0.1 (R)
1.0% Fast Red B	0.03 (Or)	0.03 (Or)	0.03 (Y)	0.03 (Y)	0.1 (Or)	0.05 (Or)	0.05 (Br)	0.03 (Or)	0.1 (R)	0.1 (R)
0.5% Fast Blue BB	0.03 (R)	0.05 (R)	0.05 (R)	0.03 (R)	0.1 (Br)	0.03 (R)	0.03 (R)	0.03 (R)	0.1 (B)	0.1 (B)
1.0% Fast Violet B	0.03 (R)	0.03 (R)	0.03 (R)	0.03 (R)	0.1 (Or)	0.03 (R)	0.03 (R)	0.03 (R)	0.1 (B)	0.1 (B)
0.2% Fast Blue B	0.03 (R)	0.05 (R)	0.05 (R)	0.05 (R)	0.1 (R)	0.03 (R)	0.03 (R)	0.05 (R)	0.1 (R)	0.1 (R)
0.2 M Magnesium chloride	0.01	0.01	0.01	0.01	0.03	0.01	0.01	—	—	—
0.2 M Magnesium chloride and 10% TEA	0.01	0.01	0.01	0.01	ND	0.01	0.01	—	—	—
0.2 M Magnesium chloride and pyridine	0.05	0.05	0.05	0.05	0.1	0.05	0.05	—	—	—

OTC: oxytetracycline; TC: tetracycline; CTC: chlortetracycline; DC: doxycycline; MINO: minocycline; DMCTC: demeclocycline; MTC: methacycline; ETC: 4-epitetracycline; ATC: anhydrotetracycline; EATC: 4-epianhydrotetracycline; Y: yellow; Or: orange; R: red; Br: brown; B: blue; ND: not detected

Table 4 Detection Limits of TCs Using RP-PTLC with Different Spray Reagents

	Detection limit (μg)									
Spray reagent	OTC	TC	CTC	DC	MINO	DMCTC	MTC	ETC	ATC	EATC
2.0% Fast Red ITR	0.05 (Y)	0.05 (Or)	0.5 (Or)	0.5 (Y)	0.1 (Y)	0.05 (Y)	0.05 (Y)	0.05 (Or)	0.03 (Br)	0.03 (Br)
1.0% Fast Red B	0.05 (Or)	0.05 (Or)	0.05 (Y)	0.05 (Y)	0.1 (Or)	0.05 (Or)	0.03 (Br)	0.05 (Or)	0.03 (Br)	0.03 (Br)
0.5% Fast Blue BB	0.03 (R)	0.03 (B)	0.03 (R)	0.03 (R)	0.1 (Br)	0.03 (V)	0.03 (V)	0.03 (B)	0.03 (G)	0.03 (G)
1.0% Fast Violet B	0.03 (R)	0.03 (R)	0.03 (R)	0.03 (R)	0.1 (Or)	0.03 (R)	0.03 (R)	0.03 (R)	0.03 (G)	0.03 (G)
0.2% Fast Blue B	0.03 (R)	0.03 (R)	0.03 (R)	0.03 (R)	0.1 (R)	0.03 (R)	0.03 (V)	0.03 (R)	0.03 (G)	0.03 (G)
0.2 M Magnesium chloride	ND	ND	ND	ND	ND	ND	ND	—	—	—
0.2 M Magnesium chloride and 10% TEA	0.01	0.01	0.01	0.01	ND	0.01	0.01	—	—	—
0.2 M Magnesium chloride and pyridine	0.05	0.05	0.05	0.05	0.1	0.05	0.05	—	—	—

OTC: oxytetracycline; TC: tetracycline; CTC: chlortetracycline; DC: doxycycline; MINO: minocycline; DMCTC: demeclocycline; MTC: methacycline; ETC: 4-epitetracycline; ATC: anhydrotetracycline; EATC: 4-epianhydrotetracyclin; ND: not detected.

Table 5 Summary of LC conditions for the Analysis of TCs

Tetracyclines	Sample	Column	Mobile phase	Extraction and cleanup	Detection (nm)	Detection Limit	Ref.
TC and PRMTC	Bulk	Cation exchange	EtOH–EDTA (pH 4.35, 4:6)	Dissolution in H_2O	250	2 μg	*(93)*
TC, CTC, ETC, EATC, and ATC	Bulk	Zipax SCX	0.3 M EDTA (pH 7.0)	Dissolution in 1 *N* HCl	429	0.04 μg	*(94)*
OTC, TC, andCTC	Urine	Anion exchange	0.005 M EDTA, 0.05 M NaCl, and 5% MeOH in H_2O 0.005 M EDTA, 0.05 M NaCl, and 30% MeOH in H_2O	Extraction with EtOAc as calcium complexes	280	4–12 μg/mL	*(95)*
OTC, TC, and CTC	Urine and plasma	μ-Bondapak C18	30% MeCN in phosphate buffer (pH 2.4) 40% MeCN in phosphate buffer (pH 2.4)	Extraction with EtOAc as calcium complexes	355	1–1.5 μg/mL	*(63)*
OTC, TC, and CTC	Urine and plasma	μ-Bondapak C18 pre-conditioned with EDTA	30% MeCN in phosphate buffer (pH 2.4) 40% MeCN in phosphate buffer (pH 2.4)	Extraction with EtOAc as calcium complexes	355	0.25–0.5 μg/mL	*(64)*

Table 5 Summary of LC conditions for the Analysis of TCs—*continued*

Tetracyclines	Sample	Column	Mobile phase	Extraction and cleanup	Detection (nm)	Detection Limit	Ref.
OTC, TC, CTC, DC, MTC, MINO, DMCTC, PRMTC, ATC, and EATC	—	Vydac TP C18	2-8% i-PrOH in 0.1 M Ammonium EDTA 1 M DEA (pH 7.3)	—	254 and 405	—	*(65)*
OTC, TC, CTC, DC, MTC, and DMCTC	Urine and serum	LiChrosorb RP-8	MeCN-0.1 M citric acid (24:76)	Extraction with EtOAc from buffered sample (pH 6.1)	350	50 ng/mL	*(47)*
TC	Plasma	LiChrosorb RP-2	MeCN-0.01 M phosphoric acid (35:65)	Ion pair extraction with TBA	357	200 ng/mL	*(66)*
OTC, TC, CTC, DC, MTC, MINO, DMCTC, PRMTC, ETC, EATC, and ATC	—	Hypersil C1 and C18	H_2O-DMF (9:1) containing 0.0005 M EDTA, 0.02 M KNO_3, 0.015 M citric acid, and 0.02 M sodium citrate	—	272	—	*(49)*
DC	Urine, plasma, and animal tissues	Nucleosil C8	MeCN-0.01 M NaH_2PO_4 (pH 2.4, 3:7)	Extraction with MeCN-0.01 M NaH_2PO_4 (pH2.4, 1:1)	270	0.5 µg/mL	*(67)*
OTC, EOTC, and AOTC	—	LiChrosorb RP-8	MeCN-0.05% TBA (2:98 or 2:8)	—	275	—	*(119)*

Table 5 Summary of LC conditions for the Analysis of TCs—*continued*

Tetracyclines	Sample	Column	Mobile phase	Extraction and cleanup	Detection (nm)	Detection Limit	Ref.
OTC, TC, CTC, DC, MTC, and DMCTC	—	LiChrosorb RP-8	MeCN-0.1 M citric acid (76:24)	—	350	—	*(48)*
TC, CTC, ETC, EATC, and ATC	Bulk	μ-Bondapak phenyl	Step gradient of 12–22% MeCNin 0.2 M phosphate buffer (pH 2.2)	Dissolution in H_2O	270	—	*(45)*
OTC, TC, CTC, DC, MTC, and DMCTC	—	LiChrosorb NH_2	MeCN-0.1 M phosphoric acid (1:9) containing 0.002 M HIBS	—	357	—	*(68)*
OTC, TC, CTC, DC, MTC, and DMCTC	—	C2, C8, and C18	MeCN-0.1 M phosphoric acid containing TBA	—	357	—	*(69)*
TC	Urine	LiChrosorb RP-8	*i*-PrOH-DEA-tetraammonium EDTA-H_2O (11:5:1:83)	Direct Injection	365	0.1 μg/mL	*(96)*
OTC, TC, and CTC	Honey	μ-Bondapak phenyl	MeCN-0.05 M phosphoric acid (1:9)	Extraction with MeCN-0.05M phosphoric acid (2:8)	272 and 357	0.1 ppm	*(70)*
TC, CTC, ETC, EATC, and ATC	Bulk	μ-Bondapak C18	MeCN-H_2O (pH 2.5, 24:76) containing 0.5% ethanolamine	Dissolution in MeOH	280	—	*(120)*

Table 5 Summary of LC conditions for the Analysis of TCs—*continued*

Tetracyclines	Sample	Column	Mobile phase	Extraction and cleanup	Detection (nm)	Detection Limit	Ref.
TC	Urine and plasma	μ-Bondapak C18	MeCN-phosphate buffer (pH 2.2, 16:84)	Direct injection after addition of TFA	357	0.05 μg/mL	(*71*)
TC, CTC, ETC, EATC, and ATC	Bulk	μ-Bondapak phenyl	Gradient elution (20–50% A) A: MeOH-H_2O-1 M phosphoric acid (75:20:5) B: MeOH-H_2O-1 M phosphoric acid (5:90:5)	Dissolution in MeOH	280	—	(*46*)
OTC, TC, CTC, DC, MTC, DMCTC, and their impurities	Bulk	LiChrosorb RP-8	MeCN-0.1 M NaH_2PO_4 (pH 8.0, 3:7) containing 0.0194 M DOA MeCN-0.05 M NaH_2PO_4 (pH 8.0, 35:65) containing 0.0097 M DOA MeCN-0.1 M NaH_2PO_4 (pH 8.0, 20:80) containing 0.0194 M DOA MeCN-0.1 M Na H_2PO4 (pH 8.0, 28:72) containing 0.0194 M DOA	Dissolution in MeOH	280	—	(*72*)

Table 5 Summary of LC conditions for the Analysis of TCs—*continued*

Tetracyclines	Sample	Column	Mobile phase	Extraction and cleanup	Detection (nm)	Detection Limit	Ref.
OTC, TC, CTC, DC, MTC, DMCTC, ETC, EATC, and ATC	Bulk	Vydac C18	Various mixtures of MeOH-0.001 M EDTA (pH 6.6)	Dissolution in H_2O	380	—	*(97)*
CTC	Premix	Vydac 201TP (C18)	H_2O-DMF (88:12) containing 0.001 M EDTA, 0.05 M citric acid, 0.013 M sodium citrate, 0.1 M KNO_3	Dissolution in 4 *N* HCl-Me_2CO-H_2O (1:13:6)	280 or 340	—	*(50)*
OTC, TC, and CTC	Meats and fishes	Wako gel DMS-10H (diethyl silica) Shimadzu PSG-100 (polystyrene)	MeCN-0.05 M phosphate buffer (pH 2.5, 1:9) MeOH-0.04 M KH_2PO_4-0.08 M EDTA (8:1:1)	Extraction with 1 *N* HCl XAD-2 cleanup	370	0.1–0.3 ppm	*(73)*
TC, CTC, ETC, EATC, and ATC	—	Cosmosil C8	MeOH-MeCN-0.01 M oxalic acid (pH 2.0, 1:1.5:5)	—	360	5 ng	(23)
		Cosmosi C18	MeOH-MeCN-0.2 M oxalic acid (pH 2.0, 1:1:3.5)		400	20 ng	*(56)*
TC, CTC, ETC, EATC, and ATC	Capsule, powder, syrup, and tablet	Nucleosil C18	MeOH-MeCN-0.2 M oxalic acid (pH 2.0, 1:1:5)	Dissolution in MeOH	400	—	

Table 5 Summary of LC conditions for the Analysis of TCs—*continued*

Tetracyclines	Sample	Column	Mobile phase	Extraction and cleanup	Detection (nm)	Detection Limit	Ref.
OTC, TC, and CTC	Animal tissues	LiChrosorb RP-8	25% aq. DMF containing 0.02 M phosphoric acid	Extraction with 5% perchloric acid C18 cartridge cleanup	340	0.1–0.2 ppm	*(74)*
OTC, TC, and CTC	Animal tissues	Nucleosil CN	MeOH-0.01M NaH_2PO_4 (pH 2.5, 75:25)	Extraction with 0.5% meta-phosphoric acid-MeOH (8:2) C18 cartridge cleanup	268	0.05–0.1 ppm	*(75)*
OTC, TC, CTC, and DC	Animal tissues	LiChrosorb RP-8	MeOH-MeCN-0.01 M oxalic acid (pH 2.0, 1:1.5:2.5)	Extraction with McIlvaine buffer (pH 4.0) containing 0.1 M EDTA C18 cartridge cleanup	350	0.05–0.1 ppm	*(24)*
DC and its impurities	Capsule, syrup, and tablet	Hamilton PRP-1	THF-0.2 M phosphate buffer (pH 8.0)-0.2 M TBA-0.1 M EDTA-H_2O (6:10:5:1:78)	Dissolution in mobile phase	254	—	*(76)*
OTC	—	Spherisorb ODS	MeCN-0.05 *N* H_2SO_4-DMF (19.5:74:6.5)	—	254	—	*(121)*
OTC and its impurities	Bulk	LiChrosorb RP-8	0.2 M ammonium oxalate-0.1 M EDTA-DMF (55:20:25)	Dissolution in 0.1 M HCl in MeOH-H_2O (15:85)	280	—	*(98)*

Table 5 Summary of LC conditions for the Analysis of TCs—*continued*

Tetracyclines	Sample	Column	Mobile phase	Extraction and cleanup	Detection (nm)	Detection Limit	Ref.
OTC, TC, CTC, DC, MTC, DMCTC, ETC, EATC, and ATC	—	Vydac RP C18	15-35% MeOH-0.001 M EDTA (pH 6.6)	—	380	—	*(99)*
OTC and TC	Fish tissues	Nucleosil C18	MeCN-0.05 M NaH_2PO_4 (pH 2.5, 35:65)	Extraction with 0.5% metaphosphoric acid-MeOH (8-2) C18 cartridge cleanup	265	0.05–0.1 ppm	*(77)*
OTC	Honey	Partsil C8	H_2O-DMF (17:3) containing EDTA (pH 3.5)	Dissolution in H_2O	355	0.5 ppm	*(100)*
OTC, EOTC, TC, and CTC	Premix	Econosphere C8	MeCN-phosphate buffer (pH 8.0) containing EDTA and DOA	Extraction with acidic MeOH	280	—	*(78)*
OTC, TC, DC, and MINO	Serum	Spheri-5 phenyl	MeOH-MeCN-TEA-0.08 M phosphate buffer (pH 2.4, 100:15:5:880)	Dilution with mobile phase containing 2% phosphoric acid	267	0.2 µg/mL	*(79)*

Table 5 Summary of LC conditions for the Analysis of TCs—*continued*

Tetracyclines	Sample	Column	Mobile phase	Extraction and cleanup	Detection (nm)	Detection Limit	Ref.
OTC, TC, CTC, DC, MTC, DMCTC, ETC, EATC, ATC, PRMTC, ICTC, and apo-OTC	Plasma	Copolymer of divinyl-benzene with styrene	MeCN-CH_2Cl_2-0.2 M acetate buffer (pH 3.6, 15:1:85) containing EDTA	Extraction with EtOAc	268 or 357	0.5 µg/mL	(*122*)
OTC, TC, and DC	Animal tissues	Polymer Lab. PLRP-S	0.01 M Phosphoric acid-MeCN-MeOH (gradient, 80:0:20 to 30:50:20)	Extraction with 1 *N* HCl liquid-liquid partition cleanup	355	0.1 ppm	(*80*)
TC, ETC, and EATC, and ATC	—	Polymer Lab. PLRP-S	*tert*-BuOH-0.2 M phosphate buffer (pH 9.0)-0.02 M BA-0.1 M EDTA (55:100:50:10)	—	254	—	(*81*)
OTC, TC, CTC, and DC	Honey	Bakerbond C8	MeOH-MeCN-0.01 M oxalic acid (1:1.5:3)	C18 cartridge pretreated with EDTA and COOH cartridge cleanup	350	0.05–0.1 ppm	(*118*)
OTC, TC, CTC, DC, MTC, DMCTC, MINO, and PRMTC	—	Chemcosorb C8	MeOH-MeCN-0.01 M oxalic acid (pH 3.0, 1:1.5:7)	—	350	—	(*36*)

Table 5 Summary of LC conditions for the Analysis of TCs—*continued*

Tetracyclines	Sample	Column	Mobile phase	Extraction and cleanup	Detection (nm)	Detection Limit	Ref.
OTC, TC, CTC, DC, MTC, DMCTC, MINO, and MINO	Honey	Chemcosorb C8	MeOH-MeCN-0.01 M oxalic acid (pH 3.0, 1:1.5:7)	C18 cartridge pretreated with EDTA and COOH cartridge cleanup	350	0.05–0.1 ppm	(*37*)
OTC, EOTC, AOTC, and apo-OTC	—	Polymer Lab. PLRP-S	*tert*-BuOH-0.2 M phosphate buffer (pH 8.0)-0.02 M TBA-0.0001 M EDTA (55:100:50:10)	—	254	—	(*82*)
OTC, TC, CTC, and DC	Animal tissues	LiChrosorb RP-8	MeOH-MeCN-0.01 M oxalic acid (1:1.5:3)	Extraction with McIlvaine buffer (pH 4.0) containing 0.1 M EDTA C18 cartridge pretreated with EDTA cleanup	350	0.01 ppm	(*38*)
OTC	Fish tissues	Hypersil ODS	0.5% $(NH_4)_2HPO_4$-0.5% DEA-MeCN-DMF (pH 2.5, 40.5:40.5:19:6)	Extraction with 1 *N* HCl-50% TCA (15-1) C18 cartridge cleanup	365	0.005 ppm	(*83*)

Table 5 Summary of LC conditions for the Analysis of TCs—*continued*

Tetracyclines	Sample	Column	Mobile phase	Extraction and cleanup	Detection (nm)	Detection Limit	Ref.
CTC	Capsule, tablet, and ointment	Zorbax C8	DMSO-$HClO_4$-H_2O (32:5:63)	Dilution with 0.01 M HCl	280	0.05%	(*125*)
OTC, TC, and CTC	Feeds	Nova-Pak C18	MeOH-MeCN-0.01 M oxalic acid (1:1.5:6.5)	Extraction with McIlvaine buffer (pH 2.0 and 4.5) containing EDTA Sephadex LH-20 cleanup	370	—	(*108*)
OTC	Fish tissues	Supelcosil LC-18DB	0.005 M phosphate buffer (pH 2.0)-MeCN-THF (81:10:9)	Extraction with phosphate buffer (pH 4.2) containing 0.1 M EDTA C8 cartridge cleanup	357	0.005 ppm	(*84*)
OTC, TC, CTC, and DC	Fish tissues	Spheri C18	MeCN-DMF-0.01 M oxalic acid (22:6:72)	Extraction with McIlvaine buffer (pH 4.0) C18 cartridge pretreated with EDTA cleanup	355	0.05–0.1 ppm	(*109*)
DC	Urine and serum	Nova-Pak C18	MeCN-AcOH-0.1 M KH_2PO_4 (75:150:125 and 65:150:125)	Extraction with McIlvaine buffer (pH 4.0) containing 0.1 M EDTA C18 cartridge cleanup	340	0.025 ppm	(*85*)

Table 5 Summary of LC conditions for the Analysis of TCs—*continued*

Tetracyclines	Sample	Column	Mobile phase	Extraction and cleanup	Detection (nm)	Detection Limit	Ref.
OTC	Fish plasma	Polymer Lab. PLRP-S	MeCN-THF-0.02 M H_3PO_4 (63:47:89)	Direct injection	350	0.1 μg/mL	*(86)*
OTC and CTC	Milk	C18 column	0.005% H_3PO_4-MeCN (1:1 and 13:7)	Amberlite XAD-2	355	0.5 ppm	*(87)*
OTC, TC, and CTC	Milk	Nova-Pak C18	0.01M oxalic acid-MeOH-MeCN gradient, (100:0:0 to 70:8:22)	Dilution with McIlvaine buffer (pH 4.0) containing 0.1 M EDTA and ultrafiltration	360	0.015–0.05 ppm	*(110)*
OTC, TC, CTC, and ECTC	Feeds	Nova-Pak C18	MeOH-MeCN-0.01M oxalic acid (1:1.5:6.5)	Extraction with McIlvaine buffer (pH 2.0 and 4.5) containing EDTA Sephadex LH-20 cleanup	370	—	*(111)*
CTC	Tissues	Polymer Lab. PLRP-S	MeCN-0.1 M glycine (125:875)	Extraction with 1M HCl-0.1M glycine Cyclohexyl cartridge cleanup	Coversion to ICTC FL; 340/420	0.02–0.05 ppm	*(123)*
OTC	Fish plasma	Cyano Spheri-5	MeOH-0.02 M oxalic acid DMF (5:95:5)	Addition of TFA	350	4 ng	*(124)*

Table 5 Summary of LC conditions for the Analysis of TCs—*continued*

Tetracyclines	Sample	Column	Mobile phase	Extraction and cleanup	Detection (nm)	Detection Limit	Ref.
OTC, TC, CTC, and DC	Animal tissues	Nova-Pak phenyl	MeOH-MeCN-0.02 M oxalic acid gradient (5:15:80 to 13:27:60)	Extraction with McIlvaine buffer (pH 4.0) containing 0.1 M EDTA C18 cartridge deactivated by silylation cleanup	355	0.005–0.01 ppm	(*61*)
OTC, TC, and CTC	Animal tissues	YMC-PACK ODS	1M Imidazole buffer-MeOH (77:23)	Extraction with McIlvaine buffer-MeOH (3:7) C18 cartridge cleanup	Formation of metal chelate with Mg FL; 380/520	0.01–0.25 ppm	(*126*)
TC, ETC, EATC, and ATC	—	Polystyrene-divinyl-benzene	*tert*-BuOH-0.2 M KH_2PO_4-0.02 M-TBA-0.01 M EDTA-H_2O (6:10:15:10:59)	—	254	—	(*88*)
CTC and its impurities	—	Polystyrene-divinyl benzene	*tert*-BuOH-1 M $HClO_4$-H_2O (5:5:10)	—	254		(*127*)
OTC and TC	Honey	Pecosphere C18	0.01 M SDS containing 0.01 M oxalic acid-MeCN (7:3)	Dilution with MeCN and H_2O	285	1.2–3.0 ppm	(*101*)

Table 5 Summary of LC conditions for the Analysis of TCs—*continued*

Tetracyclines	Sample	Column	Mobile phase	Extraction and cleanup	Detection (nm)	Detection Limit	Ref.
OTC and TC	Milk	Nucleosil C18 pre-treated with CTC	MeCN-0.02 M H_3PO_4 (24:76)	Extraction with TBA into CH_2Cl_2	355	0.01 ppm	(*89*)
OTC, TC, and CTC	Milk	Micro Pak ODS	MeCN-0.01 M oxalic acid (3:7)	Matrix solid phase dispersion (C18 with EDTA and oxalic acid)	365	0.1 ppm	(*116*)
OTC, TC, and CTC	Salmon muscle	LiChroCART RP-18	MeOH-MeCN-0.01 M oxalic acid (10:17:73)	Extraction with McIlvaine buffer C18 cartrudge cleanup	355	0.08–0.05 ppm	(*115*)
OTC	Fish tissues	Micro Pak ODS	MeOH-MeCN-0.01 M oxalic acid (2.5:27.5:70)	Matrix solid phase dispersion C18 (with EDTA and oxalic acid)	365	0.05 ppm	(*117*)
OTC, TC, and CTC	Animal tissues	LiChrosorb RP-8	MeCN-0.01 M oxalic acid (1:1)	Extraction with succinate buffer after formation of metal chelate with copper, XAD-2 cleanup	350	0.01–0.05 ppm	(*113*)

Table 5 Summary of LC conditions for the Analysis of TCs—*continued*

Tetracyclines	Sample	Column	Mobile phase	Extraction and cleanup	Detection (nm)	Detection Limit	Ref.
OTC and CTC	Honey	Asahipak Gel ODP-50	OTC; MeOH-0.1 M glycine buffer (1:20) containing SHS CTC: MeOH-0.1 M glycine buffer (1:5)	C18 cartridge and ion pair extraction cleanup	Formation of metal chelates FL; 370/505	0.005–0.02 ppm	(*102*)
OTC, TC, and CTC	Milk	Nova Pak C18	MeOH-MeCN-0.01 M oxalic acid (30:50:30 or 0:60:40)	C18 cartridge cleanup	Particle beam MS	0.1 ppm	(*103*)
OTC, TC, CTC, and DC	Animal tissues	LiChrosorb RP-8	MeOH-MeCN-0.01 M oxalic acid (pH 2.0) (1:1.5:2.5)	Extraction with McIlvaine buffer (pH 4.0) containing 0.1M EDTA C18 cartridge cleanup	350	0.05–0.1 ppm	(*112*)
OTC	Blood and serum	Polymer Lab. PLRP-S	MeCN-0.02 M phosphoric acid (23:77) containing SHS	Automatic dialysis	350	0.05 ppm	(*90*)
MTC, EMTC, OTC, DC, EOTC, and ADMTC	—	Polymer Lab. PLRP-S	2-Methyl-2-PrOH-0.2 M phosphate buffer-0.01 M EDTA-H_2O (2.5:10:10:77.5)	—	254	—	(*91*)
OTC	Salmon muscle	Ultrasphere ODS	MeOH-0.02 M phosphate buffer (pH 2.25) (60:190)	Extraction with phosphate buffer Sephadex G-25 cleanup	365 and 254	0.05 ppm	(*92*)

Table 5 Summary of LC conditions for the Analysis of TCs—*continued*

Tetracyclines	Sample	Column	Mobile phase	Extraction and cleanup	Detection (nm)	Detection Limit	Ref.
OTC	Animal serum	Hisep shielded hydrophobic phase	MeOH-0.2 M oxalic acid (pH 7.0) (1:9)	Direct injection	360	0.05 ppm	(*128*)
OTC, TC, and CTC	Animal tissues	Nova Pak C18	Phosphate citric buffer-MeCN (72:28) containing 0.005 M TBA	Extraction with McIlvaine buffer (pH 4.0) containing 0.1M EDTA C18 cartridge cleanup	365	0.01 ppm	(*62*)
OTC	Salmon muscle	Ultrasphere ODS	MeCN-THF-0.025 M oxalic acid (225:25:750) containing 0.01 M SOS	Extraction with 1% metaphosphoric acid	355	0.05 ppm	(*129*)
OTC, TC, CTC, DC, MINO, MTC, and DMCTC	Milk	Polymer Lab. PLRP-S	MeOH-MeCN-0.01M oxalic acid gradient (0:0:100 to 8:22:70)	Metal chelate affinity column cleanup	355	0.005 ppm	(*114*)
OTC, TC, and CTC	Milk	Polymer Lab. PLRP-S	0.05 M Oxalate buffer (pH 2.0) containing 0.005 M octanesulfonate-MeCN gradient (80:20 to 62:38)	Deproteinization by HCl and MeCN	365	0.005 ppm	(*130*)

Table 5 Summary of LC conditions for the Analysis of TCs—*continued*

Tetracyclines	Sample	Column	Mobile phase	Extraction and cleanup	Detection (nm)	Detection Limit	Ref.
OTC, TC, and CTC	—	Polymer Lab. PLRP-S Supelco LC-18 Supelco LC-18-DB	Mixture of MeOH, MeCN, DMF, and 0.05 M buffer at various ratios	—	355	—	*(131)*
OTC, TC, CTC, and DC	Milk	Bakerbond C8	MeOH-MeCN-0.01 M oxalic acid (pH 2.0) (1:1.5:2.5)	Extraction with McIlvaine buffer (pH 4.0) containing 0.1M EDTA C18 cartridge cleanup	350	0.01 ppm	*(60)*
OTC, TC, CTC, and DC	Honey	Inertsil phenyl	MeOH-MeCN-0.005 M TFA (2:2:11)	C18 cartridge pretreated with EDTA cleanup	Frit FABMS	0.1 ppm	*(107)*
OTC, TC, and CTC	Animal tissues	LiChrosorb RP18	14% i-PrOH in 0.05 M aq. DEA containing 0.001 M EDTA (pH 7.3)	Extraction with 0.02 *N*-HCl-0.01 M EDTA after coversion to anhydroform reextraction with $MeCl_2$	280	—	*(146)*
MINO and its metabolites	Serum and urine	C18	Gradient of 10–40% MeCN in 0.2% aq. formic acid	Mixture with 0.025 mM EDTA	352	—	*(147)*

Table 5 Summary of LC conditions for the Analysis of TCs—*continued*

Tetracyclines	Sample	Column	Mobile phase	Extraction and cleanup	Detection (nm)	Detection Limit	Ref.
OTC, TC, CTC, DC, and MINO	—	Vydac 201H554 C18	MeCN-0.01M EDTA (pH 6.6) (27:73) MeCN-0.1% TFA (24:76)	—	Eupropium phosphores cence	1–10 ng	(*148*)
OTC	Plasma	LiChrosorb RP8	MeOH-MeCN-0.01 M oxalic acid (10:15:75)	Extraction with EtOAc-i-PrOH (6:0.5)	357	5 ng/mL	(*149*)
MINO	Bulk	Chemospher C8 Rosil C8 Hypersil C8	DMF-0.2 M ammonium oxalate-0.1 M EDTA (25:55:20)	Dissolution in 0.01 N HCl	280		(*150*)
		LiChrosorb C8 Particil C8 Zorbax C8	DMF-0.2 M ammonium oxalate-0.1 M EDTA-H_2O (15:55:20:10)				
		Nucleosil C8	DMF-0.2 M ammonium oxalate-0.1 M EDTA-H_2O (10:55:20:15)				
DC	Serum	Ultrabase C18	MeCN-phosphoric acid (pH 2.5) (28:72)	Extraction with EtOAc from buffered sample (pH 6.1)	350	25 ng/mL	(*151*)

Table 5 Summary of LC conditions for the Analysis of TCs—*continued*

Tetracyclines	Sample	Column	Mobile phase	Extraction and cleanup	Detection (nm)	Detection Limit	Ref.
OTC, TC, CTC, DC, DMCTC, and MINO	Animal tissues	Chromospher C8	MeCN-0.01 Moxalic acid (pH 2.0) (2:8)	Metal chelate affinity column cleanup	365	10 ng/g	*(152)*

HIBS: 1-hydroxy-3,5-diisobutylbenzenesulfonic acid; DOA: N, N-dimethyloctylamine, MTC: methacycline, DMCTC: demeclocycline, PRMTC: rolitetracycline, TEA: triethanolamine, AOTC: anhydrooxytetracycline, TCA: trichloroacetic acid, ICTC: isochlortetracycline, TFA: trifluoroacetic acid, DMF: N, N-dimethylformamide, ATC: anhydrotetracycline, EATC: 4-epianhyrotetracycline, ETC: 4-epitetracycline, DMSO: dimethyl sulfoxide, MS: mass spectrometry, MINO: minocycline, ADMTC: 2-acetyl-2-decarboxamidometacycline, EOTC: 4-epioxytetracycline, FL: fluorescence ex/em, TBA: tetracutylammonium, THF: tetrahydrofuran, SOS: sodium octane sulfonate, SHS: sodium 1-heptane sulfonate, DEA: diethanolamine, FABMS: fast atom bombardment mass spectrometry, and EDTA: disodium ethylenediaminetetraacetate.

Separations are shown in Figure 6 under optimal conditions *(23)*. In this section RP–LC separation of TCs using a mobile phase containing oxalic acid is described in detail.

In the previous section, although the Rf values of OTC and TC, and of ETC and TC in RP–TLC, were similar. Using the C8 TLC plate gave good resolution between OTC and TC, and good separation was shown between ETC and TC on the C18 plate. Because capacity factors [$k'=(t_r - t_0)/t_0$, where t_r is the retention time of the sample peak and t_0 is the retention time for a nonretained peak] of OTC and TC and of ETC and TC were similar in RP–LC, the resolution factors [$Rs=2(t_1 - t_2)/(w_1+w_2)$, where t_1 and t_2 are retention times and w_1 and w_2 are the corresponding peak widths at the baseline] between OTC and TC and between ETC and TC were estimated using Cosmosil C8 and C18 columns. The Cosmosil C8 and C18 columns had methanol–acetonitrile–0.01 M aqueous oxalic acid solution (pH 2.0) (1:1.5:5) as a mobile phase to compare C8 and C18 modified silica gel columns for their suitability in analyzing TCs. *Rs* values between OTC and TC are 1.46 and 1.18 using the C8 and C18 columns, where the corresponding *Rs* values between ETC and TC are 1.20 and 1.35, respectively. Therefore, the C8 column is suitable for analyzing the parent TCs and the C18 column is suitable for analyzing impurities of TC.

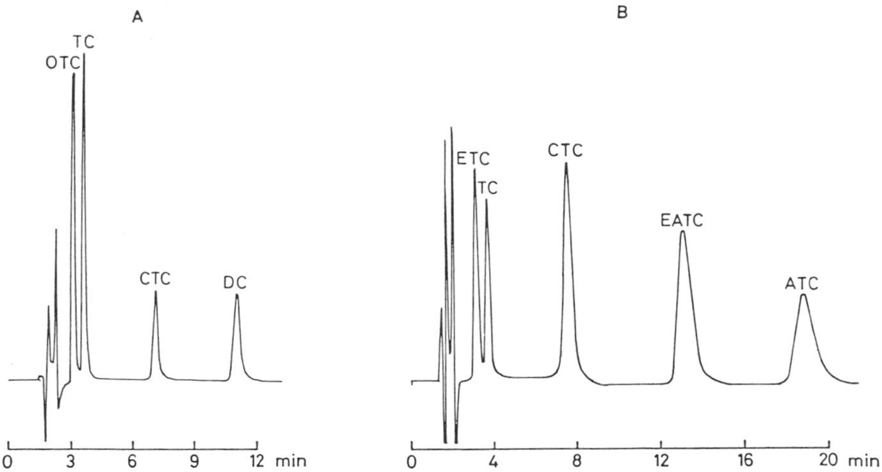

Figure 6 LC separation of TCs
LC conditions: (A) Column: Cosmosil 5C8 (15 cm × 4.6 mm i.d.); mobile phase: CH_3OH-CH_3CN-0.01 M oxalic acid (aq. pH 2.0) (1:1.5:5); detection: 360 nm; flow rate: 1.0 mL/min (B) Column: Cosmosil 5C18 (15 cm × 4.6 mm i.d.); mobile phase: CH_3OH-CH_3CN-0.2 M oxalic acid (aq. pH 2.0) (1:1:3.5); detection: 400 nm; flow rate: 1.0 mL/min
SOURCE: **Reproduced by permssion of Elsevier Science. See reference *(23)*.**

The influence of oxalic acid concentration was examined using methanol–acetonitrile–aqueous oxalic acid solution (pH 2.0) in ratios of 1:1.5:5 and 1:1:3.5 as mobile phases for the parent TCs and the impurities of TC, respectively. The asymmetry factor (*As*, which is the ratio of the rear to the front lengths of the peak along a line parallel to and 10 % distant from its base) improved with increasing acid concentration. Good *As* values are obtained above 0.01 and 0.2 M for the parent TCs and the impurities of TC, respectively. Values of *k ′* are also related to the acid concentration. When 0.01 M aqueous oxalic acid solution is used for the parent TCs, the *k ′* values are sufficient to separate the TCs, and the *Rs* value between the OTC and the TC peaks is 1.46. Using 0.2 M aqueous oxalic acid solution, suitable retention times are shown for the impurities of TC and the *Rs* value between ETC and TC is 1.29. Therefore, 0.01 M aqueous oxalic acid solution is used for the parent TCs and 0.2 M is used for the impurities of TC.

The influence of the pH of the aqueous oxalic acid solution in the mobile phase on *k ′* and *As* values was investigated using methanol–acetonitrile–0.01 M aqueous oxalic acid solution (1:1.5:5) and methanol–acetonitrile–0.2 M aqueous oxalic acid solution (1:1:3.5) as mobile phases for the parent TCs and the impurities of TC, respectively. CTC and DC were completely separated below pH 3.0, but they overlapped above pH 4.0. For the impurities of TC, the *k ′* values of EATC and ATC were too large above pH 2.5, and above pH 3.5 oxalic acid precipitated from the mobile phase. All the TCs had the best *As* values at pH 2.0.

Although various proportions of methanol or acetonitrile with 0.01 or 0.2 M aqueous oxalic acid solution (pH 2.0) were tested as the mobile phase, unsatisfactory results were obtained suggesting that a mixture of methanol and acetonitrile is required to separate TCs. Mixtures of methanol and acetonitrile at various ratios were tested using the mobile phases (methanol–acetonitrile)–0.01 M aqueous oxalic acid solution (pH 2.0) (1:2) for the parent TCs, and (methanol–acetonitrile)–0.2 M aqueous oxalic acid solution (pH 2.0) (2:3.5) for the impurities of TC. For the parent TCs, ratios of 1:2, 1:1.5, and 1:1 for the methanol—acetonitrile portion of the mobile phase gave good separation. However, with the 1:1 mixture, the retention times of TCs were too long, and the 1:1.5 mixture, gave a better resolution between OTC and TC than did the 1:2 mixture. For the impurities of TC, a 1:1 mixture of methanol and acetonitrile gave good separation, but with a 2:1 mixture and with methanol alone the retention times were too long. Thus, the 1:1.5 and 1:1 mixtures were used for the parent TCs and the impurities of TC, respectively.

Various ratios of the organic solvent (methanol–acetonitrile, 1:1.5 and 1:1) and aqueous oxalic acid solution (0.01 and 0.2 M, pH 2.0) were tested as mobile phases. For the parent TCs, a poor separation between OTC and TC was obtained using methanol–acetonitrile–0.01 M aqueous oxalic acid solution (pH 2.0) (1:1.5:2.5). When methanol–acetonitrile–0.01 M aqueous oxalic acid solution (pH 2.0) (1:1.5:7.5) was used, the best separation between OTC and TC was achieved, but the retention times of DC and CTC were too long. A good separation among the parent TCs was obtained on a C8 column using methanol–acetonitrile–0.01 M aqueous oxalic acid solution (pH 2.0) (1:1.5:5). For the impurities of TC, the *k'* values of EATC and ATC were too large with methanol–acetonitrile–0.2 M aqueous oxalic acid solution (pH 2.0) (1:1:3) and the resolution between ETC and TC was poor. Methanol–acetonitrile–0.2 M aqueous oxalic acid solution (pH 2.0) (1:1:3.5) is recommended as a mobile phase to separate the impurities on a C18 column.

Large differences in the chromatographic behavior of analytes have been observed using packing materials from different suppliers. As shown in Figure 7, mobile phases containing oxalic acid made it possible to separate TCs on various different column packing materials. These results show that even if packing materials from different suppliers are used, variation of the ratios of organic solvent and aqueous oxalic acid solution in the mobile phase enables good separation.

Figure 8 illustrates good separation of the eight parent TCs obtained under modified LC conditions (*36*). However, it was not possible to identify PRMTC on the chromatogram using the usual solvents for sample preparation such as methanol and water, because PRMTC decomposed rapidly to give TC (*52–54*). The simultaneous determination of the eight TCs including PRMTC is possible using tetrahydrofuran, dimethyl sulfoxide, and the mobile phase as sample preparation solvents and it is effective for the analysis of PRMTC in pharamaceutical preparation (*36*). However, if the LC method is applied to residue analysis, it is expected that PRMTC will be rapidly converted into TC in vivo or during the analytical procedure. Therefore, we introduce the separation of seven TCs other than PRMTC.

To obtain optimal separation conditions for the seven TCs, various combinations of methanol, acetonitrile, and the aqueous oxalic acid solution were examined. Methanol–acetonitrile–0.01 M aqueous oxalic acid solution (pH 2.0) (1:1.5:7) gave suitable *k'* and *Rs* values for the TCs except for MINO, which was not retained on the column using this mobile phase. De Leenheer and Nelis (*48*) reported that MINO is one of the most lipophilic TCs at neutral pH, but its *k'* at lower pH is too small. This is caused by the dimethylamino group at the C6 position in MINO, as well as *k'* being a function of the pH of the mobile phase. Therefore, the influence of the pH of the mobile phase on *k'* was investigated using methanol–acetonitrile–0.01 M aqueous oxalic acid solution (1:1.5:7). As shown in Figure 9, the *k'* value of MINO increased with increasing mobile phase pH and was suitable for analysis above pH 3.0. Because good separations among the seven TCs were obtained at pH 3.0, methanol–acetonitrile–0.01 M aqueous oxalic acid solution (1:1.5:7) at pH 3.0 is recommended as the mobile phase for a C8 column. The calibration graphs prepared by the peak-height method for determining TCs showed linear relationships between 1 and 10 ng.

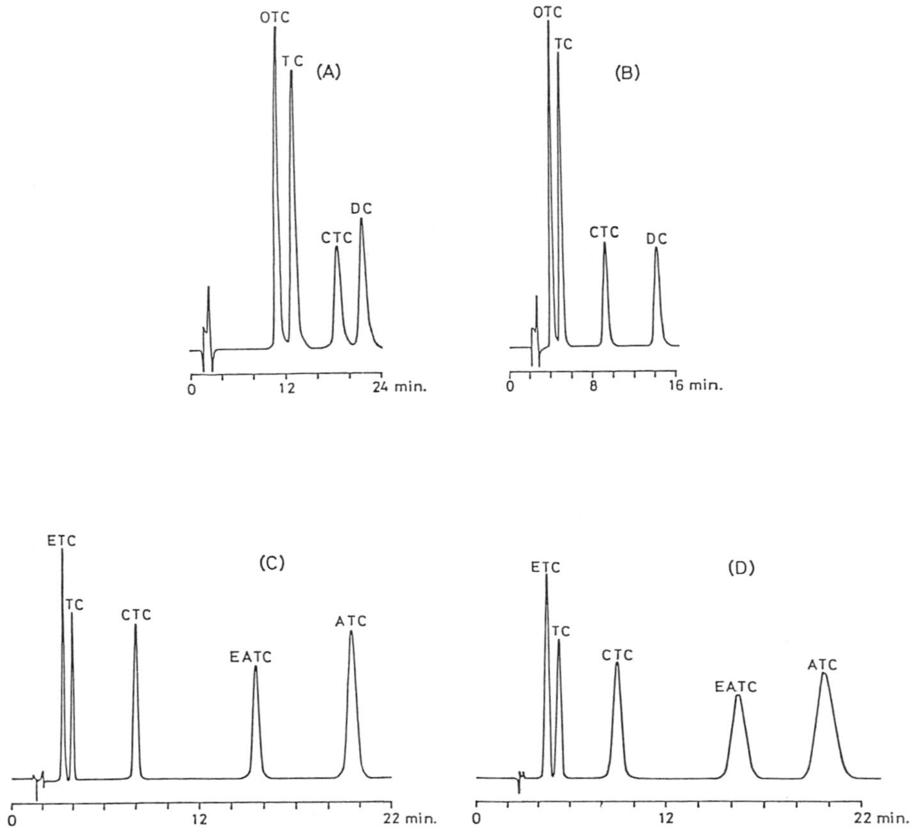

Figure 7 LC separation of TCs using various packing materials from different suppliers
Columns and mobile phases: (A) Column: LiChrosorb PR-8 (250 × 4.0 mm i.d., 10 μm); mobile phase: CH_3OH-CH_3CN-0.01 M oxalic acid (aq. pH 2.0) (1:1.5:2). (B) Column: Nucleosil 5C8 (150 × 4.6 mm i.d., 5 μm); mobile phase: CH_3OH-CH_3CN-0.01 M oxalic acid (aq. pH 2.0) (1:1.5:7.5). (C) Column: Nucleosil 5C18 (150 × 4.6 mm i.d., 5 μm); mobile phase: CH_3OH-CH_3CN-0.2 M oxalic acid (aq. pH 2.0) (1:1:4). (D) Column: μBondpak C18 (300 × 3.9 mm i.d., 10 μm); mobile phase: CH_3OH-CH_3CN-0.2 M oxalic acid (aq. pH 2.0) (1:1:4). Flow rate: 1 mL/min. Detection: (A) and (B): 360 nm, (C) and (D): 400nm.

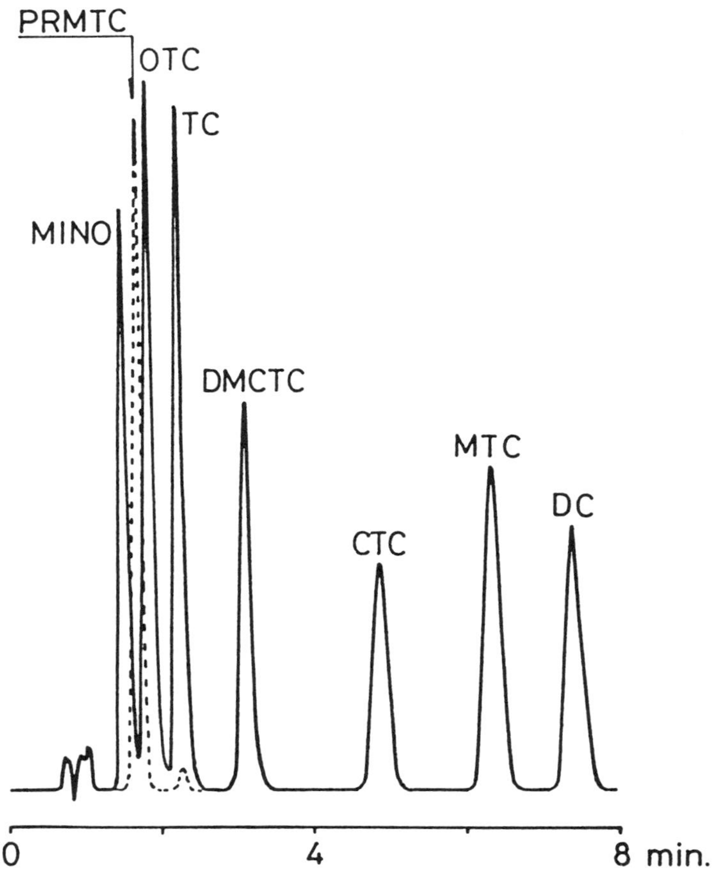

Figure 8 LC separation of the eight TCs. Column: Chemcosorb 3C8 (3 μm, 75 × 4.6 mm i.d.). Mobile phase: CH_3OH-CH_3CN-0.01 M oxalic acid (aq. pH 3.0) (1:1.5:7). Flow rate: 1 mL/min. Detection: 350 nm

- - - - PRMTC ——— Mixture of seven TCs

SOURCE : Reproduced by permission of Elsevier Science. See reference *(36)*.

Optimization of High-Performance Liquid Chromatographic Conditions Using a Volatile Mobile Phase

The existence of residual silanol groups and metals in LC column packing materials influences the chromatography of silanol and metal sensitive compounds. To avoid this influence, triethylamine and oxalic acid are added to the mobile phase, or a polystyrene-divinylbenzene copolymer (PSDVB) is used as a column packing material (Table 5). Most of the previously reported LC conditions

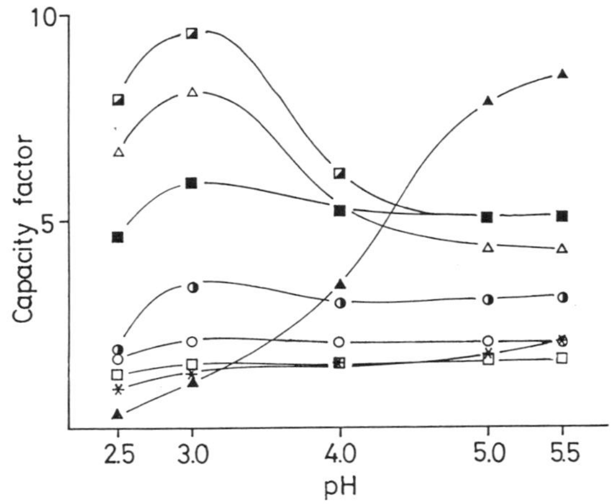

Figure 9 Influence of pH of mobile phase on capacity factor. Key: □ : OTC, O : TC, ■ : CTC, ◪ : DC, ▲ : MINO, Δ : MTC, ◑ : DMCTC, ✱ : PRMTC. Column: Chemcosorb 3C8 (3 μm, 75 × 4.6 i.d.). Mobile phase: CH_3OH-CH_3CN-0.01 M oxalic acid (aq.) (1:1.5:7). Flow rate: 1 mL/min. Detection: 350 nm. SOURCE: **Reproduced by permission of Elsevier Science. See reference *(36)*.**

for TCs using both chemically modified silica gel and PSDVB as column packing materials are based on mobile phases containing such nonvolatile compounds as oxalic and citric acids, EDTA, and ion-pair reagent to control peak tailing. However, mobile phases containing nonvolatile compounds, when used in LC–MS, have caused clogging at the interface and a buildup of deposits in the ion source, so that the LC–MS cannot be operated for a prolonged period (*103*, *104*). Mobile phases containing such nonvolatile buffers, while necessary for the chromatography of the TCs, are unsuitable for routine use in LC–MS analyses.

Recently, well end-capped alkyl-bonded silica gels have been synthesized from 99.99% pure silica gel, and have been successfully applied to separating silanol and metal sensitive compounds without the above additives (*105*, *106*). In this section, the separation of TCs on this type of column packing using a volatile mobile phase is described (*107*).

Five types of well end-capped alkyl-bonded chromatographic packings synthesized from 99.99% pure silica gel are commercially available, including Inertsil C18, Inertsil C8, Inertsil Ph, TSK-GEL C18-80TS, and TSK-GEL C8-80TS. As previously described, using a C8 conventional column provides better resolution between OTC and TC than a C18 column. To compare the suitability of the new C8, C18, and Ph columns (5μm, 250 × 4.0 mm, i.d.) for analyzing TCs, the antibiotics were separated using methanol–acetonitrile–0.01 M aqueous trifluoroacetic acid (TFA) solution (1:1:5) as a mobile phase. Peak shape and resolution were monitored as performance indicators. On all columns *As* factors for TCs were between 1.00 and 1.05, indicating no tailing and no leading peaks. For the separation of TCs, although excellent separations were obtained on all columns between TC and CTC, and between CTC and DC, different chromatographic behavior was observed in the separation of OTC and TC. The *Rs* values between OTC and TC on the C8 and Ph columns showed better resolution than those on the C18 columns. Therefore, C8 and Ph columns for separating the parent TCs were selected and optimal mobile phases were investigated using mixtures of methanol, acetonitrile, and 0.01 M TFA solution at various ratios. The best separations were obtained when methanol–acetonitrile–0.01 M TFA solution was used at ratios of 3:3:14, 2:3:15, and 6:19:75 on TSK GEL C8-80TS, Inertsil C8, and Inertsil Ph, respectively, as shown in Figure 10. These volatile mobile phases can be used in LC–MS without clogging problems at the interface (*107*).

These LC methods introduced above have been applied to the analysis of impurities in pharmaceutical preparations and to residual TCs in milk, honey, and animal tissues in combination with various cleanup procedures described below.

Extraction and Cleanup

Because TCs are apt to bind to proteins in biological samples (*1, 18, 19*), strong acid and acidic deproteinizing agents are suitable for extracting TCs from biological samples (*30, 73–75, 77, 78, 80, 83, 94*). However, TCs are decomposed to the anhydro-form under acidic conditions (< pH 2.0) and converted to their epimers by extended heating. Therefore, extraction has more recently been achieved using mild acidic solvent containing EDTA (0.1 M EDTA–McIlvaine buffer, pH 4.0) (*21, 24, 37, 38, 59–62, 85, 107–112*).

For cleanup procedures, Amberlite XAD-2 (*30, 73, 113*) and chelate extraction (*4, 6, 95, 113, 114*) have been employed, however, these methods require complicated procedures. To avoid potential problems during extraction and cleanup caused by the undesirable properties of TCs, pretreating the C18 cartridges with EDTA (*21, 37, 38, 59, 60, 67, 107, 109*) or silylation reagent (*61*), adding oxalic acid to the eluent (*24, 103, 112, 115*), and the matrix solid phase dispersion using C18 adsorbent with EDTA and oxalic acid (*116, 117*) have been tried. These techniques have allowed sufficient separations of TCs and coextractives in food samples and a wide range of detection limits to be obtained.

In this section, practical extraction and cleanup methods using C18 and weak cation exchange cartridges for LC and TLC with simple detection of TCs in foods including fish, animal tissues, egg, milk, and honey are introduced.

Fish, Animal Tissues, Eggs, and Milk

Liquid chromatography. TCs form chelate complexes with metal ions and are capable of irreversible adsorption on alkyl-bonded reversed-phase packing, so that they are not eluted from C18 cartridges with methanol, acetonitrile, or ethanol.

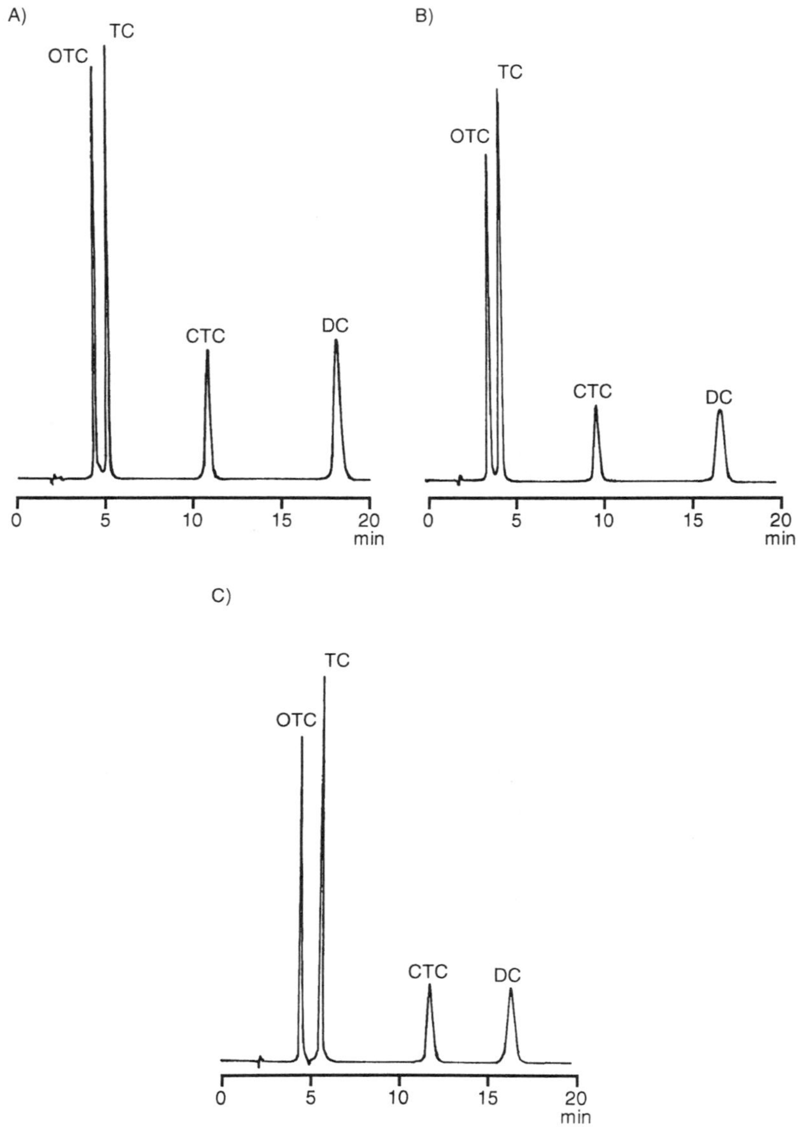

Figure 10 LC separation of TCs using a volatile mobile phase. (A) Column: TSK GEL C8-80TS (150 × 4.6 mm, i.d.). Mobile phase: Methanol-acetonitrile-0.01 M TFA (3:3:14). (B) Inertsil C8 (150 × 4.6 mm, i.d.). Mobile phase: Methanol-acetonitrile-0.01 M TFA (2:3:15). (C) Inertsil Ph (150 × 4.6 mm, i.d.). Mobile phase: Methanol-acetonitrile-0.01 M TFA (6:19:75). Flow rate: 1 mL/min. Detection: 350 nm.

The main area of investigation has been the elution of TCs from the cartridge. We previously suggested that only a mobile phase containing oxalic acid gives excellent separation of TCs by RP–TLC and RP–LC using conventional stationary phases. Because the separations using C18 cartridges and RP columns are based on the same principle, it was assumed that TCs could be eluted from a C18 cartridge with methanol containing oxalic acid, and the required concentration and volume of methanolic oxalic acid solution were examined. All TCs were completely recovered from C18 cartridges using 10 mL of 0.01 M methanolic oxalic acid solution.

The suitability of various solvents for applying TCs onto C18 cartridges was investigated. Using the solvents listed in Table 6, TCs were applied to Bond Elut C18 and Baker 10 C18 cartridges and then eluted with 10 mL of 0.01 M methanolic oxalic acid solution. Table 6 indicates that satisfactory results were obtained by using only mild acidic solvents; McIlvaine buffer (pH 4.0), 0.1 M EDTA–McIlvaine buffer (pH 4.0), 1% aqueous trichloroacetic acid solution, and 1% aqueous

Table 6 Solvent Comparison for Applying TCs on C18 Cartridges

	Recovery, %[a]							
	Bond Elut C18				Baker 10 C18			
Applying solvent	OTC	TC	CTC	DC	OTC	TC	CTC	DC
0.1M Na_2 EDTA-McIlvaine buffer (pH 4.0)	96	96	96	98	100	100	98	98
McIlvaine buffer (pH 4.0)	98	106	98	93	97	100	100	94
5% Metaphosphoric acid (aq.)	100	100	99	95	96	89	90	87
5% Trichloroacetic acid (aq.)	94	95	94	92	92	83	99	90
5% Phenol (aq.)	63	69	75	64	35	74	82	77
1N Hydrochloric acid (aq.)	63	56	60	55	58	44	62	53
5% Perchloric acid (aq.)	71	62	66	68	57	42	57	52

OTC: oxytetracycline; TC: tetracycline; CTC:chlortetracycline; DC: doxycycline.
[a]Recovery from 60 mL of solvent containing TCs each (5 μg). Mean of three replicates
SOURCE: Reproduced by permission of Elsevier Science. See reference (24).

metaphosphoric acid solution. These results are due to the instability of TCs and the packing materials of the cartridge in strong acidic solutions.

In order to apply these results to the analysis of TCs in animal liver, it was necessary to test whether or not TCs were retained on the C18 cartridge in the presence of liver coextractives. An extract prepared from 5 g of beef liver using 0.1 M EDTA–McIlvaine buffer (pH 4.0) containing TCs was applied on Bond Elut C18 and Baker 10 C18 cartridges. The TCs were then eluted with 10 mL of 0.01 M methanolic oxalic acid solution. The recoveries of all TCs from both C18 cartridges were almost 100%. To investigate the suitability of various extractants for analyzing TCs in liver, TCs from liver fortified at 1.0 ppm were extracted in three steps using 20, 20, and 10 mL of the extractants shown in Table 7. The results demonstrate that 0.1 M EDTA–McIlvaine buffer (pH 4.0) and McIlvaine buffer (pH 4.0) are most suitable. Because the presence of EDTA improves the recoveries of TCs from liver homogenate (19), 0.1 M EDTA–McIlvaine buffer (pH 4.0) was selected. In order to investigate the applicability of this method to various foods, recovery trials were conducted. Table 8 shows that good recoveries and coefficients of variation were obtained with no interfering peaks. Detection limits were 0.05 ppm for OTC and TC and 0.1 ppm for CTC and DC (Figure 11).

Therefore, the most suitable analytical method for residual TCs in food using LC is as follows: animal liver (5 g) is blended three times with 20, 20, and 10 mL of 0.1 M EDTA–McIlvaine buffer (pH 4.0) and centrifuged (24). The supernatant is applied on a C18 cartridge, the TCs are eluted with 10 mL of 0.01 M methanolic oxalic acid solution, and then are determined by LC (Scheme 1).

Thin-layer chromatography with spray reagents. Because TLC can be carried out in all laboratories without the need for special equipment, applying the above

Table 7 Comparison of Extractants

	Recovery, %[a]			
Extractant	OTC	TC	CTC	DC
0.1M Na_2EDTA-McIlvaine buffer (pH 4.0)	87	87	80	68
McIlvaine buffer (pH 4.0)	88	86	80	68
5% Trichloroacetic acid (aq.)	76	77	57	40
5% Metaphosphoric acid (aq.)	73	75	64	50

OTC: oxytetracycline; TC: tetracycline; CTC: chlortetracycline; DC: doxycycline
[a]Recovery of TCs from 5 g of homogenized beef liver fortified at the level of 1.0 ppm using three step extractions 20, 20, and 10 mL
Mean of three replicates

Table 8 Recovery of TCs from Fortified Foods

Sample	*n*	Recovery, % (C.V., %)[a]			
		OTC	TC	CTC	DC
Beef liver	4	88 (2.4)	88 (1.1)	80 (2.9)	68 (2.0)
Pork liver	4	91 (1.1)	88 (1.6)	89 (2.6)	69 (3.3)
Chicken liver	5	86 (1.9)	81 (1.7)	81 (2.0)	70 (0.7)
Beef kidney	5	91 (2.7)	86 (1.0)	92 (2.6)	81 (2.9)
Pork kidney	5	89 (1.2)	86 (2.3)	86 (2.1)	83 (2.2)
Beef muscle	5	92 (2.5)	88 (2.7)	93 (2.1)	85 (3.5)
Pork muscle	5	93 (1.4)	90 (1.6)	95 (1.4)	86 (2.2)
Chicken muscle	5	91 (0.6)	90 (0.7)	93 (1.3)	84 (0.4)
Milk	5	87 (0.8)	83 (0.9)	87 (0.6)	84 (1.4)
Egg	4	82 (0.7)	80 (0.8)	82 (1.6)	86 (2.1)
Yellow tail muscle	5	91 (3.5)	88 (2.5)	93 (2.9)	83 (0.9)
Eel muscle	5	94 (1.1)	89 (0.8)	88 (3.1)	85 (1.4)

OTC: oxytetracycline; TC: tetracycline; CTC: chlortetracycline; DC: doxycycline
[a]Recovery of TCs from 5 g of homogenized foods fortified at the level of 1.0 ppm, analyzed by LC
SOURCE: Reproduced by permission of Elsevier Science. See reference (24)

C18 cartridge cleanup with oxalic acid has been attempted. Although methanolic oxalic acid was successfully used as an eluent for TCs from a C18 cartridge, it was impossible to concentrate the eluate because of the decomposition of TCs caused by the oxalic acid during evaporation. For this reason, it could not be applied to TLC analysis. Because pretreating the TLC plates with EDTA is quite effective in controlling the undesirable properties of TCs, pretreating C18 cartridges with EDTA was tried. TCs were completely recovered from C18 cartridges pretreated with EDTA by various organic solvents such as methanol, acetonitrile, ethanol, and ethyl acetate. To find a suitable eluent from C18 cartridge for TLC analysis of TCs, the recoveries of TCs from beef liver were investigated using various solvents as eluents. After adding TCs to beef liver and extracting with 0.1 M EDTA–McIlvaine buffer (pH 4.0), the extract was applied to a Baker 10 C18 cartridge that had been pretreated with saturated aqueous EDTA. The cartridge was washed with 20 mL of water, TCs were eluted, and they were determined by TLC–UV densitometry. The two-step elution shown in Scheme 2 was quite effective. The best recoveries were obtained by eluting with 10 mL of ethyl acetate followed by 20 mL of methanol–ethyl acetate (5:95) and no interfering peaks

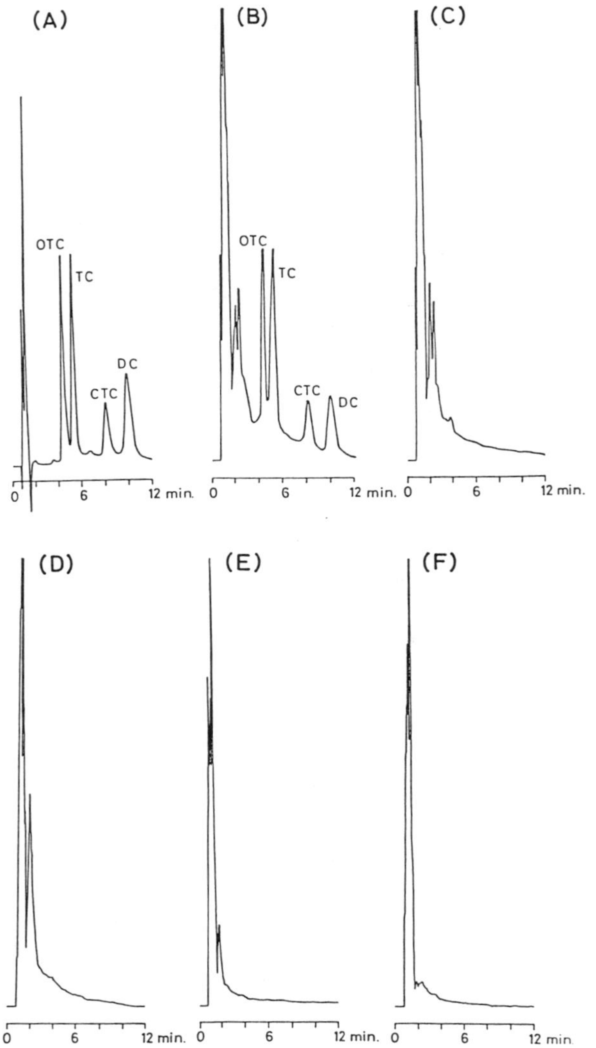

Figure 11 Typical high-performance liquid chromatograms of various food extracts. (A) Standard of TCs (50 ng) (B) Fortified beef liver extract (1.0 ppm). (C) Beef liver extract. (D) Beef kidney extract. (E) Eel muscle extract. (F) Milk extract. LC conditions: column, LiChrosorb RP-8 (10 μm, 250 × 4.0 mm, i.d.); mobile phase, methanol-acetonitrile-0.01 M oxalic acid (aq., pH 2.0) (1:1.5:2); detection, 350 nm.

SOURCE: **Reproduced by permission of Elsevier Science. See reference *(24)*.**

Sample (5 g)

blend three times using 20, 20, and 10 mL of 0.1 M Na_2EDTA-McIlvaine buffer (pH 4.0)

centrifuge 4000 r.p.m., 10 min

Supernatant

filtration

Filtrate

Baker 10 C18

wash with H_2O (20 mL)

elute with 0.01 M $(COOH)_2$-CH_3OH (10 mL)

Eluent

HPLC

Scheme 1 Analytical procedure for residual TCs in food using LC

appeared on densitometric profiles as seen in Figure 12. Overall recoveries are shown in Table 9.

For screening TCs in animal tissues, a combination of the Baker 10 C18 cleanup and TLC detection system with diazonium salt described above was attempted. However, because some interfering substances from the liver caused a color reaction with diazonium salt, it was impossible to discriminate between TCs below 0.3 ppm and the interfering substances on the TLC plate. Satisfactory detection limits for the TCs could not be obtained using the spray reagents.

Sample (5 g)

blend with 0.1 M Na_2EDTA-McIlvaine buffer (pH 4.0) three times (20, 20, and 10 mL)
centrifuge (4000 r.p.m., 5 min)

Supernatant

filtration

Filtrate

Baker 10 C18 (pretreated with sat. aq. Na_2EDTA)

wash with water (20 mL)
aspirate to dry the cartridge (5 min)
elute with ethyl acetate (10 mL) followed by 5 % methanol-ethyl acetate (20 mL)

Eluate

add isopropanol (5 mL)
evaporate under reduced pressure at 30°C

Residue

dissolve in methanol (0.1 mL)

RP-TLC

Scheme 2 Screening method for residual TCs in foods using TLC

TCs are converted into anhydro-TCs by acid treatment followed by heating, and iso-TCs by alkaline treatment (*14*). Further, the degradation products of TCs produce different colors from their parent compounds when reacted with diazonium salts, and the color intensity of the degradation products is frequently superior to that of the parent compounds (*57*). If these reactions were applied to the detection of TCs, a more sensitive and specific detection method should be possible. To generate the degradation products, hydrochloric acid and aqueous ammonia were used as spray reagents prior to the color development. After developing TCs on C8 TLC plates the TCs were converted to their degradation products by spraying with 2 M hydrochloric acid and 28% aqueous ammonia. The

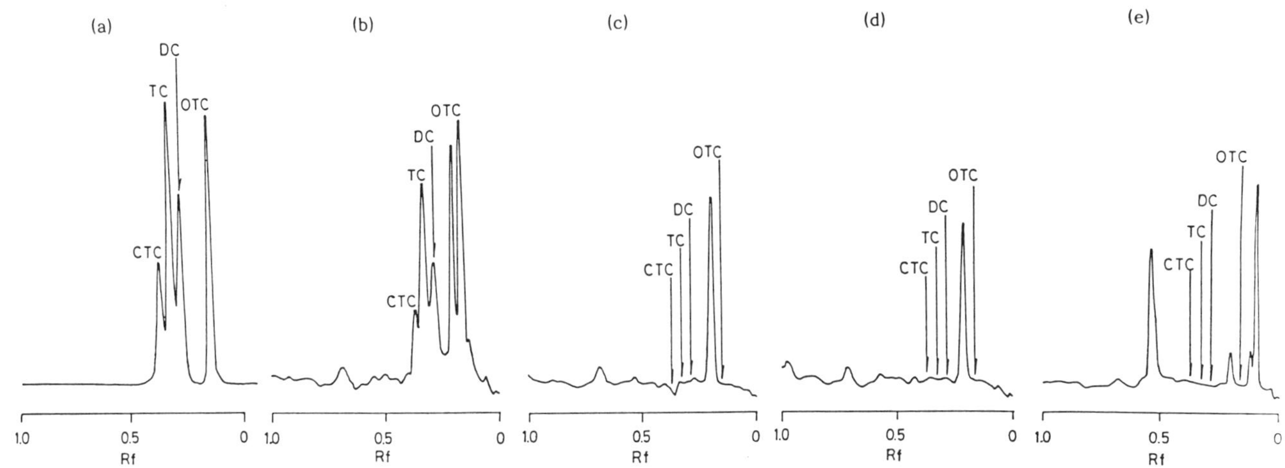

Figure 12 Typical UV-densitometric profiles of beef tissues
(A) Standard of TCs (0.5 μg). (B) Fortified beef liver (1.0 ppm). (C) Beef liver. (D) Beef kidney. (E) Beef muscle
TLC conditions: plate, C8-TLC; solvent system, methanol-acetonitrile-0.5 M oxalic acid (aq., pH 2.0) (1:1:4); detection, 350 nm.
SOURCE: Reproduced by permission of Elsevier Science. See reference (38).

Table 9 Recovery of TCs from Fortified Animal Tissues

		Recovery, % (C.V., %)[a]			
Sample	*n*	OTC	TC	CTC	DC
Beef liver	4	93 (1.6)	71 (3.5)	64 (3.7)	58 (1.0)
Beef kidney	4	92 (2.2)	76 (3.3)	67 (6.2)	71 (1.5)
Beef muscle	4	92 (4.6)	84 (2.5)	81 (3.3)	87 (3.6)
Pork liver	4	91 (5.0)	72 (4.9)	65 (8.9)	56 (1.9)
Pork kidney	4	85 (4.2)	79 (8.1)	67 (9.8)	72 (9.9)
Pork muscle	4	90 (4.4)	82 (6.7)	76 (7.1)	80 (9.0)

OTC: oxytetracycline; TC: tetracycline; CTC: chlortetracycline; DC: doxycycline
[a]Recovery of TCs from 5 g of homogenized animal tissues fortified at the level of 1.0 ppm, analysed by LC
SOURCE: Reproduced by permission of Elseview Science. See reference (*38*).

degradation products were detected by spraying with an aqueous solution of diazonium salt. More sensitive and specific colors of TCs were given with the following order of treatments: spraying with 2 M hydrochloric acid, heating at 150°C for 1 min, and spraying with 28% aqueous ammonia, 0.5% aqueous Fast Blue BB Salt and pyridine. The colors of OTC, TC, CTC, and DC were dark blue, deep green, dark red, and red, respectively, and their detection limits were 0.03 µg on a C8 TLC plate. The practical detection limits of TCs in beef liver, kidney, and muscle using a combination of the above C18 cartridge cleanup and this detection system were 0.1 ppm (38).

Therefore, this method combining cleanup on C18 cartridges pretreated with EDTA and detection with spray reagents after RP–TLC is recommended as a screening method for residual TCs in animal tissues (Scheme 2).

Honey

Liquid Chromatography

The above C18 cartridge cleanup procedures did not give satisfactory results for the residual analysis of TCs in honey. The chromatogram showed many peaks arising from substances in honey that interfered with the analysis of TCs. Using a C18 cartridge alone for both extraction and cleanup was not always sufficient for the residual analysis of TCs in honey. A more effective cleanup operation that can separate TCs from impurities is essential. We intended to develop a double cartridge system employing reversed-phase and ion-exchange cartridges for extraction and cleanup. The strategy was designed to meet the following conditions: (*a*) eluting TCs from C18 cartridge with ethyl acetate containing methanol, and (*b*) eluting TCs from COOH cartridge with the LC mobile phase.

Cation exchange cartridges were expected to be effective as cleanup cartridges because TCs contain a dimethylamino group. Baker 10 COOH and SO_3 were examined. First, the suitability of the mobile phase, methanol–acetonitrile–0.01 M aqueous oxalic acid solution (pH 3.0) (2:3:16) as the eluent for cleanup cartridges was investigated. Baker 10 SO_3 cartridges gave poor recoveries, but all TCs were completely eluted from Baker 10 COOH cartridges with the LC mobile phase.

As described above, a mixture of methanol and ethyl acetate was effective for eluting TCs in animal tissues from a Baker 10 C18 cartridge pretreated with saturated aqueous Na_2EDTA. The required concentration of methanol and the volume of the mixture were examined and all TCs in 25 g of honey were completely recovered from the C18 cartridge by 50 mL of 10% methanol–ethyl acetate.

On the basis of the preliminary studies, a precise analytical method for TCs in honey is proposed in Scheme 3. To evaluate its capability, the overall recoveries of TCs, coefficients of variation, and detection limits were evaluated and are seen in Table 10, and typical chromatograms are seen in Figure 13. These results indicate that the LC method, in combination with Baker 10 C18 and COOH cartridge cleanup, is applicable and recommended for the practical analysis of seven TCs in honey with good recovery, higher accuracy, and better detection limits (*37*, *118*).

Thin-Layer Chromatography with Spray Reagents

Good separation among MINO, CTC, DC, MTC, and OTC and among MINO, OTC, TC, DMCTC, and CTC were obtained in NP– and RP–TLC, respectively. Because their combination enabled us to identify the seven TCs, both TLC systems were used to detect TCs with spray reagents. The magnesium chloride-TEA method reported by Szab et al. (*43*) was chosen to visualize the TCs. The detection limits of TCs on both plates were investigated with the spray reagents (Tables 3 and 4). The formation of fluorescent spots under UV light (360 nm) after spraying the NP–TLC plate with 0.2 M magnesium chloride was the most sensitive (MINO, 0.03 µg; others, 0.01 µg). Although using this spray reagent followed by 10% methanolic TEA could not detect MINO on both plates, this method gives enough sensitivity for other TCs (0.01 µg). Because other combinations of TLC and spray reagents did not give satisfactory results, the following combinations were selected: 0.2 M aqueous magnesium chloride and NP–TLC for detecting MINO,

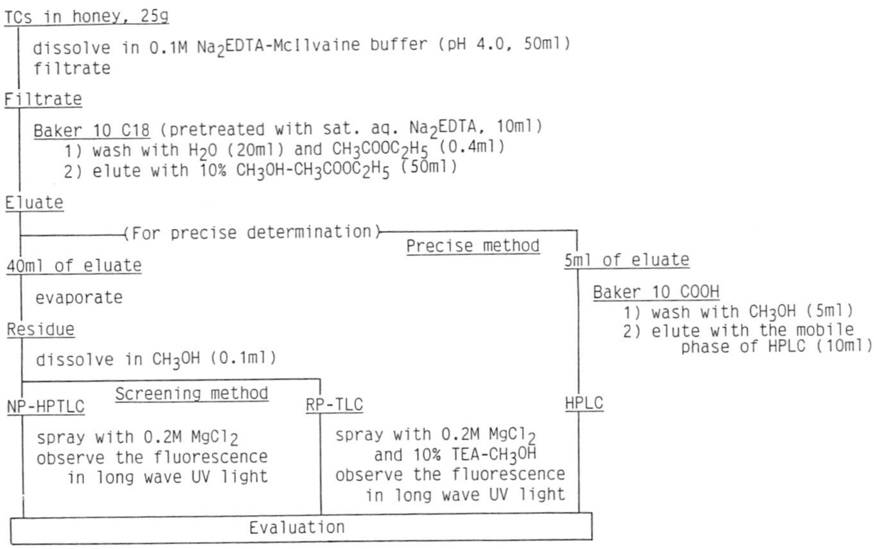

Scheme 3 Systematic analytical method for residual TCs in honey
SOURC E: Reproduced by permission of Elsevier Scienc. See reference *(37)*.

Table 10 Recovery of TCs from Fortified Honey

	MINO	OTC	TC	DMCTC	CTC	MTC	DC
Recovery (%)[a]	86	93	92	88	99	84	97
C.V. (%)	2.1	0.9	2.5	2.9	2.7	4.3	2.7
Detection limit (ppm)	0.02	0.02	0.02	0.02	0.05	0.05	0.05

MINO: minocycline; OTC: oxytetracycline; TC: tetracycline; DMCTC: demeclocycline; CTC: chlortetracycline; MTC: methacycline; DC: doxycycline

[a]Recovery of TCs from 25 g of honey fortified at the level of 1.0 ppm, analyzed by LC. Results of six replicates.

SOURCE: Reproduced by permission of Elsevier Science. See reference *(37)*.

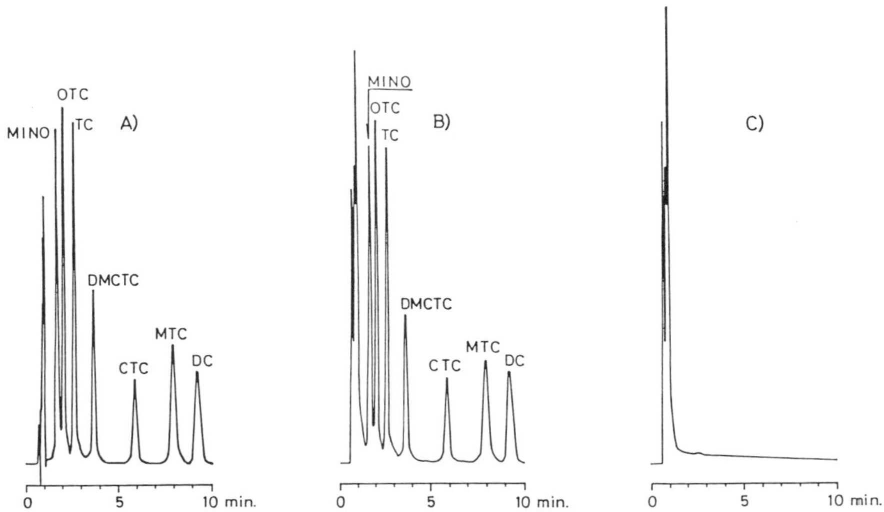

Figure 13 Typical high-performance liquid chromatograms of honey extracts. (A) Standard of TCs (10 ng), (B) fortified honey (1.0 ppm), (C) honey. LC conditions: column, Chemcosorb 3C8 (3 μm, 75 × 4.6 mm, i.d.); moblie phase, methanol-acetonitrile-0.01 M oxalic acid (aq., pH 3.0) (1:1.5:7); detection, 350nm.
SOURCE : **Reproduced by permission of Elsevier Science. See reference (*37*).**

CTC, DC, MTC, and OTC, and 0.2 M aqueous magnesium chloride followed by 10% methanolic TEA and RP–TLC for detecting OTC, TC, DMCTC, and CTC.

When TLC methods are applied to the analysis of TCs in honey, the final sample solution is concentrated because of the low sensitivity. However, the final sample solution in the double cartridge system (Baker 10 C18/COOH) contains oxalic acid for precise quantitative analysis so it cannot be concentrated. Further, direct elution from the C18 cartridge with 10% methanol–ethyl acetate without any washing of the cartridge gave many interfering spots from honey on the TLC plates. Therefore, washing the C18 cartridge with 0.4 mL of ethyl acetate before eluting the TCs allowed TCs at a level of 0.1 ppm in honey to be detected on the NP-TLC plate with 0.2 M aqueous magnesium chloride. Further, except for MINO, they could be detected at a level of 0.05 ppm on the RP–TLC plate with 0.2 M aqueous magnesium chloride followed by 10% methanolic TEA. Washing the C18 cartridge with 0.4 mL of ethyl acetate is essential to detect residual TCs in honey with good sensitivity. Thus, a simple screening method for residual TCs in honey has been established. Furthermore, as shown in Scheme 3, a simultaneous analytical system has been established by a combination of the above TLC and LC methods (*37*).

Mass Spectrometry

Any determination method can be subject to interferences, so when a drug residue is detected at trace levels in food, confirmation is usually required before regulatory action can be taken. Supporting evidence for a positive finding may be obtained by reinjecting the extract on a column with a different surface chemistry, reanalyzing with a second method, or characterizing the analyte by photodiode array detection. However, to confirm the presence of a residue, mass spectral evidence is unsurpassed because it provides direct structural characterization of the analyte. Any of these approaches may serve to identify a false positive or provide evidence consistent with the presence of a particular drug residue, but only mass spectral evidence has the potential to provide absolute confirmation.

The quality of mass spectral evidence can range from conclusive to merely supportive. In cases where an instrument is not pushed to its detection limits, it may be possible to confirm a residue by a full scan analysis and then compare the full spectrum of the analyte in the sample extract to that of a chemical standard. More commonly, only selected ions are monitored to detect lower levels. There is general agreement that the minimum criteria for confirmation include the monitoring of four ions (*132*, *133*), although only three ions may suffice if they include the molecular ion and two structurally significant fragments (*133*, *134*). The retention time and relative ion intensities found in the sample analysis must be comparable to those measured for a chemical standard.

There are a number of reports that cover a variety of techniques for applying mass spectrometry (MS) to the analysis of TC residues in food (*59*, *60*, *103*, *107*, *135–137*). For analytes such as the TCs, which lack the volatility needed for GC–MS, other chromatographic means have been used to separate and introduce the analyte into the ion source. Few analysts in a routine regulatory environment have access to the full range of modern MS instrumentation, so the choice of technique may be dictated by availability rather than relative merits. To allow analysts to make the best use of available instrumentation, the different MS techniques that have been used for TC analysis are described, recommendations are made, and speculations on future developments in this field are made.

Tandem Mass Spectrometry

The earliest report using MS for the analysis of a tetracycline residue in food appeared in 1985 when Traldi et al. (*135*) used the collision-activated decomposition mass-analyzed ion kinetic energy spectrometry (CAD MIKES) technique on a ZAB 2F mass spectrometer (VG Analytical, Manchester, UK) to quantitatively determine and confirm OTC residues in bovine milk and meat. Crude extracts spiked with an internal standard were introduced directly into the ion source and the CAD MIKE spectrum of the molecular ion showed 18 product ions. Quantitation was by single ion monitoring of the most abundant product ion, *m/z* 442 $[M\text{-}H_2O]^+$, and had a detection limit of 1 ppb. This detection limit is much lower than can be achieved by any of the other MS methods described here. When this study was published, the practical application of the technique was limited because the instrumentation was uncommon outside of research laboratories. However, as MS–MS technology becomes more affordable and available such methods will come into more routine use.

Thin-Layer Chromatography–Fast Atom Bombardment Mass Spectrometry

As described earlier, TCs can be separated on a C8 TLC plate using methanol–acetonitrile–0.5 M oxalic acid. The R_f for each parent TC in bovine muscle extract (cleaned up as in Scheme 2) reproducibly matches that for a chemical standard (*59*). Thus, for a suspect positive sample it is possible to locate the analyte spot on a developed plate by running a chemical standard in parallel and locating it with a UV lamp. Once the spot has been located and a matrix applied, the plate is loaded into a holder in the fast atom bombardment (FAB) source for MS analysis. The best matrix for TCs is thioglycerol, because it releases the TCs from the stationary phase and does not produce any interfering cluster ions.

If analyzed in this way, the sensitivity is very poor (5 µg/spot) because the spot on the plate is too diffuse and spreads further when the matrix is applied. To enhance the sensitivity, a simple condensation technique was developed to focus an analyte spot (*138*) that reduces the detection limit 50-fold to 0.1 µg. A rectangular area including the analyte spot is cut and a trapezoid is drawn around the spot. The stationary phase outside the trapezoid is scraped off and a small volume of methanol is deposited along the base of the trapezoid. In 30 s the analyte has been reconcentrated into a 0.5 × 2 mm zone at the top of the trapezoid.

Standards and samples were analysed on a JEOL (JEOL, Tokyo, Japan) JMS-AX505W magnetic sector mass spectrometer with a TLC–FAB ion source (*59*). The FAB gun was operated with xenon gas at 5 kV and an acceleration voltage of 3 kV. The spectra of OTC, TC, and CTC display intense $[M+H]^+$ and two fragment ions which have been identified as $[M+H–NH_3]^+$ and $[M+H–NH_3–H_2O]^+$ (*139*). The spectrum of DC shows only $[M+H]^+$ and $[M+H–NH_3]^+$. In bovine liver and kidney at concentrations of 0.1 ppm, $[M+H]^+$ is seen, and in some cases one fragment ion is also detectable. In bovine muscle at 0.1 ppm the protonated molecular ion and all the fragment ions are seen, except in the case of TC which shows only $[M+H]^+$. At higher levels in incurred samples, all of the ions can be seen and easily distinguished from the background. This technique has also been applied to milk analysis (*60*), in which, as Figure 14 shows, $[M+H]^+$ could be seen at 50 ppb for all TCs as well as two fragment ions for TC and one fragment ion for DC.

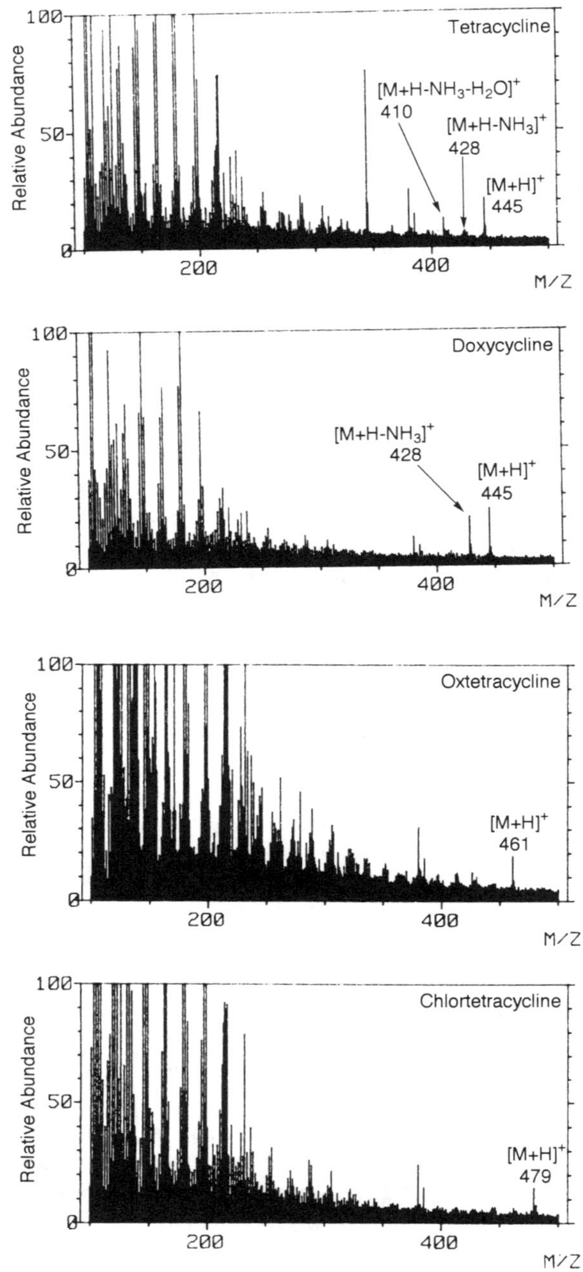

Figure 14 TLC–FAB mass spectra of the TCs in milk at 0.05 ppm

Liquid Chromatography–Mass Spectrometry

Compared to TLC–FAB-MS, LC–MS has the added convenience of on-line operation, the ability to record selected ion chromatograms, and can perform background corrections. There are a number of LC–MS interfaces available, but only three have been applied to TC analysis in food. In a comparative study, Voyksner et al. (*140*) examined thermospray (TSP) and particle beam (PB) interfaces for a number of different drugs. TSP was the recommended interface for β-lactams, cephapirin and cytidine based on the generation of several structurally significant fragment ions. However, for the TCs very limited fragmentation was seen, whereas in PB LC–MS with methane negative ion chemical ionization (NCI), a minimum of four relatively intense fragment ions for each TC were seen. This suggested that TSP could provide supporting evidence for a suspect positive finding, whereas a method for TCs using PB had the potential to meet the minimum criteria for a confirmatory procedure. The recent application of a third technique, frit FAB LC–MS (also known as continuous-flow FAB) to the analysis of TCs will also be described.

Particle beam liquid chromatography–mass spectrometry. PB LC–MS in NCI mode has been successfully employed by Kijak et al. (*103*) to confirm residues of

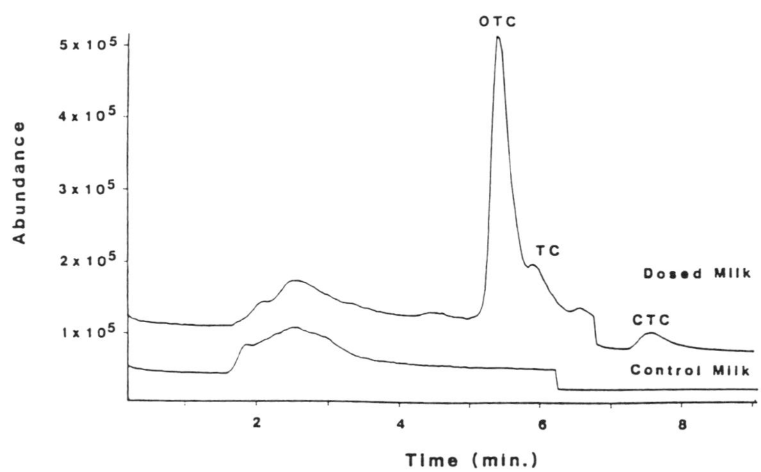

Figure 15 NCI PB LC–MS total ion chromatograms (four ions for each TC) of control milk and dosed milk containing 0.13 ppn CTC, 0.13 ppm TC and 0.46 ppm OTC. Column: Novapak C18 (150 × 3.9 mm i.d.). Mobile phase: Acetonitrile-0.05 M oxalic acid-methanol (50:30:20)
SOURCE: Reproduced by permission of John Wiley & Sons Limite. See reference (*103*).

OTC, TC, and CTC in milk as seen in Figure 15. In this study, milk filtrates were prepared by ultrafiltration and cleaned up on C18 cartridges with conditions similar to those in Scheme 1. Final elution was with 10 mL of 0.01 M oxalic acid in methanol, and the extract was concentrated to 0.2 mL under a stream of nitrogen at 45°C. LC separation was carried out on a Novapak C18 column (150 × 3.9 mm i.d.) with acetonitrile–0.05 M oxalic acid–methanol (50:30:20). On a 5988A single quadrupole mass spectrometer (Hewlett Packard Co., Palo Alto, CA, USA) with PB interface operating in NCI mode, four ions were monitored for each TC as shown in Table 11. Each of the three TCs studied could be confirmed using four ions at 100 ppb. Some intra- and inter-day variability was reported in the abundances of some ions relative to base peak. The interface also clogged due to the oxalic acid from the concentrated extract, so a switching valve was installed post-column to divert the flow to waste during the first 3 min of each run. Although

Table 11 Fragment Ions in the LC–MS Spectra of the Tetracyclines

		m/z (%RI)		
Compound	Mol. wt.	FAB[a]	PB NCI[b]	+TSP
TC	444	445 (100)	444 (100)	445 (100)
		428 (20)	426 (78)	427 (34)
		410 (64)	401 (10)	
			383 (15)	
CTC	478	479 (100)	478 (62)	479 (100)
		462 (41)	460 (25)	435 (15)
		444 (44)	442 (100)	
			434 (6)	
OTC	460	461 (100)	460 (100)	461 (100)
		444 (18)	442 (35)	418 (38)
		426 (45)	416 (3)	
			399 (15)	
DC	444	445 (78)	ND[c]	ND
		428 (100)		

RI: relative intensity; FAB: fast atom bombardment; PB NCI: particle beam negative ion chemical ionization; TSP: thermospray; TC: tetracycline; CTC: chlortetracycline; OTC: oxytetracycline; DC: doxycycline; ND: not determined

[a]From reference 139.

[b]Average %RI values calculated from data in reference 103.

[c]Not done.

this eliminated the clogging, it seemed to exacerbate the variation in relative ion abundances. It may be possible to improve the performance of this method by using the C18 cartridge cleanup described in Scheme 2 to eliminate the nonvolatile constituent and thereby dispense with the switching valve and allow the final extract to concentrate without risking degradation in the presence of acid. Oxalic acid can also be eliminated from the mobile phase while improving resolution by using one of the LC systems described in Figure 10, or below for TSP LC–MS.

Thermospray liquid chromatography–mass spectrometry. Both positive ion TSP LC–MS and negative ion TSP LC–MS with discharge ionization have been used for detecting TCs in fermentation broths (*104*). Preliminary reports of positive ion TSP LC–MS used to detect TCs in food (*113*, *136*, *137*) and a report on detecting doxycycline in urine and serum (*141*) are available. As shown in Table 11, the TSP spectra of TCs display an intense $[M+H]^+$, with one relatively intense fragment ion, resulting presumably from the loss of water ($[M+H–H_2O]^+$, *m/z* 427 for TC) or cyanic acid ($[M+H\text{-}HOCN]^+$, *m/z* 418 for OTC), or from the cleavage of the carboxyamide group from ring A ($[M+H\text{-}CONH_2]^+$, *m/z* 435 for CTC).

The TSP spectra and chromatograms were generated on a VG Trio 2 single quadrupole mass spectrometer with VG Plasmaspray/Thermospray interface operated in positive ion Thermospray mode (VG Analytical Ltd., Manchester, UK). To reduce tissue coextractives interference we ran a gradient on an Inertsil C8 (GL Science, Tokyo, Japan) column (5 μm, 4.0 × 250 mm) from 100% 0.1 M ammonium formate (pH 3, adjusted with trifluoroacetic acid) to acetonitrile–0.1 M ammonium formate (pH 3) (30:70) over 10 min, then held for 10 min. The relatively high buffer concentration assists ion production, which occurs by evaporation from highly charged microdroplets in the spray cone. With a flow rate of 1 mL/min, as shown in Figure 16, OTC, TC, and CTC were fully resolved and eluted at 11.8, 13.1, and 17.3 min, respectively. Vaporizer and source temperatures were 185 and 190°C respectively and repeller voltage 130 V. Optimal temperatures in any TSP application vary with flow rate, mobile phase composition, source and interface condition, and interface design. As deposits slowly accumulate and the condition of the vaporizer degrades, the optimum vaporizer temperature will change. If the optimum temperature drifts only a few degrees from the operating temperature, it will significantly affect ion production. Keeping instrumental parameters at optimum is a regular and frequent element of TSP analysis.

Under optimal conditions with single ion monitoring of $[M+H]^+$, the detection limits in bovine and porcine muscle and kidney prepared as in Scheme 1, are 0.1 ppm for OTC and TC and 0.3 ppm for CTC. For TC, the fragment ion at 427 *m/z* can be seen at 0.2 ppm, as shown in Figure 17. The fragment ions for OTC and CTC can be detected at higher levels.

Frit fast atom bombardment liquid chromatography–mass spectrometry. We have recently applied frit FAB LC–MS to residual TC analysis in honey (*107*).

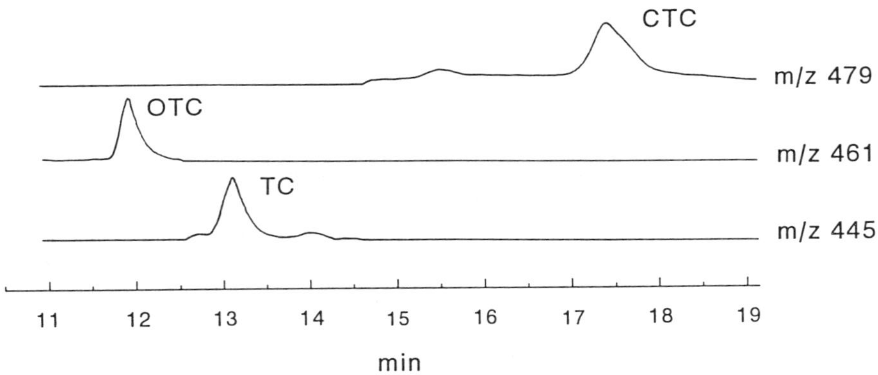

Figure 16 Single ion monitoring of CTC, OTC and TC standards by TSP LC–MS. Column: Inertsil C8 (5 μm, 250 × 4.0 mm i.d.); mobile phase: acetonitrile-0.1 M ammonium formate (pH 3), gradient from (0–100) to (30–70) over 10 min and held for 10 min

This interface allows LC to be linked to a FAB source for on-line continuous FAB ionization. Unlike the PB and TSP interfaces, which are able to handle a normal flow rate of 1 mL/min, frit FAB LC–MS has an optimum flow rate range 1–5 μL/min. A narrow bore column to reduce flow rate and narrow the analyte bonds must be used with a post column splitter *(142)* to divert the bulk of the LC effluent. An Inertsil Ph (5μm, 250 × 2.1 mm) column operating at a 0.1 mL/min

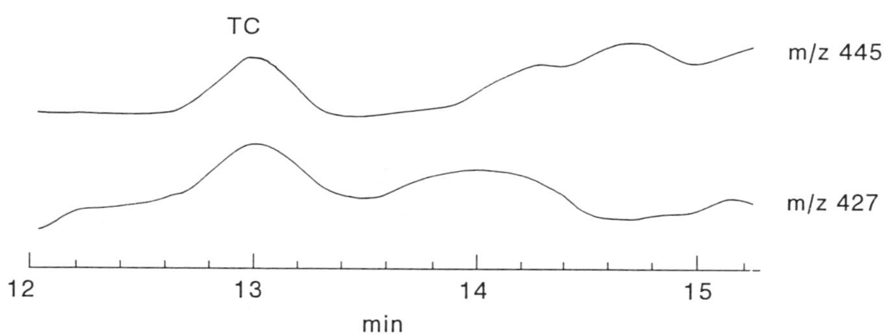

Figure 17 TSP LC—MS selected ion chromatograms of incurred bovine muscle containing 0.19 pp TC. Column: Inertsil C8 (5 μm, 250 × 4.0 mm); Mobile phase: Acetonitrile-0.1 M ammonium formate (pH 3)

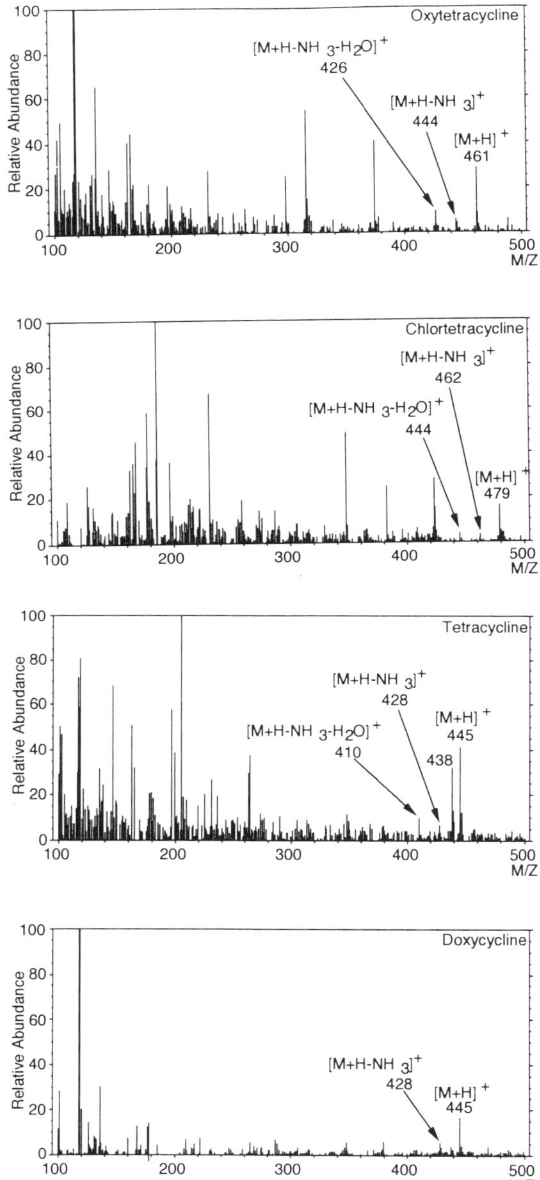

Figure 18 Frit FAB LC/mass spectra of the TCs in honey at 0.2 ppm

flow rate and a 1:20 split ratio gives a 5 μL/min flow rate, separates the TCs, and gives good sensitivity with a mobile phase of methanol–acetonitrile–0.005 M aqueous trifluoroacetic acid (2:2:11) containing 1% thioglycerol.

Analyses were conducted on a JMS-AX505W (JEOL Ltd., Tokyo, Japan) magnetic sector mass spectrometer with an ion source temperature of 60°C and a neutral xenon beam as the primary beam FAB ionization. Acceleration voltages for the primary and secondary beams were 3 and 5 kV, respectively. Scanning range was from 100 *m/z* to 1500 *m/z* at a cycle time of 6.5 s. The characteristic ions shown in Table 11 are the same as described earlier for TLC–FABMS: $[M+H]^+$, $[M+H–NH_3]^+$, and $[M+H–NH_3–H_2O]^+$.

Honey samples were cleaned up for the analysis using a C18 cartridge pretreated with ethylenediaminetetraacetic acid (EDTA) (*21*). The final extract was dissolved in water rather than methanol or mobile phase because it resulted in better LC resolution. When honey is spiked at 0.2 ppm with CTC, DC, OTC, and TC the full spectrum recorded at the top of each peak clearly shows the protonated molecular ion and both fragment ions, as seen in Figure 18. A comparison with blank honey shows that there are no interfering peaks. The detection limit for protonated molecular ion of CTC, DC, OTC, and TC is estimated to be 0.1 ppm. The mass chromatograms of an incurred honey sample containing 0.24 ppm CTC are shown in Figure 19.

The success of a confirmation method routinely used in a regulatory laboratory is measured against five performance criteria: specificity, sensitivity, reproducibility, ruggedness, and practicality. In their current state of development none of

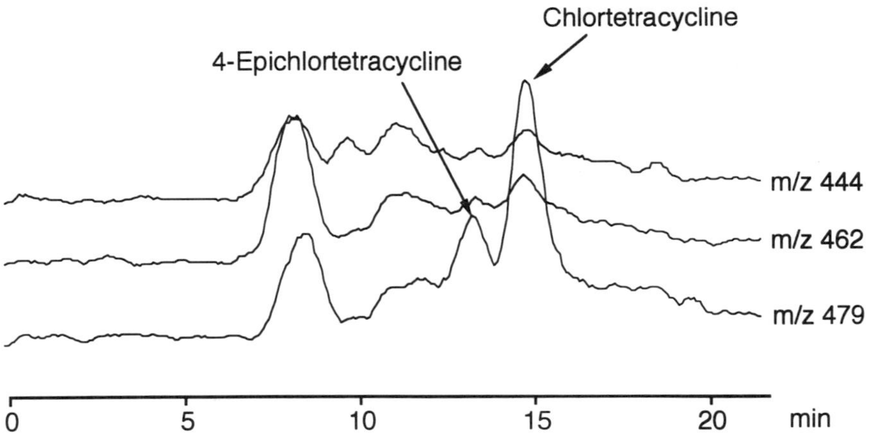

Figure 19 Frit FAB LC/mass chromatograms for an incurred honey sample containing 0.24 ppm CTC

the methods described fully satisfy these measures at the 0.1 ppm level, but overall the recommended methods for confirming TC residues in foods are frit FAB LC–MS and PB NCI LC–MS, both of which are capable of meeting the minimum criteria for a confirmatory method. TLC–FABMS can also meet these criteria, but as an off-line technique it is less convenient to use. TSP LC–MS has similar sensitivity to the other techniques, but lacks sufficient fragmentation.

Incremental improvement can be expected in all of these methods as the techniques are refined, but substantial advances in the development of confirmatory methods for TC residues can also be anticipated in the future as alternate interface technologies are exploited and tandem MS becomes more widely available and affordable. Practical considerations in a routine laboratory demand that more universal interface applications be implemented. Atmospheric pressure interfaces (*143*) may be able to satisfy this need and simultaneously address sensitivity, reproducibility, and ruggedness. As one example, the ionspray interface has been compared to TSP and PB PCI for a number of drugs including oxytetracycline, and gave approximately a 100-fold improvement in sensitivity (*144*). The suitability of this interface in combination with tandem MS for analyzing antibiotic residues has been demonstrated by quantitating and confirming erythromycin A and metabolite residues in salmon tissue at low ppb levels (*145*).

Conclusion

OTC, TC, CTC, and DC are commonly applied to food-producing animals including honeybees at therapeutic levels to treat disease and at subtherapeutic levels for promoting growth and enhancing feed conversion. Their cost effectiveness guarantees continued widespread use, so regulatory agencies must be prepared to monitor for the presence of residues of TCs in the food supply. Many chemical methods for analyzing TCs in food have been reported over the last 20 years, but were plagued by persistent problems due to protein binding, chelation with metal ion contaminants, and binding to silanol groups in the stationary phase. Only since 1985, with the discovery that the use of oxalic acid and EDTA at each step (extraction, cleanup, and separation) could block chelation and binding, have simple and effective methods available. Now, with the commercial availability of RP packing materials made from high purity silica gel, excellent chromatographic results can be achieved without using oxalic acid, EDTA, or other nonvolatile components in the mobile phase. This was an important advance for the development of LC–MS confirmation methods, because nonvolatile constituents in the mobile phase can clog the interface and contaminate the ion source.

In this chapter, we have described recent developments in the chemical analysis of TCs and recommended practical methods for analyzing TC residues in food. The recommended methods range from simple TLC screening procedures, which require no instrumentation or specialized equipment, to LC determinative methods on isocratic systems with UV detection, and finally to confirmation methods with specialized mass spectrometers and interface technologies. Methodology is now available to allow regulatory laboratories to tailor their testing programs to satisfy their needs while also considering available equipment and expertise.

The refinement of all of these methods will continue as new insights and new analytical technology are applied to the problem. Unlike the screening and determination of TC residues, which have reached a mature stage of development after two decades of research, confirmation by LC–MS is still in its infancy. Major advances can be anticipated in this area, since interface technology is evolving rapidly and promising interfaces have yet to be applied to the confirmation of TC residues in food. From the perspective of a regulatory laboratory, monitoring a wide range of nonvolatile drug residues in food and striving to meet generally accepted confirmation criteria, it is apparent that the technology that will find long-term success will consist of a universal, sensitive, and rugged soft-ionization technique coupled to a tandem MS.

Acknowledgments

We wish to thank Professor M. Suzuki, Faculty of Pharmacy, Meijo University, for his useful discussion and suggestions.

References

1. B. M. Dugger, *Ann. N. Y. Acad. Sci.*, **51**: 177, 1948.
2. *Fed. Regist.*, Part III, **39**: (22), 3935 (1974).
3. *British Pharmacopoeia 1988*, vol. I (London: Her Majesty's Stationary Office, 1988), 562.
4. *European Pharmacopoeia, Part II*, 2nd ed. (France: Maisonneuve S. A., 1993), 211.
5. G. F. Flimter, *J. Am. Med. Assoc.*, **184**: 111, 1963.
6. M. Fluop and A. Drapkin, *N. Engl. J. Med.*, **19**: 986, 1965.
7. L. I. Ehrlich and H. S. Stein, *Pediatrics*, **31**: 339, 1963.
8. S. R. Sulkowski and J. R. Haserick, *J. Am. Med. Assoc.*, **189**: 178, 1964.
9. L. A. Mitscher, *The Chemistry of the Tetracycline Antibiotics.*, 1st ed. (New York, NY: Marcel Dekker, 1978), 52.
10. L. A. Mitscher, *The Chemistry of the Tetracycline Antibiotics,* 1st ed. (New York, NY: Marcel Dekker, 1978), 91.
11. L. A. Mitscher, *The Chemistry of the Tetracycline Antibiotics,* 1st ed. (New York, NY: Marcel Dekker, 1978), 110.

12. K. W. Kohn, *Anal. Chem.*, **33**: 862, 1961.
13. H. Piger and C. Schlatter, *Analyst*, **101**: 808, 1976.
14. L. A. Mitscher, *The Chemistry of the Tetracycline Antibiotics,* 1st ed., (New York, NY: Marcel Dekker, 1978), 123.
15. M. Ishidate and T. Sakaguchi, *Chem. Pharm. Bull.*, **6**: 1, 1958.
16. J. Y. Lee and W. Evrett., *J. Am. Chem. Soc.*, **103**: 5221, 1981.
17. S. Ishino, T. Sakaguchi, I. Morimoto, and T. Okutsu, *Yakugaku Zasshi*, **101**: 118, 1981.
18. H. Zia and J. C. Price, *J. Pharm. Sci.*, **65**: 226, 1976.
19. C. Bogert and A. M. Kroon, *J. Pharm. Sci.*, **70**: 186, 1981.
20. H. Oka, K. Uno, K. -I. Harada, Y. Kaneyama, and M. Suzuki, *J. Chromatogr.*, **260**: 457, 1983.
21. H. Oka, K. Uno, K. -I. Harada, and M. Suzuki, *Yakugaku Zasshi*, **103**: 531, 1983.
22. H. Oka, K. Uno, K. -I. Harada, and M. Suzuki, *J. Chromatogr.*, **284**: 227, 1984.
23. H. Oka, K. Uno, K. -I. Harada, K. Yasaka, and M. Suzuki, *J. Chromatogr.*, **298**: 435, 1984.
24. H. Oka, H. Matsumoto, K. Uno, K. -I. Harada, S. Kadowaki, and M. Suzuki, *J. Chromatogr.*, **325**: 265, 1985.
25. P. P. Ascione, J. B. Zagar, and G. P. Chrekian, *J. Pharm. Sci.*, **56**: 1393, 1967.
26. A. A. Fernandez, V. T. Noceda, and E. S. Carrera, *J. Pharm. Sci.*, **58**: 443, 1969.
27. N. D. Gyanchandani, I. J. McGilveray, and D. W. Hughes, *J. Pharm. Sci.*, **59**: 224, 1970.
28. I. C. Dijkhuis and M. R. Brommet, *J. Pharm. Sci.*, **59**: 558, 1970.
29. C. Radecka and W. L. Wilson, *J. Chromatogr.*, **57**: 297, 1971.
30. J. J. Ryan and J. A. Dupont, *J. Assoc. Off. Anal. Chem.*, **57**: 828, 1974.
31. R. Bossuyt, R. V. Renterghem, and G. Waes, *J. Chromatogr.*, **124**: 37, 1976.
32. C. D. Gatsonis and C. A. Ageloudis, *Pharmazie*, **37**: 649, 1982.
33. G. J. Kapadia and G. S. Rao, *J. Pharm. Sci.*, **53**: 223, 1964.
34. Y. Nishida, E. Tsuchiya, and S. Toyoshima, *Yakugaku Zasshi*, **87**: 516, 1967.
35. U. Valcavi, A. Brandt, G. B. Corsi, F. Minoja, and G. Pascucci, *J. Antibiotics*, **34**: 34, 1981.
36. H. Oka, Y. Ikai, N. Kawamura, K. Uno, M. Yamada, K. -I. Harada, M. Uchiyama, H. Asukabe, and M. Suzuki, *J. Chromatogr.*, **393**: 285, 1987.
37. H. Oka, Y. Ikai, N. Kawamura, K. Uno, M. Yamada, K. -I. Harada, and M. Suzuki, *J. Chromatogr.*, **400**: 253, 1987.
38. Y. Ikai, H. Oka, N. Kawamura, M. Yamada, K. I. Harada, and M. Suzuki, *J. Chromatogr.*, **411**: 313, 1987.

39. W. Naidong, S. Geelen, E. Roets, and J. Hoogmartens, *J. Pharm. Biomed. Anal.*, **8**: 891, 1990.
40. W. Naidong, C. Hauglustaine, E. Roets, and J. Hoogmartens, *J. Planar. Chromatogr.*, **4**: 63, 1991.
41. D. L. Simmons, H. S. L. Woo, C. M. Koorengevel, and P. Seers, *J. Pharm. Sci.*, **55**: 1313, 1966.
42. E. Ragazzi and G. Veronese, *J. Chromatogr.*, **132**: 105, 1977.
43. A. Szabó, M. K. Nagy, and E. Tömörkény, *J. Chromatogr.*, **151**: 256, 1978.
44. A. I. H. Omer, E. A. G. Kariem, and R. B. Salama, *J. Chromatogr.*, **205**: 486, 1981.
45. N. Muhammad and J. A. Bodnar, *J. Pharm. Sci.*, **69**: 928, 1980.
46. K. Dihuidi, E. Roets, J. Hoogmartens, and H. Vanderhaeghe, *J. Chromatogr.*, **246**: 350, 1982.
47. A. P. De Leenheer and H. J. C. F. Nelis, *J. Pharm. Sci.*, **68**: 999, 1979.
48. H. J. C. F. Nelis and A. P. De Leenheer, *J. Chromatogr.*, **195**: 35, 1980.
49. J. H. Knox and J. Jurand, *J. Chromatogr.*, **186**: 763, 1979.
50. H. R. Howell, L. L. Rhodig, and A. D. Sigler, *J. Assoc. Off. Anal. Chem.*, **67**: 572, 1984.
51. L. A. Mitscher, *The Chemistry of the Tetracycline Antibiotics,* 1st ed. (New York, NY: Marcel Dekker, 1978), 58.
52. D. Hughes, W. L. Wilson, A. G. Butterfield, and J. Pound, *J. Pharm. Pharmacol.*, **26**: 79, 1974.
53. R. Böcker and C. J. Estler, *Arzneim. Forsch.*, **29**: 1690, 1979.
54. R. Böcker and C. J. Estler, *Arzneim. Forsch.*, **29**: 1693, 1979.
55. E. Stahl, *Thin-Layer Chromatography,* 2nd ed. (Heidelberg, Germany: Springer-Verlag, 1969), 573.
56. H. Oka and M. Suzuki, *J. Chromatogr.*, **314**: 303, 1984.
57. H. Oka, K. Uno, K. -I. Harada, M. Hayashi, and M. Suzuki, *J. Chromatogr.*, **295**: 129, 1983.
58. J. S. Kang and S. Ebel, *J. Planar. Chromatogr.*, **2**: 434, 1989.
59. H. Oka, Y. Ikai, J. Hayakawa, K. Masuda, K. -I. Harada, M. Suzuki, V. Martz, and J. D. MacNeil, *J. Agric. Food Chem.*, **41**: 410, 1993.
60. H. Oka, Y. Ikai, J. Hayakawa, K. Masuda, K. -I. Harada, and M. Suzuki, *J. AOAC Int.*, **77**: 891, 1994.
61. E. J. Mulders and D. V. De Lagemaat, *J. Pharm. Biomed. Anal.*, **7**: 1829, 1989.
62. J. R. Walsh, L. V. Walker, and J. J. Webber, *J. Chromatogr.*, **596**: 211, 1992.
63. J. P. Sharma, E. G. Perkins, and R. F. Bevill, *J. Chromatogr.*, **134**: 441, 1977.
64. J. P. Sharma and R. F. Bevill, *J. Chromatogr.*, **166**: 213, 1978.
65. G. D. Mack, and R. B. Ashworth, *J. Chromatogr. Sci.*, **16**: 93, 1978.

66. S. Eksborg, H. Ehrsson, and U. Lönroth, *J. Chromatogr.*, **185**: 583, 1979.
67. R. Böcker, *J. Chromatogr.*, **187**: 439, 1980.
68. S. Eksborg, *J. Chromatogr.*, **208**: 78, 1981.
69. S. Eksborg and B. Ekqvist, *J. Chromatogr.*, **209**: 161, 1981.
70. U. Jürgens, *Z. Lebensm. Unters. Forsch.*, **173**: 356, 1981.
71. J. Hermansson, *J. Chromatogr.*, **232**: 385, 1982.
72. J. Hermansson and M. Anderson, *J. Pharm. Sci.*, **71**: 222, 1982.
73. Y. Onji, M. Uno, and K. Tanigawa, *J. Assoc. Off. Anal. Chem.*, **67**: 1135, 1984.
74. H. Terada, M. Asanoma, and Y. Sakabe, *Eisei Kagaku*, **30**: 138, 1984.
75. Y. Hoshino, M. Horie, N. Nose, and H. Iwasaki, *J. Food Hyg. Soc. Jpn.*, **25**: 430, 1984.
76. K. Dihuidi, M. J. Kucharski, E. Roets, J. Hoogmartens, and H. Vanderhaeghe, *J. Chromatogr.*, **325**: 413, 1985.
77. M. Horie, Y. Hoshino, N. Nose, H. Iwasaki, and H. Nakazawa, *Eisei Kagaku*, **31**: 371, 1985.
78. G. S. Chappell and J. E. Houglum, *J. Assoc. Off. Anal. Chem.*, **69**: 28, 1986.
79. K. Tyczkowska and A. L. Aronson, *J. Assoc. Off. Anal. Chem.*, **69**: 760, 1986.
80. W. A. Moats, *J. Chromatogr.*, **358**: 253, 1986.
81. K. Wolfs, E. Roets, J. Hoogmartens, and H. Vanderhaeghe, *J. Chromatogr.*, **358**: 444, 1986.
82. N. H. Khan, E. Roets, J. Hoogmartens, and H. Vanderhaeghe, *J. Chromatogr.*, **405**: 229, 1987.
83. I. Nordlander, H. Johnsson, and B. Osterdahl, *Food Addit. Contam.*, **4**: 291, 1987.
84. A. Rogstad, V. Hormazabal, and M. Yndestad, *J. Liq. Chromatogr.*, **11**: 2337, 1988.
85. M. E. Sheridan and G. S. Clarke, *J. Chromatogr.*, **434**: 253, 1988.
86. K. Eivindvik and K. E. Rasmussen, *J. Liq. Chromatogr.*, **12**: 3061, 1989.
87. V . L. Ellerbroek, *Arch. Lebensmittelhyg.*, **40**: 7, 1989.
88. N. H. Khan, P. Wera, E. Roets, and J. Hoogmartens, *J. Liq. Chromatogr.*, **13**: 1351, 1990.
89. D. J. Fletouris, J. E. Psomas, and N. A. Botsoglou, *J. Agric. Food Chem.*, **38**: 1913, 1990.
90. T. Agasøster and K. E. Rasmussen, *J. Chromatogr.*, **570**: 99, 1991.
91. W. Naidong, K. Verresen, E. Roets, and J. Hoogmartens, *J. Chromatogr.*, **586**: 61, 1991.
92. R. G. Aoyama, K. M. McErlane, and H. Erber, *J. Chromatogr.*, **588**: 181, 1991.
93. A. G. Butterfield, D. W. Hughes, W. L. Wilson, and N. J. Pound, *J. Pharm. Sci.*, **64**: 316, 1975.

94. R. F. Lindauer, D. M. Cohen, and K. P. Munnelly, *Anal. Chem.*, **48**: 1731, 1976.
95. J. P. Sharma, G. D. Koritz, E. G. Perkins, and R. F. Bevill, *J. Pharm. Sci.*, **66**; 1319, 1977.
96. B. G. Charles, J. J. Cole, and P. J. Ravenscroft, *J. Chromatogr.*, **222**: 152, 1981.
97. A. Aszalos, C. Hanneke, M. J. Hayden, and J. Crawford, *Chromatographia,* **15**: 367, 1982.
98. W. N. Barnes, A. Ray, and l. J. Bates, *J. Chromatogr.*, **347**: 173, 1985.
99. A. Aszalos, *Chromatographia,* **20**: 313, 1985.
100. P. Sporns, S. Kwan, and L. A. Roth, *J. Food Prot.*, **49**: 383, 1986.
101. T. Galeano-Diaz, A. Guiberteau-Cabanillas, and F. Salinas, *Anal. Lett.*, **23**: 607, 1990.
102. K. Nakaya, Y. Kobayashi, and N. Tanahashi, *J. Food Hyg. Soc. Jpn.*, **32**: 43, 1991.
103. P. J. Kijak, M. G. Leadbetter, M. H. Thomas, and E. A. Thompson, *Biol. Mass Spectrom.*, **20**: 789, 1991.
104. G. B. Kenion, G. T. Carter, J. Ashraf, M. M. Siegel, and D. B. Boeder, *Liquid Chromatography Mass/Spectrometry*, ACS Symposium Series 420 (Washington, D.C.: American Chemical Society, 1990), 140.
105. M. Ohhira, F. Ohmura, and T. Hanai, *J. Liq. Chromatogr.*, **12**: 1065, 1989.
106. M. Horie, K. Saito, T. Nose, M. Tera, and H. Nakazawa, *J. Liq. Chromatogr.*, **16**: 1463, 1993.
107. H. Oka, Y. Ikai, J. Hayakawa, K. -I. Harada, M. Suzuki, R. Himei, M. Horie, H. Nakazawa, and J. D. MacNeil, *J. Agric. Food Chem.* **42**: 2215, 1994.
108. E. Martinez and W. Shimoda, *J. Assoc. Off. Anal. Chem.*, **71**: 477, 1988.
109. H. Björklund, *J. Chromatogr.*, **432**: 381, 1988.
110. M. Thomas, *J. Assoc. Off. Anal. Chem.*, **72**: 564, 1989.
111. E. Martinez and W. Shimoda, *J. Assoc. Off. Anal. Chem.*, **72**: 848, 1989.
112. H. Oka, Y. Ikai, N. Kawamura, and J. Hayakawa, *J. Assoc. Off. Anal. Chem.*, **74**: 894, 1991.
113. W. H. H. Farrington, J. Tarbin, J. Bygrave, and G. Shearer, *Food Addit. Contam.*, **8**: 55, 1991.
114. M. C. Carson, *J. AOAC Int.*, **76**: 329, 1993.
115. G. J. Reimer and L. M. Young, *J. Assoc. Off. Anal. Chem.*, **72**: 813, 1989.
116. A. R. Long, L. C. Hsieh, M. S. Malbrough, C. R. Short, and S. A. Barker, *J. Assoc. Off. Anal. Chem.*, **73**: 379, 1990.
117. A. R. Long, L. C. Hsieh, M. S. Malbrough, C. R. Short, and S. A. Barker, *J. Assoc. Off. Anal. Chem.*, **73**: 864, 1990.
118. H. Oka, Y. Ikai, N. Kawamura, K. Uno, M. Yamada, K. -I. Harada, M. Uchiyama, H. Asukabe, Y. Mori, and M. Suzuki, *J. Chromatogr.*, **389**: 417, 1987.

119. D. Mourot, B. Delepine, J. Boisseau, and G. Gayot, *J. Chromatogr.*, **190**: 486, 1980.
120. J. Y. C. Hon and L. R. Murry, *J. Liq. Chromatogr.*, **5**: 1973, 1982.
121. J. Torel, J. Cillard, P. Cillard, and M. Vie, *J. Chromatogr.*, **330**: 425, 1985.
122. H. J. E. M. Reeuwijk and U. R. Tjaden, *J. Chromatogr.*, **353**: 339, 1986.
123. W. J. Blanchflower, R. J. McCracken, and D. A. Rice, *Analyst*, **114**: 421, 1989.
124. B. Iversen, A. Aanesrud, and A. K. Kolstad, *J. Chromatogr.*, **493**: 217, 1989.
125. N. H. Khan, E. Roets, J. Hoogmartens, and H. Vanderhaeghe, *J. Pharm Biomed. Anal.*, **7**: 339, 1989.
126. Y. Yoneda, N. Furushiro, and R. Andoh, *J. Food Hyg. Soc. Jpn.*, **30**; 42, 1989.
127. W. Naidong, E. Roets, and J. Hoogmartens, *Chromatographia,* **30**: 105, 1990.
128. R. Ueno, K. Uno, and T. Aoki, *J. Chromatogr.*, **573**: 333, 1992.
129. G. Carignan, K. Carrier, and S. Sved, *J. AOAC Int.*, **76**: 325, 1993.
130. C. R. White, W. A. Moats, and K. L. Kotula, *J. AOAC Int.*, **76**: 549, 1993.
131. C. R. White, W. A. Moats, and K. L. Kotula, *J. Liq. Chromatogr.*, **16**: 2873, 1993.
132. W. G. de Ruig, R. W. Stephany, and G. Dijkstra, *J. Assoc. Off. Anal. Chem.*, **72**: 487, 1989.
133. T. Cairns and E. G. Siegmund, *Crit. Rev. Food Sci.*, **30**: 397, 1991.
134. J. Sphon, *J. Assoc. Off. Anal. Chem.*, **61**: 1247, 1978.
135. P. Traldi, S. Daolio, B. Pelli, R. Maffei Facino, and M. Carini, *Biomed. Mass Spectrom.*, **12**: 493, 1985.
136. W. H. H. Farrington, J. Tarbin, and J. Bygrave, *Proceedings EuroResidue Conference*, ed. N. Haagsma, A. Ruiter, and P. B. Czedik-Eysenberg (Utrecht, The Netherlands: University of Utrecht, 1990) 179.
137. S. Porter, *Proceedings EuroResidue II Conference*, ed. N. Haagsma, A. Ruiter, and P.B. Czedik-Eysenberg (Utrecht, The Netherlands: University of Utrecht, 1993), 533.
138. H. Oka, Y. Ikai, F. Kondo, N. Kawamura, J. Hayakawa, K. -I. Harada, K. Masuda, and M. Suzuki, *Rapid Commun. Mass Spectrom.*, **6**: 89, 1992.
139. K. -I. Harada, K. Masuda, M. Suzuki, H. Oka, Y. Ikai, and J. Hayakawa, *Org. Mass Spectrom.*, **28**: 1512: 1993.
140. R. D. Voyksner, C. S. Smith, and P. C. Knox, *Biomed. Environ. Mass Spectrom.*, **19**: 523, 1990.
141. J. L. Riond, K. Tyczkowska, and J. E. Riviere, *Am. J. Vet. Res.*, **50**: 1329, 1989.
142. Y. Ikai, H. Oka, J. Hayakawa, K. -I. Harada, and M. Suzuki, *Mass Spectrosc.*, **39**: 199, 1991.

143. E. C. Huang, T. Wachs, J. J. Conboy, and J. D. Henion, *Anal. Chem.*, **62**: 713A, 1990.
144. S. Pleasance, R. Bailey, P. Blay, D. North, and M. A. Quilliam, *Proceedings 39th ASMS Conference on Mass Spectrometry and Allied Topics*, Nashville, TN, May, 19-24, 1991, 1362.
145. S. Pleasance, J. Kelly, M. D. Leblanc, M. A. Quilliam, R. K. Boyd, D. D. Kitts, K. McErlane, M. R. Bailey, and D. North, *Biol. Mass Spectrom.*, **21**: 675, 1992.
146. R. B. Ashworth, *J. Assoc. Off. Anal. Chem.*, **68**: 1013, 1985.
147. R. H. Böcker, R. Peter, G. Machbert, and W. Bauer, *J. Chromatogr.*, **568**: 363, 1991.
148. J. X. Duggan, *J. Liq. Chromatogr.*, **14**: 2499, 1991.
149. H. J. Nelis, J. Vandenbranden, A. de Kruif, F. Belpaire, and A. P. De Leenheer, *J. Pharm. Sci.*, **81**: 1216, 1992.
150. W. Naidong, K. Vermeulen, I. Quintens, E. Roets, and J. Hoogmartens, *Chromatographia*, **33**: 560, 1992.
151. I. Gastearena, M. C. Dio-Viéiez, E. Segura, M. M. Goñi, M. J. Renedo, and D. Fos, *Chromatographia*, **35**: 524, 1993.
152. J. M. Degroodt, B. Wyhowski de Bukanski, and S. Srebrnik, *J. Liq. Chromatogr.*, **16**: 3515, 1993.

Chapter 11

Chemical Analysis of Peptide Antibiotics

Yoshitomo Ikai
Aichi Prefectural Institute of Public Health, Japan

Contents

Introduction

Many kinds of peptide antibiotics have been used as veterinary medicines and feed additives. Bacitracin, enramycin, virginiamycin, and thiopeptin are most commonly used for growth promotion, improvement of feed efficiency, and control of animal diseases. These antibiotics are usually administered to animals in feeds at low concentration, and produce residues in tissues at very low or undetectable levels. Bioassays, including bioautography, have been primarily used for analyzing residues of these antibiotics in livestock products, while chemical assays have rarely been applied.

Chemical assays are rarely used in analyzing livestock products for several reasons. Procedures required for the cleanup of residues of peptide antibiotics in biological matrices are usually complicated and time-consuming, because many interfering substances that are structurally similar to these analytes are contained in the sample matrices. Detection limits (0.05 ppm or lower) required to meet regulatory requirements for most antibiotics cannot be achieved by the usual UV detection system for liquid chromatographic (LC) analysis, because these compounds lack suitable chromophore groups in their structures. Furthermore, most of these peptide antibiotics consist of two or more active components, the content of each component is not constant, and the isolated pure components are either not commercially available or are very expensive. A major reason chemical methods are rarely applied to analyzing samples for peptide antibiotics is that it is difficult to obtain working standards that are essential in chemical analysis.

However, chemical methods, in comparison with bioassays, can generally offer advantages both in speed and in the potential for identification and quantification of target analytes. In addition, recent developments in chemical analysis techniques and apparatus provide opportunities for determining peptide antibiotics at the same detection levels that are obtained by bioassay. If both bioassay and chemical assay techniques are jointly applied to the residue analysis of antibiotics in foods, more reliable results may be expected. To achieve this objective, reliable, sensitive chemical analysis methods are required for all peptide antibiotics used for treatment of food-producing animals.

Recent developments in chemical analysis techniques are described, including the preparation of working standards of the above peptide antibiotics, and a practical approach to the analysis of their residues in agricultural products is introduced.

Chemical and Biological Properties

Bacitracin

Bacitracin (BC) is a linear-ring peptide antibiotic produced by *Bacillus subtilis* and *B. licheniformis* that was discovered by Johnson et al. in 1945 (*1*). Commercial formulations of BC contain a complex of many components classified into BC-A, -B, -C, -D, -E, -F, and -G, with the main component being BC-A. BC-F is a degradation product of BC-A that shows nephrotoxicity. Although BC-A, -B, and -C are microbiologically active, BC-F exhibits no activity. The structures of BC-A and -F are known, but other minor components have never been elucidated. Figure 1 shows the structures, chemical formulas, and molecular weights of BC-A and -F. BC is active against Gram-positive organisms, and its manganese and zinc salts are mainly used as feed additives for growth promotion. BC is soluble in polar solvents, such as water and alcohols, but is practically insoluble in acetone, ether, and chloroform. It easily forms complexes with heavy metals such as copper, and it is stable in dry form and acid solutions, but not in alkaline solutions. In particular,

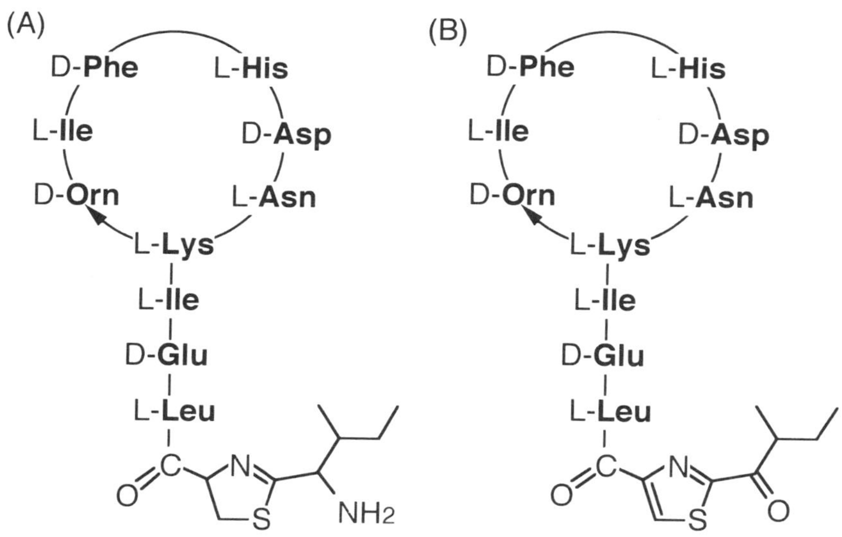

Figure 1 Structures of bacitracin components
(A) Bacitracin A, $C_{66}H_{103}N_{17}O_{16}S$, MW: 1421
(B) Bacitracin F, $C_{66}H_{100}N_{16}O_{17}S$, MW: 1418

BC-A easily decomposes to BC-F in weak alkaline solution. BC-A has a weak absorption maximum near 250 nm, but this is not usable for residue analysis because of a lack of sensitivity.

Colistin

Colistin (CL) is a linear-ring peptide antibiotic produced by *B. polimyxa* var. *colistinus* that was found by Koyama et al. (*2*) in 1950. It is a member of the polymyxin family of antibiotics, and has as its main components CL-A and -B. Their structures, chemical formulas and molecular weights are shown in Figure 2. The ratio between CL-A and -B is not constant, and small amounts of similar components are present as minor constituents. Among the peptide antibiotics used as feed additives, only CL, which is usually administered as the sulfate, has activity against Gram-negative organisms. CL sulfate is soluble in water, slightly soluble in methanol, and practically insoluble in acetone and ether. It is stable both in dry form and in water solution. CL components do not have any specific fluorophore

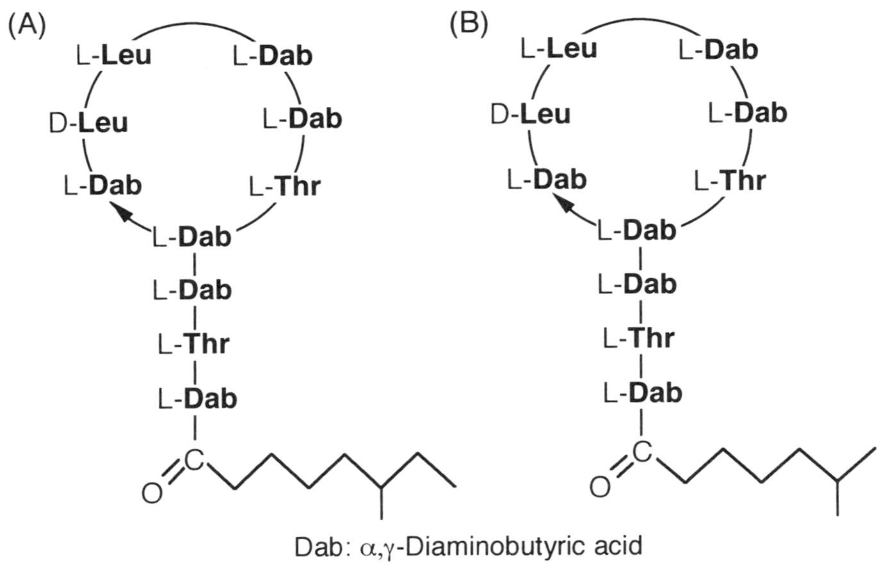

Figure 2 Structures of colistin components
(A) Colistin A, $C_{53}H_{100}N_{16}O_{13}$, MW: 1168
(B) Colistin B, $C_{52}H_{98}N_{16}O_{13}$, MW: 1154

and UV chromophore, so detection by high-performance liquid chromatography (LC) at residue levels of interest is difficult without including a suitable derivatization step in an analytical method.

Virginiamycin

Virginiamycin (VGM) is a mixture of peptolides (macrocyclic lactones containing a peptide part) produced by *Streptomyces virginiae* that was isolated by Somer et al. (*3*) in 1955. From the strain of *S. virginiae,* the following two types of components with antimicrobial activity are isolated: VGM-M_1 and -M_2, called factor M, and VGM-S_1, -S_2, -S_3, -S_4, and -S_5, called factor S. These two types of components exhibit a natural synergism. Commercially available VGM is prepared by mixing the isolated components of VGM-M_1 and -S_1 at a ratio of 75 and 5%, respectively. Their structures, chemical formulas, and molecular weights are shown in Figure 3. VGM is primarily effective against Gram-positive bacteria, and is used as a feed additive to promote growth and to prevent infection. It is soluble in alcohols, acetone, chloroform and benzene, but is practically insoluble in water and acidic water. Although it is stable in dry form and in neutral and acidic solution, it is deactivated rapidly in alkaline solution above pH 9.5. Both components of VGM have fluorophores and the fluorescence of VGM-S_1 is intense enough to provide adequate detection sensitivity. While that is not the case for VGM-M_1, it does have a strong UV adsorption near the wavelength of 220 nm and may be used instead.

Enramycin

Enramycin (ER) is a linear-ring peptide antibiotic produced by *Streptmyces fungicidicus* that was isolated by Higashiide et al. (*4*) in 1967. ER consists of two major components named ER-A and -B, and their structures are shown in Figure 4 together with their chemical formulas and molecular weights. It is active against Gram-positive and Acid-fast bacteria, and its monohydrochloride is usually administered to animals as a feed additive. ER hydrochloride is soluble in diluted hydrochloric acid, DMF, and pyridine, slightly soluble in water and methanol, and practically insoluble in ethanol, acetone, ether, and chloroform. Because ER is an amphoteric substance, the solubility in water depends on pH. It is highly stable in dry form and in aqueous solution with a pH of 3.5–7.5, but it is deactivated outside this pH range. ER has an absorption maximum near 230 nm, but the intensity is not entirely sufficient for residue analysis.

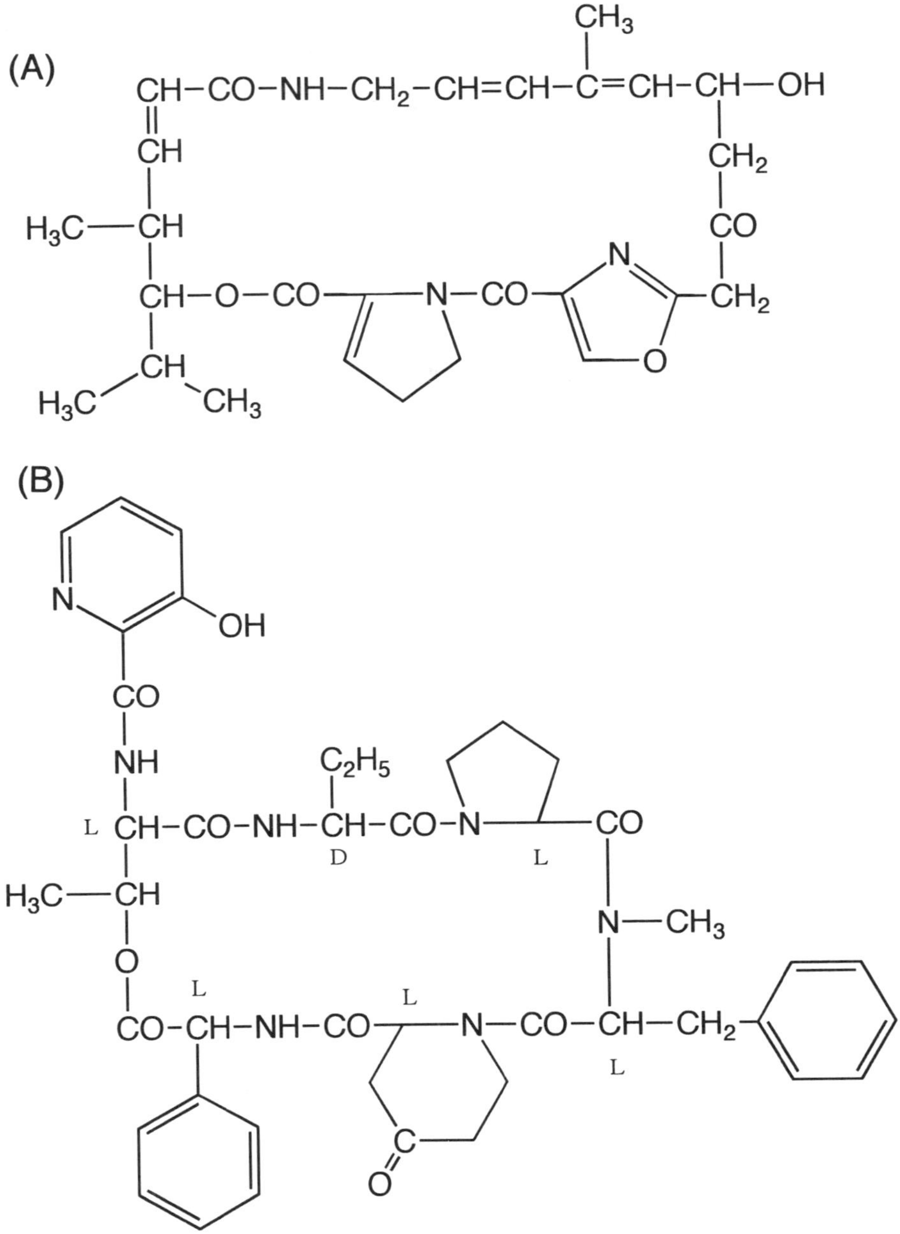

Figure 3 Structures of virginiamycin components
(A) Virginiamycin M_1, $C_{28}H_{35}N_3O_7$, MW: 525
(B) Virginiamycin S_1, $C_{43}H_{49}N_7O_{10}$, MW: 823

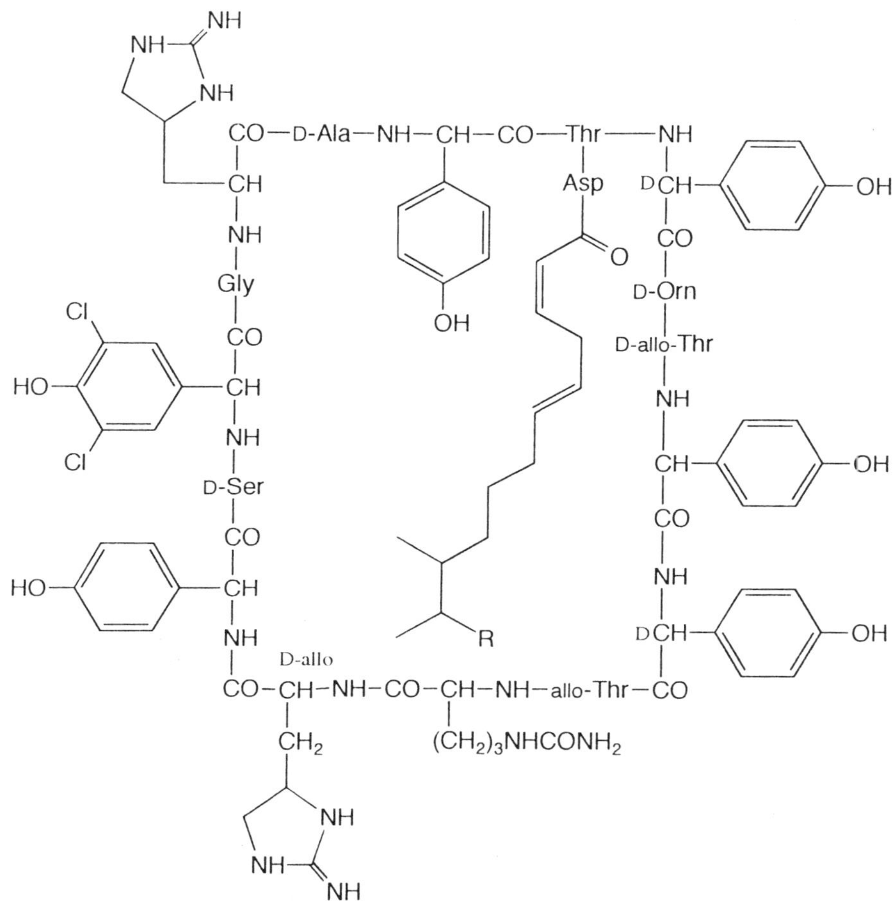

Figure 4 Structures of enramycin components
Enramycin A, R=CH_3, $C_{107}H_{138}Cl_2N_{26}O_{31}$, MW: 2352
Enramycin B, R=C_2H_5, $C_{108}H_{140}Cl_2N_{26}O_{31}$, MW: 2366

Thiopeptin

Thiopeptin (TPT) is a mixture of TPT-A_1, -A_2, -A_3, -A_4 and -B produced by *Nocardia tateyamenis*. The majority of TPT is TPT-B, and the TPT components are characterized by the presence of sulfur in their structures. Figure 5 shows the structure, chemical formula, and molecular weight of TPT-B. TPT is active against Gram-positive bacteria and is used only as a feed additive. It is soluble in chloroform, pyridine, dioxane, and DMF, it is slightly soluble in alcohols, acetone and ethyl acetate, and is practically insoluble in hexane, ether, and water. It is stable

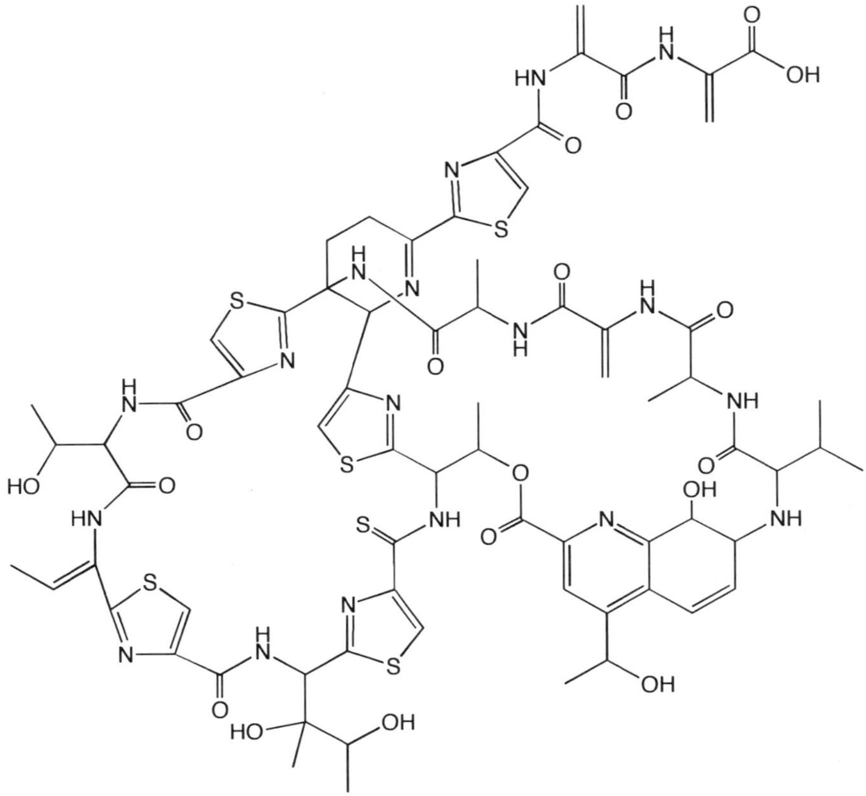

Figure 5 Structure of thiopeptin B
$C_{72}H_{90}N_{18}O_{22}S_6$, MW: 1750

dry and in aqueous solution from pH 4–6, but it is deactivated outside this pH range.

Separation

Because peptide antibiotics are nonvolatile and thermally unstable, thin-layer chromatography (TLC) and LC are more suitable separation techniques for

Table 1 TLC Conditions for the Separation of Peptide Antibiotics

	TLC plate	Solvent system	Rf	Detection	Ref.
BC	Silica gel	MeOH	0.23	bioassay	(*5*)
	Silica gel	*n*-BuOH-MeOH (1:1)	0.2	bioassay	(*5*)
	Silica gel	chloroform-MeOH (1:1)	0.15	bioassay	(*5*)
	Kieselgel G	*n*-BuOH-AcOH-water-pyridine (30:22:38:6)	0.61	ninhydrin	(*6*)
	Kieselgel G	*n*-BuOH-water-pyridine-AcOH-EtOH (60:10:6:15:5)	0.66	ninhydrin	(*6*)
	Silica gel 60	EtOH-Water-25% NH_4OH aq. soln. (8:1:1)	0.45	bioassay	(*7*)
	Silica gel TLC	chloroform-MeOH-0.75% AcOH aq. soln. (30:20:4)	0.30–0.60	ninhydrin	(*10*)
	RP-8	MeOH-0.1 M K_2SO_4 aq. soln. (pH 2.0) (7:3) containing 1.0% TEA	0.18–0.62	ninhydrin	(*10*)
CL	Kieselgel G	Acetone-water-AcOH-2 N NH_4OH aq. soln. (15:5:1:2)	0.95	ninhydrin	(*11, 12*)
	Silica gel 60 H silanized	Acetone-0.1 N HCl aq. soln. (25:75) containing 1% NaCl	0.21–0.32	ninhydrin	(*13*)
	RP–18	Acetone-2 M NaCl aq. soln. (1:1)	0.46–0.58	ninhydrin, fluorescamine	(*14*)

MeOH: methanol; BuOH: butanol; AcOH: acetic acid; EtOH: ethanol; TEA: triethylamine

analyzing these compounds. Tables 1 and 2 show the reported separation conditions using TLC and LC, respectively. The peptide antibiotics contain many components, so analysts should consider the minor and major components when optimizing the separation conditions. If all components are not successfully resolved, an accurate determination of the main components cannot be expected and examining the purity of working standards is difficult. Therefore, separation techniques for separating each component of the peptide antibiotics that can be recommended will be presented.

Thin-Layer Chromatography

Bacitracin. Some researchers have tried to separate BC components using silica gel (*5–7*), ion exchange resin (*8*), and reversed-phase (RP)-type silica gel (*9*) plates. However, separating only the main component, BC-A, was discussed, and there have been no reports of separating minor components. In addition, some

Table 2 LC Conditions for the Separation of Peptide Antibiotics

	Column	Mobile phase	Detection (nm)	Ref.
BC	μ-Bondapak C18 (300 × 4.6 mm i.d.)	MeOH-MeCN-0.01M phosphate buf. (pH 4.5), gradient, 1.0 mL/min	UV (254)	(*16*)
	Supel Cosil LC-8 (5μm, 150 × 3 mm i.d.)	7.1% MeOH, 23.6% MeCN in phosphate-EDTA buf., adjusted pH 6.8, 2.0 mL/min	UV (214)	(*17*)
	Chromatorex Ph (5μm, 250 × 4.0 mm i.d.)	MeCN-0.01 M TFA aq. soln., gradient, 1.0 mL/min	UV (234)	(*18*)
	Capcell Pak C18 (5μm, 150 × 4.6 mm i.d.)	MeOH-0.04M Na_2HPO_4 aq. soln. (6:4), 1.3 mL/min	UV (234)	(*20*)
	YMC basic (C8) (250 × 4.6 mm i.d.)	MeOH-0.05 M phosphate buf. (pH 6.5), gradient, 1.0 mL/min	UV (215)	(*19*)
CL	μ-Bondapak C18 (300 × 4.6 mm i.d.)	MeOH-MeCN-0.02 M phosphate buf. (pH 2.0), gradient, 1.0 mL/min	moving-wire-FID	(*16*)
	Nucleosil 5C18 (200 × 4 mm i.d.)	22.5% MeCN in 0.005 M tertrate buf. (pH 3.0) containing 0.005 M 1-butanesulfonate and 0.05 M Na_2SO_4, 1.0 mL/min	UV (220)	(*21*)
	Nucleosil 5C18 (200 × 0.5 mm i.d.)	MeOH-0.2 M KCl/HCl buf. (pH 2.0) (1:1), 0.01 mL/min	UV (210)	(22)
	Ultrasphere ion-pair (5 μm, 250 × 4.6 mm i.d.)	MeCN-water (23:77) containing 0.1 M phosphate buf. (pH 3.0), 1.0 mL/min	UV (185)	(23)
	Nucleosil 5C18 (150 × 4.6 mm i.d.)	22% MeCN, 0.023 M H_3PO_4, 0.01 M AcOH and 0.05 M Na_2SO_4 aq. soln. adjusted pH 2.5 by TEA, 0.9 mL/min	UV (220)	(*24*)
	Inertsil Ph (5μm, 250 × 4.0 mm i.d.)	MeCN-0.01 M TFA aq. soln. (24:76), 1.0 mL/min	UV (220)	(*14*)

Table 2 LC Conditions for the Separation of Peptide Antibiotics—*continued*

	Column	Mobile phase	Detection (nm)	Ref.
VGM	Hypersil ODS (5μm, 250 × 4.6 mm i.d.)	MeOH-MeCN-0.001 M H_3PO_4 aq. soln. (pH 2.12) (44:10:46), 1.0 mL/min	UV (254), FL (EX 295, Em >389)	(*25*)
	TSKgel ODS-120T (5μm, 150 × 4.6 mm i.d.)	MeOH-MeCN-0.015 M NaH_2PO_4 aq. soln.-THF (43:22:34:1), 0.5 mL/min	FL (EX 311, Em 427)	(*27*)
	Supercosil LC-18-DB (5μm, 150 × 4.6 mm i.d.)	MeCN-0.01 M $NH_4H_2PO_4$ aq. soln., gradient, 1.0 mL/min	UV (254)	(*30*)
	Kaseisorb LC-ODS-300-5 (250 × 4.6 mm i.d.)	MeCN-water (35:65), 0.7 mL/min	UV (235)	(*31*)
ER	Nucleosil 5C18 (150 × 4.6 mm i.d.)	MeCN-0.05 M NaH_2PO_4 aq. soln. (35:65), 0.5 mL/min	UV (230)	(*26*)

BC: bacitracin; MeOH: methanol, MeCN: acetonitrile; EDTA: ethylenediaminetetraacetic acid; TFA: trifluoroacetic acid; CL: colistin; AcOH: acetic acid; TEA: triethylamine; VGM: virginiamycin; THF: tetrahydrofrane; ER: enramycin.

of the conditions reported are difficult to apply in an actual analysis because of poor resolution. To provide the best possible separation of BC components using TLC, the following separation conditions were proposed (*10*). In these conditions, two different types of TLC plates, silica gel high-performance thin-layer chromatography (TLC) and RP C8-modified silica gel high-performance TLC plates were used in a complementary fashion. Chloroform–methanol–0.75% aqueous acetic acid (30:20:4) was used as a solvent system for the silica gel high-performance TLC plate, and methanol–0.1 M aqueous potassium sulfate, pH 2.0 (7:3), containing 1.0% of triethylamine (TEA), was used as a solvent system for the C8 TLC plate. In the solvent system for silica gel TLC, separation depended on the concentration of acetic acid. Because the shapes of the spots on the C8 plate were improved by increasing the concentration of potassium sulfate and TEA in the solvent system, these two solutes are considered to be important modifiers for separating BC on the RP TLC plate. As seen in Figure 6, eight and ten spots were observed by ninhydrin detection on the silica gel TLC and C8 TLC plates. These separation conditions were considered to be usable not only for residue analysis but also for examining the purity of isolated components.

Colistin. A number of reports have been published on the separation of colistin (*5, 9, 11–13*). Among these, only Thomas et al. (*13*) separated CL into three spots using a silanized silica gel plate with acetone–0.1 *N* hydrochloric acid (25:75) containing 1% sodium chloride as the solvent system. However, while the TLC plate used is not popular now, a similar or improved separation of CL may be expected by using conventional RP-type C8 and C18 TLC plates. As a result of optimization using several organic solvents (methanol, ethanol, acetonitrile, and acetone) and aqueous solutions of salts (sodium chloride and sodium sulfate) in the solvent system, a satisfactory separation of CL was obtained on C18 TLC plates

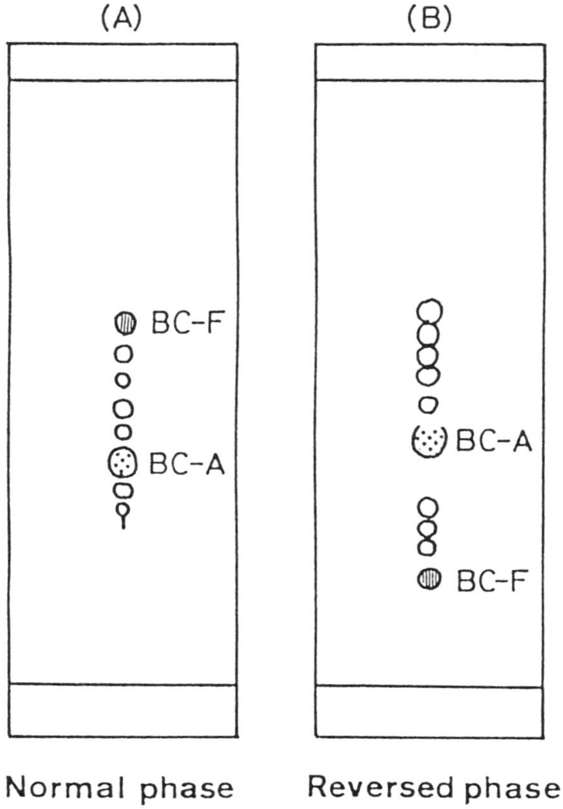

Figure 6 Separation of BC components on TLC
(A) Silica gel TLC, (B) RP-8 TLC
Separation conditions: see Table 1.
SOURCE : Reproduced by permission of *Journal of Chromatography*. See reference (*10*).

using acetone–2 M aqueous sodium chloride solution (1:1) as shown in Figure 7 (*14*). In this solvent system, the concentration of sodium chloride was the most important factor for the separation. The spots of CL components did not separate without salt in the solvent system because of extreme tailing, and the shapes of the spots were improved by increasing the concentration of the salt. The tailing has been attributed to the influence of silanol groups on the TLC stationary phase, and the salt added to the solvent system can reduce the interaction between the silanols and the components of CL. However, salt could not be used at concentrations in excess of 2.5 M in the solvent system, because the salt solution was not miscible with acetone at that concentration. A concentration of 2.0 M sodium chloride aqueous solution was therefore used in the solvent system. Using these separation conditions in both residue analyses and for quality checks of working standards is recommended.

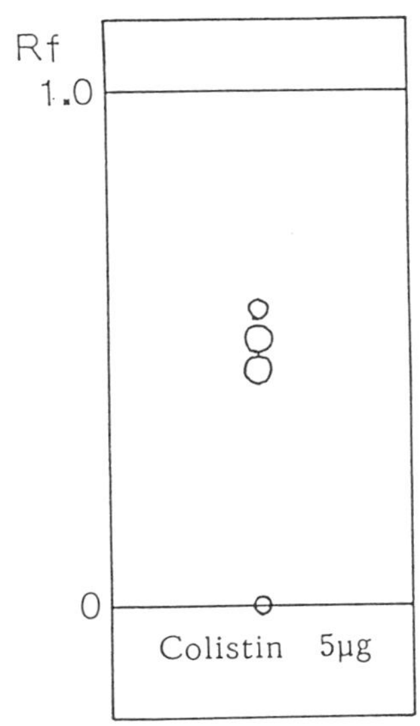

Figure 7 Separation of CL components on TLC.
Separation conditions: see Table 1.
Detection procedures: Visualization method–(a) spray 0.5% ninhydrin-*n*-butanol solution; (b) heat the plate at 120°C for 1 min. Fluorescence method–(a) spray 0.5 M borate buffer (pH 10); (b) spray 0.2 % fluorescamine-acetone solution; (c) observe the fluorescence under UV-light (365 nm).

High-Performance Liquid Chromatography

Bacitracin. For the LC separation of BC components, gradient elution has commonly been employed (*16*, *18*, *19*). However, isocratic separation is more desirable for routine residue analyses and the preparative separation of BC components, because it is simple, rapid, and requires a less expensive LC system. To separate the BC components under isocratic conditions, some conventional reversed-phase LC columns were tested in combination with various mobile phases, but satisfactory results were not obtained in any trial because of peak tailing. The tailing may be caused by the interaction between the amino group of BC and the residual silanol groups on the column. Some silanol-free stationary phases, such as capsule and polymer type LC columns, were therefore tested and a satisfactory separation of BC was obtained using a combination of Capcell Pak (Shiseido, Tokyo, Japan) C18 LC column (capsule type) and methanol–0.04 M Na_2HPO_4 aqueous solution (6:4) as the mobile phase (*20*). As shown in Figure 8, a commercial BC complex could be separated into 22 peaks within 25 min. In this LC system, the pH value of the aqueous solution in the mobile phase was one of the most effective parameters in controlling the separation of the components. The separation of BC components was improved by increasing the pH value of aqueous solution in the mobile phase (Figure 9), with the best separation near pH 9.0. Because 0.04 M Na_2HPO_4 aqueous solution is at pH 9.4, it was used in the mobile phase without further pH adjustment. These results suggest that the selection of a silanol-free LC column and use of an alkaline mobile phase are key points in obtaining a satisfactory separation of BC components under isocratic LC conditions.

Colistin. Some reports have been published concerning the separation of polymyxin antibiotics including CL (*14*, *16*, *21–24*), and most of them describe satisfactory separations using a C18 LC column. A highly concentrated salt, such as sodium sulfate, is required in the mobile phase in order to obtain good separation. This is attributed to the presence of five amino groups present in the structures of CL-A and -B, which may interact with the silanol groups of the column, thereby affecting the separation. Adding highly concentrated salts to the mobile phase controls these interactions. However, too high a salt concentration in the mobile phase shortens the column life and causes various problems with the LC apparatus, such as clogging of the connection tubing. Furthermore, a mobile phase containing nonvolatile salts is not applicable to LC–MS analysis. Therefore, I investigated separation conditions using volatile materials only in the mobile phase.

Because a combination of acetonitrile and trifluoroacetic acid (TFA) aqueous solution is frequently used as a mobile phase for separating peptide compounds, this combination was applied to CL. Satisfactory separation was not obtained using conventional LC columns, but wide-pore ODS and phenyl type LC columns gave excellent separation of CL components, as observed in Figure 10. These results

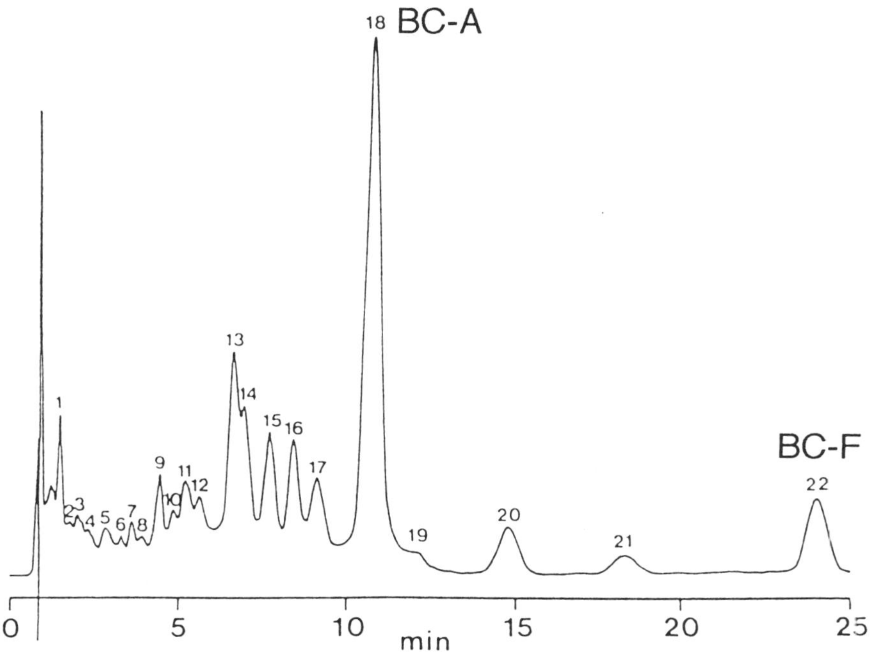

Figure 8 Separation of BC under isocratic LC conditions.
Conditions: Column, Capcell Pak C18 (5 μm, 150 × 4.6 mm i.d.); mobile phase, methanol-0.04 M Na_2HPO_4 aq. soln. (6:4); flow rate, 1.3 mL/min; detection, 234 nm.
SOURCE: Reproduced by permission of *Journal of Chromatography*. See reference (*20*).

suggest that wide-pore or phenyl type LC columns are suitable for separating CL components, because they do not require any salt in the mobile phase.

Virginiamycin and Enramycin. There are reports concerning the separation of VGM and ER components using LC (*25–27, 30, 31*) that show that the components of each antibiotic are well separated using conventional ODS columns and mobile phases as seen in Figures 11 and 12. Because the components of VGM and ER have no primary amino group in their structures, the influence of residual silanol groups on the LC column is less than in the cases of BC and CL. As a result, there are no special requirements for specific LC columns or mobile phases to separate VGM and ER components.

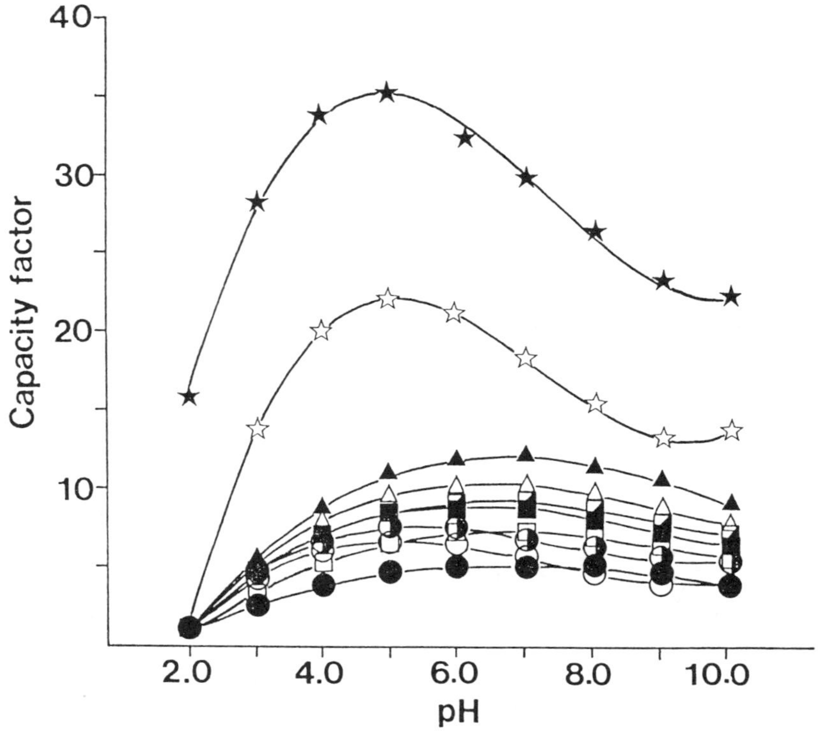

Figure 9 Influence of pH values of mobile phase on the separation of BC components.
Conditions: Column, Capcell Pak C18 (5μ m, 150 × 4.6 mm i.d.); mobile phase, methanol-0.04 M Na_2HPO_4 aq. soln. (6:4); flow rate, 1.3 mL/min; detection, 234 nm.
○: peak 11; ●: peak 12; ◑: peak 13; □: peak 14; ■: peak 15; ◩: peak 16; △: peak 17; ▲: peak 18 (BC-A); ☆: peak 20; ★: peak 22 (BC-F); No. of peaks, see Fig. 8.

SOURCE : **Reproduced by permission of** *Journal of Chromatography*. **See reference (*20*).**

Detection

Fluorometric Detection on TLC

Generally, visualization of the spot with ninhydrin is used to detect peptides and amino acids on TLC plates, as shown in Table 1. This technique is simple, rapid, and reliable, but it is not suitable for residue analyses of the peptide antibiotics because of a lack of sensitivity. To provide adequate sensitivity for the detection of the peptides and amino acids at residue levels on the TLC, fluorometric derivatization reagents for the amino group, especially fluorescamine, are most

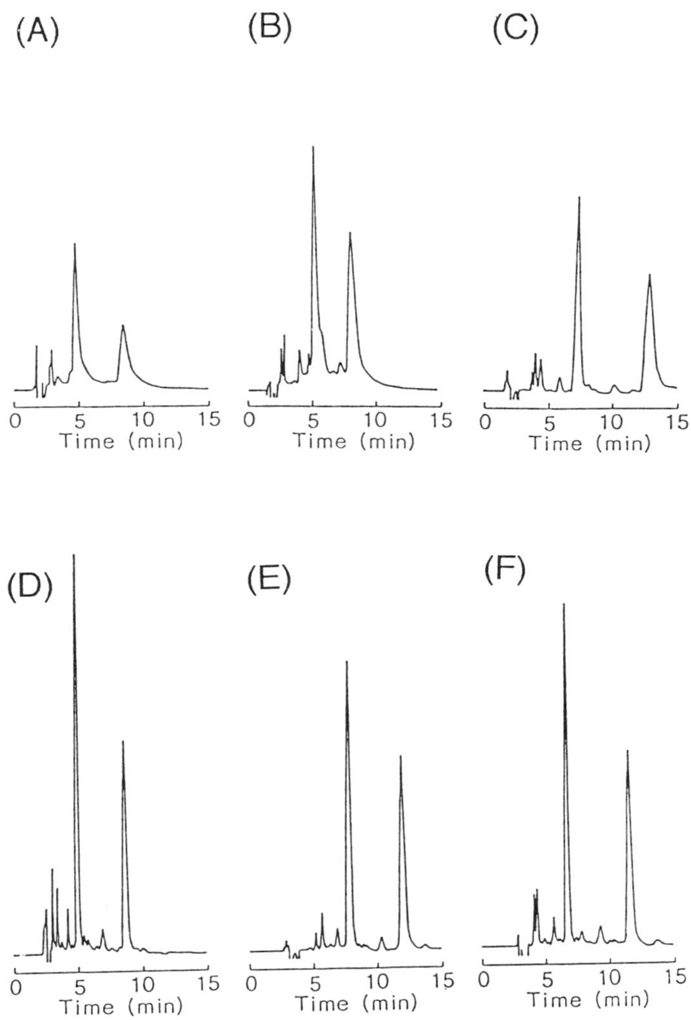

Figure 10 Separation of CL using various LC columns.
Conditions: Mobile phase, mixture of acetonitrile (a) and 0.01 M TFA aq. soln. (b); flow rate, 1.0 mL/min; detection; 220 nm. (A): Nucleosil 5C18 (5 μm, 150 × 4.6 mm i.d.); mobile phase ratio a:b, 23:77. (B): LiChrosorb RP-18 (5 μm, 150 × 4.6 mm i.d.); mobile phase ratio a:b, 30:70. (C): LiChrosorb RP-Select B (5 μm, 250 × 4.6 mm i.d.); mobile phase ratio a:b, 24:76. (D): Kaseisorb ODS-300-5 (wide pore, 250 × 4.6 mm i.d.); mobile phase ratio a:b, 20:80. (E): Nucleosil 7 Ph (7 μm, 250 × 4.6 mm i.d.); mobile phase ratio a:b, 20:80. (F): Inertsil 5 Ph (5 μm, 250 × 4.0 mm i.d.); mobile phase ratio a:b, 25:75.

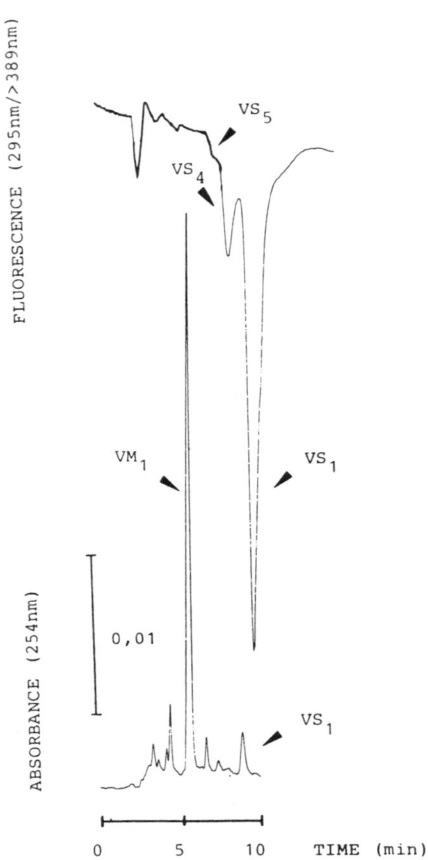

Figure 11 Separation of virginiamycin.
Conditions: Column, Hypersil ODS (5 μm, 250 × 4.6 mm i.d.); mobile phase, acetonitrile-methanol-phosphoric acid aq. soln. (pH 2.12, about 1 mM) (44:10:46); flow rate, 1.0 mL/min; UV detection, 254 nm; FL detection, Ex=295 nm and Em > 389 nm.
SOURCE : **Reproduced by permission of** *Journal of Chromatography*. **See reference (25).**

frequently used. Fluorescamine has the ability to derivatize both primary and secondary amino groups, it reacts quickly without heating, and it produces a reaction product with an intense fluorescence. This method has been applied to the detection of colistin components separated on a TLC plate, and the derivatization procedures are shown in Figure 7, together with the separation. CL-A and -B in 10 ng of commercial CL complex were successfully detected, although the detection limit of the ninhydrin method was 1 μg. This is a satisfactory detection

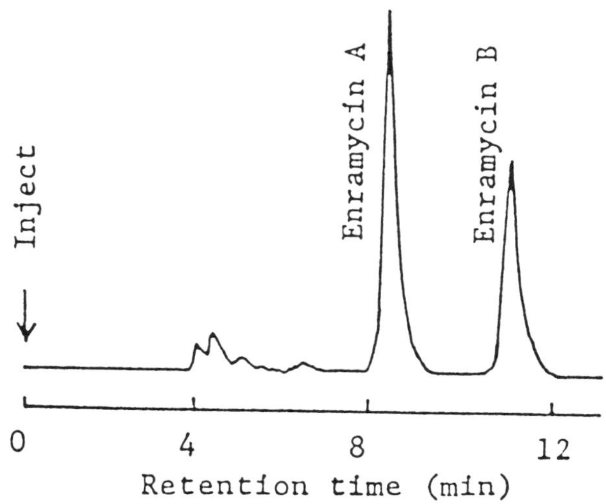

Figure 12 Separation of enramycin.
Conditions: Column, Nucleosil 5C18 (5 μm, 250 × 4.6 mm i.d.); mobile phase, acetonitrile-0.05 M NaH_2PO_4 (pH 4.0) (35:65); flow rate, 0.5 mL/min; detection, 230 nm.
SOURCE : **Reproduced by permission of** *Journal of the Hygienic Society of Japan.* **See reference (*26*).**

limit for residue analysis. Because this fluorometric method is simple and sensitive, its use for detecting not only CL but other peptide antibiotics as well is recommended.

UV Detection in Liquid Chromatography

This is a most popular detection method for LC analysis, and it is frequently used to analyze pharmaceutical preparations and residues when the analyte has a comparatively strong UV chromophore. UV detection has been used for the residue analysis of VGM (*30, 31*) and ER (*26*) in meat and for the determination of BC in feed (*17*). However, satisfactory results could not be obtained when attempts were made to apply this detection method to the residue analysis of BC and CL in food, because of a lack of sensitivity and also causes interference from the sample matrix.

Fluorescence Detection in Liquid Chromatography

This is one of the most effective detection methods for residue analysis because of its sensitivity and selectivity. However, the application of this method is limited to analytes in which specific fluorophores exist in the structure, like VGM (*27*), or to those analytes that fluoresce with a suitable derivatizing agent. Derivatization with *o*-phthalaldehyde (OPA) is frequently used for the fluorescent determination of amino compounds such as peptides and amino acids. To apply the OPA derivatization method to the determination of LC-separated BC and CL components, various reaction parameters were optimized. Although this derivatization is applicable to both pre- and postcolumn systems, a postcolumn system was chosen to simplify the preparation of sample solution. Satisfactory chromatograms of BC-A, and CL-A and -B are obtained. Figure 13 shows the postcolumn derivatization LC system used and operating conditions, and the resulting LC profiles of BC-A, while CL-A and -B are shown in Figure 14. The detection limits of each component were 1 ng, although those using UV detection were over 10 ng. This postcolumn florescence derivatization LC system is simple, rapid, and

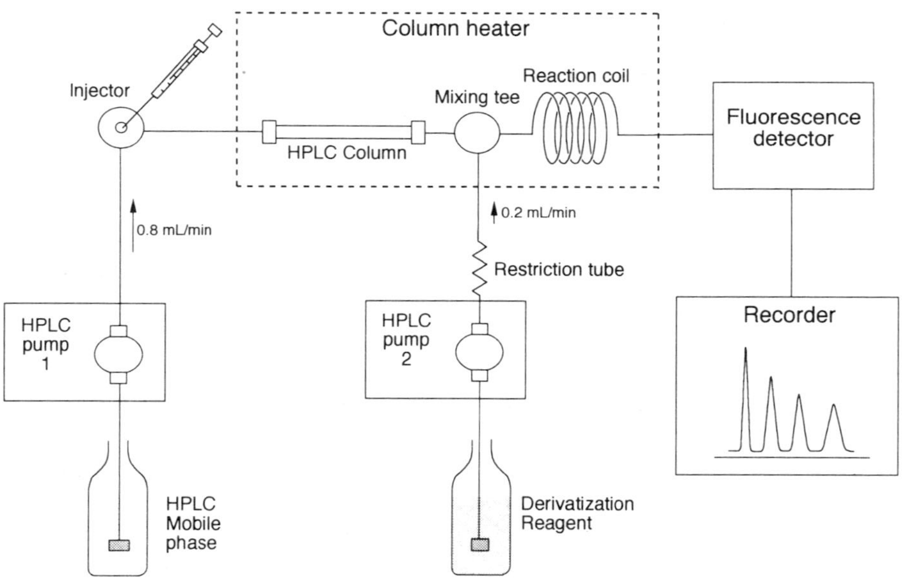

Figure 13 Post-column fluorescence derivatization LC system
Conditions: Flow rate of mobile phase, 0.8 mL/min; flow rate of derivatization reagent, 0.2 mL/min; reaction coil, 50 cm × 0.25 mm i.d.; reaction temp, 30°C; detection, Ex=340 nm, Em=455 nm. Preparation of derivatization reagents: OPA (1.6 g) was dissolved in 20 mL of ethanol. The OPA solution, 15 mL of 2-mercaptoethanol and 30 mL of 10% Brij-35 aqueous solution were mixed and filled up to 1000 mL with 0.04 M borate buffer (pH 10.0)

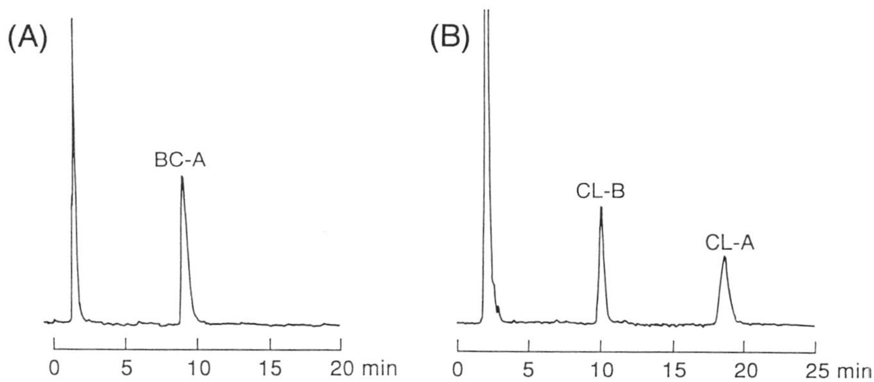

Figure 14 LC profiles of BC-A, CL-A and -B.
(A) BC-A. Conditions: Column, Inertsil Ph (5 μm, 150 × 4.6 mm i.d.); mobile phase, acetonitrile-0.01 M TFA aq. soln. (26:74); sample size, 10 ng; other conditions, see Figure 13. (B) CL-A and -B. Conditions: Mobile phase, acetonitrile-0.017 M TFA aq. soln. (20:80); sample size, 10 ng (each); other conditions, same as above.

sensitive, making it applicable to the residue analysis of not only BC and CL but also other peptide antibiotics.

Preparation of Working Standard

Peptide antibiotics consist of two or more components, and because the isolated pure components are either not commercially available or are very expensive, analysts must isolate the components to obtain working standards. The previously reported isolation conditions for peptide antibiotics are listed in Table 3. Most workers have used preparative LC or medium-pressure liquid chromatography (MPLC) to isolate the components. BC-A, CL-A and -B, and VGM-M_1 and -S_1 have been successfully isolated using LC (*20*, *22*, *24*) and MPLC (*25*).

Although LC and MPLC can achieve high resolution with simple operation, they have several disadvantages including the quantity of the sample that my be applied is limited and sample may be lost by irreversible adsorption on the stationary phase. On the other hand, to efficiently isolate the main components from BC, CL, and VGM complex, high-speed countercurrent chromatography (HSCCC) (*14*, *32–34*) and droplet countercurrent chromatography (DCCC) (*29*) have been utilized. Figure 15 shows the elution curve of BC components separated by preparative HSCCC using a chloroform–ethanol–methanol–water mixture (5:3:3:4) as the solvent system (*34*). BC-A was successfully isolated from BC

Table 3 Preparative Conditions of Peptide Antibiotics

Apparatus		Operation conditions	Monitor (nm)	Ref.
LC	BC	Column: Capcell Pak C18 (5 μm, 150 × 4.6 mm i.d.), Mobile phase: MeOH-0.05 M Na_2SO_4 aq. soln. (6:4), Flow rate: 1.3 mL/min	UV (234)	*(20)*
	BC	Column: YMC basic (C8, 200Å, 5 μm, 250 × 21 mm i.d.), Mobile phase: MeOH-0.05 M KH_2PO_4 aq. soln. (pH 6.5) (59:41), Flow rate: 20 mL/min	UV (215)	*(19)*
	CL	Column: Amberlite XAD-2 (100-200 mesh, 550 × 16 mm i.d.), Mobile phase: MeOH-0.2 M NaCl-0.01 M HCl, gradient, Flow rate: 0.5 mL/min	VIS (570)	*(22)*
	CL	Column: octadecyldimethylsilyl substituted LiChrosorb Si 100 (10 μm, 250 × 40 mm i.d.), Mobile phase: 23% MeCN in 0.023 M H_3PO_4, 0.01 M AcOH and 0.05 M Na_2SO_4 aq. soln. adjusted pH 2.5 by TEA, Flow rate: 20 mL/min	UV (365)	*(24)*
	VGM	Column: TSKgel ODS 120T (5μm, 150 × 4.6 mm i.d.), Mobile phase: MeOH-MeCN-0.015 M NaH_2PO_4 aq. soln.-THF (43:22:34:1), Flow rate: 0.5 mL/min	FL (Ex 311, Em 427)	*(27)*
MPLC	VGM	Column: LiChroprep 60 (25-40 μm, 300 × 23 mm i.d.), Mobile phase: chloroform-MeOH-AcOH (100:12:12)	UV (276)	*(25)*
HSCCC	BC	Solvent system: chloroform-EtOH-MeOH-water (5:3:3:4), Mobile phase: lower layer, Flow rate: 3.0 mL/min, Revolution speed: 800 rpm	UV (254)	*(34)*
	CL	Solvent system: *n*-BuOH-0.04 M TFA aq. soln. (1:1), Mobile phase: upper layer, Flow rate: 2.0 mL/min, Revolution speed: 800 rpm	UV (220)	*(14)*
DCCC	VGM	Solvent system: benzene-chloroform-MeOH-water (26:14:24:6), Mobile phase: upper layer, Flow rate: 24 mL/h	UV (300)	*(29)*

BC: bacitracin; MeOH: methanol; CL: colistin; MeCN: acetonitrile; AcOH: acetic acid; TEA: triethylamine; VGM: virginiamycin; THF: tetrahydrofrane; MPLC: medium-pressure liquid chromatography; HSCCC:high-speed countercurrent chromatography; EtOH: ethanol; BuOH: butanol; TFA: trifluoroacetic acid; DCCC: droplet countercurrent chromatgraphy complex under these conditions. In addition, HSCCC was applied to the isolation of CL-A and -B from a commercial CL complex using *n*-butanol–0.04M TFA aqueous solution (1:1) as a solvent system (*14*). Because the purity of each isolated component was over 90% for BC-A and CL-A and -B from the LC chromatogram, they were usable as working standards for chemical analyses. This apparatus is somewhat specific to the Japanese Laboratories of Public Health, but it is useful for the preparative separation of complex samples because of its excellent separation ability, its lack of requirement for salt in the solvent system, its absence of

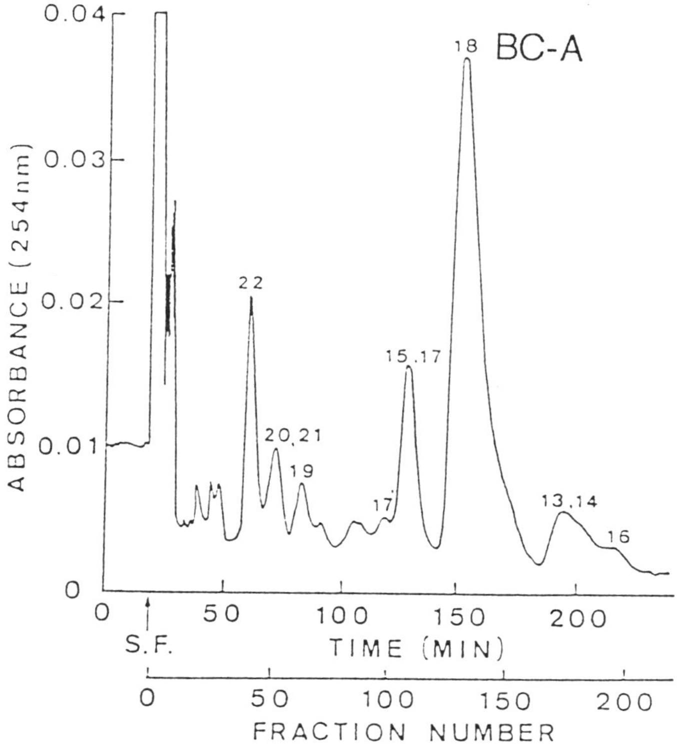

Figure 15 HSCCC elution curve of commercial BC complex.
Conditions: Apparatus, HSCCC-1A (Shimadzu, Kyoto, Japan); revolution speed, 800 rpm; solvent system, chloroform-ethanol-methanol-water (5:3:3:4); mobile phase, lower layer; flow rate of mobile phase, 3.0 mL/min; detection, 254 nm; sample size, 50 mg.
SOURCE : Reproduced by permission of *Journal of Chromatography*. See reference (*34*).

irreversible adsorption on the stationary phase, and its capability for sample loading. Separation conditions for HSCCC and DCCC are listed in Table 3, together with preparative LC conditions.

Extraction and Cleanup from Agricultural Products

There are very few published reports concerning the chemical analysis of residual peptide antibiotics in food (*26, 27, 30, 31*), none of which include BC and CL. Table 4 lists the extraction solvents, the cleanup and determination techniques used, and the recovery data of each method, including one under development in the Aichi Prefectrual Institute of Public Health for residues of BC and CL (Y. Ikai, unpublished data).

Bacitracin and Colistin

Gallagher et al. (*17*) reported a quantitative LC method for BC in feed using a C18 prepacked cartridge in the cleanup procedure that provided excellent recoveries. A possible starting point in establishing a simple and rapid cleanup method for residual BC in livestock products using a C18 prepacked cartridge. Based on this, the following procedure was planned. BC is extracted from the sample by homogenization with extraction solvent followed by centrifugation and filtration, and the sample extract is then directly applied to the cartridge. After washing the cartridge with suitable solvent, the BC is eluted with a suitable eluent and the eluate is directly injected into the LC system without any additional treatment.

To adapt this procedure to the actual analysis of BC, various parameters were optimized. Because BC is soluble in water and alcohols and is stable in acid solution, aqueous solutions of acids (sulfuric acid, hydrochloric acid, perchloric acid, and TFA) and methanol were examined as extraction solvents. From this a 0.5% sulfuric acid aqueous solution was selected because it gave the best recoveries and gave the fewest interfering peaks on the chromatogram. Bond Elut C18 (500 mg) was the cartridge used and was pretreated with disodium ethylenediaminetetraacetic acid (Na_2EDTA) aqueous solution to avoid the formation of complexes of BC with metals. As a washing solvent for the cartridge, 25% acetonitrile–0.1M TFA aqueous solution was used, because it was most effective in removing interfering substances from the sample. For the elution of BC from the cartridge, LC mobile phase was used so that the eluate may be directly injected into the chromatograph. Scheme 1 illustrates the analytical procedure based on the results of these experiments. BC-A fortified in beef muscle at 1 ppm was recovered using this procedure and was determined using the postcolumn flourescence derivatization LC system. LC chromatograms of fortified and blank beef muscles

Table 4 Extraction and Cleanup Conditions of Peptide Antibiotics in Feed and Foods

	Sample	Extraction and cleanup	Determination	Recovery	Detection limit	Ref.
BC	feed	Pretreatment: defat with petroleum ether, Extraction: MeOH-MeCN-phosphate buf., Cleanup: Bond Elut C18 cartridge	UV–LC	96–101%	—	(*17*)
	beef muscle	Extraction: 0.5% sulfuric acid aq. soln., Cleanup: Bond Elut C18 cartridge	FL–LC	BC-A: 63.7%	BC-A: 0.2 ppm	—[a]
CL	beef muscle	Extraction: 0.5% sulfuric acid aq. soln., Cleanup: Bond Elut C18 cartridge	FL–LC	CL-A: 64.0%, CL-B: 64.9%	CL-A: 0.2 ppm, CL-B: 0.2 ppm	—[a]
VGM	beef, pork, and chicken meat	Extraction: MeCN, Cleanup: liquid-liquid partition	FL–LC	M_1: 98.9%, S_1: 99.5%	M_1: 0.1 ppm, S_1: 0.01 ppm	(*27*)
	swine tissue	Extraction: MeOH-0.2 M $NH_4H_2PO_4$ aq. soln., Cleanup: liquid-liquid partition	UV–LC	M_1: 83.3–98.2%	—	(*30*)
	pork and chicken meat	Extraction: MeOH-3% phosphotungstic acid aq. soln., Cleanup: liquid-liquid partition, Sep-Pak silica cartridge	UV–LC	M_1: 93.6–95.2%	M_1: 0.05 ppm	(*31*)
ER	beef, pork, and chicken meat	Pretreatment: heating in boiling water, Extraction: acetone-water, Cleanup: liquid-liquid partition	UV–LC	54.0–62.4%	0.2 ppm	(*26*)

BC: bacitracin; MeOH: methanol; MeCN: acetonitrile; CL: colistin; VGM: virginiamycin; ER: enramycin

[a] Y. Ikai, unpublished data.

are shown in Figure 16. The average recovery of four replicates and the detection limit of BC-A were 63.7% (CV, 7.0%) and 0.2 ppm, respectively.

The analytical procedure for BC was modified for analyzing CL and the resulting procedure is shown together with that of BC in Scheme 1. Recoveries of CL-A and -B from beef muscle were 64.0% and 64.9% (CV, 3.0% and 3.1%),

Sample (5 g)
- add 20 ml of 0.5 % sulfuric acid aq. soln.
- homogenize for 30 sec
- centrifuge (8000 r.p.m., 10 min)
- filtrate (filter: No. 4C)

(repeat)

Filtrate

BOND ELUT C18 (500 mg)*
- wash with 5 ml of water
- wash with 5 ml of washing solvent**
- elute with 10 ml of effluent***

Eluate

HPLC analysis

Scheme 1 Cleanup procedures of residual BC and CL in meat.
Conditions for BC: * Pretreat with 5 mL of methanol, 5 mL of water followed by 5 ml of 2 M Na_2EDTA aq. soln. ** 25 % Acetonitrile and 0.1 M TFA in water.* Acetonitrile-0.01 M TFA aq. soln. (26:74). Conditions for CL: * Pre-treat with 5 mL of methanol and 5 mL of water. ** 15 % Acetonitrile and 0.1 M TFA in water. *** Acetonitrile-0.017 M TFA aq. soln. (20:80).**

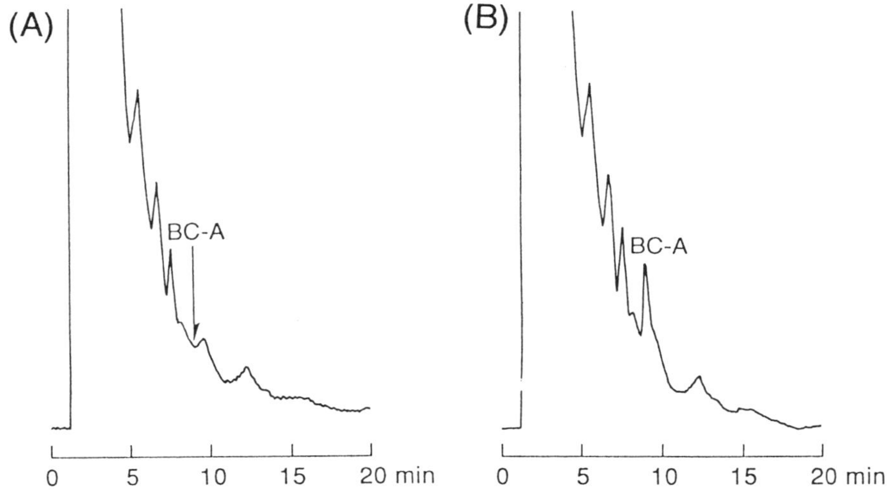

Figure 16 Typical LC profiles of beef meat cleaned up by method for BC. (A) Blank. (B) Addition of BC-A at the level of 1.0 ppm.

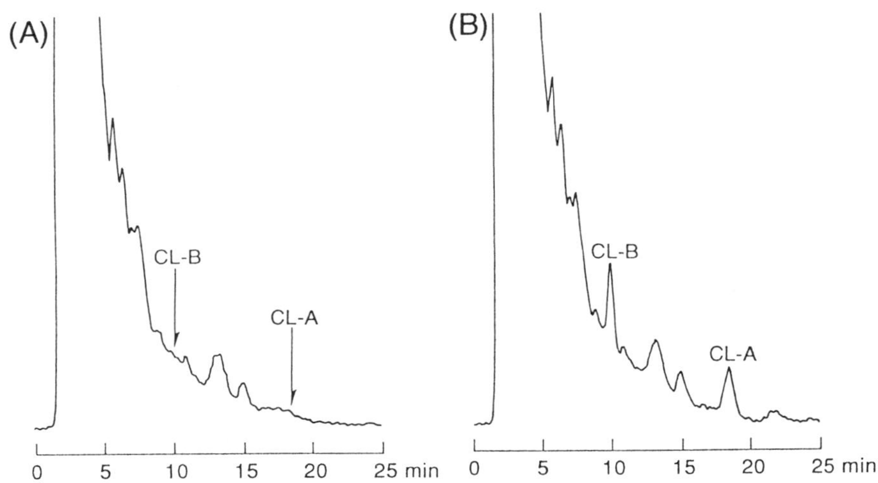

Figure 17 Typical LC profiles of beef meat cleaned up by method for CL. (A) Blank. (B) Addition of CL-A and -B at the level of 1.0 ppm.

respectively, at a fortification level of 1 ppm, with a detection limit of 0.2 ppm for both components. The LC chromatograms of fortified and blank beef muscle samples are shown in Figure 17.

These analytical methods are very simple and rapid, because they do not require any time-consuming procedures such as concentrating the extract by rotary evaporator, liquid–liquid partitioning using separatory funnels, or column chromatography. The required time from sampling to determination is approximately 1 h. Because Japanese official guidelines for the acceptable minimum recovery and detection limit, in the analysis of residual antimicrobial agents in food require 70% and 0.05 ppm, respectively, the recoveries and detection limits obtained are somewhat inadequate. The low recovery is attributed to the precipitation of the analyte together with the sample matrix at the extraction process, and the unsatisfactory detection limits are due to interfering peaks on the chromatograms. Future work is intended to modify the procedures by improving the extraction solvent and developing additional cleanup steps to obtain satisfactory recoveries and detection limits.

Virginiamycin

Nagase et al. (*27*) Moats et al. (*30*) and Saito et al. (*31*) have reported LC analytical methods for determining residual VGM in animal tissues. In these reports, VGM was extracted by acetonitrile or mixtures of methanol and aqueous solution of deproteinizing agents, followed by cleanup based on liquid–liquid partition. An additional cleanup using prepacked cartridges was adapted to the procedure that provided an excellent cleanup effect on the chromatogram (*31*). Sufficient recoveries (over 80%) were obtained in all these methods (*27*, *30*, *31*), with separations performed on an ODS column. UV detection was used for determining VGM-M_1 (*30, 31*) and a satisfactory detection limit (below 0.05 ppm) was achieved. For the determination of VGM-S_1, fluorescence detection was successfully used (*27*). Based on the performance of these methods, the additional cleanup step and the determination conditions are key points for the chemical analysis of residual VGM in livestock products.

Enramycin

Only one report has been published concerning the residue analysis of EN in meat by Horie et al. (*26*). After pretreating the sample by heating in boiling water, ER was extracted with acetone–water under acidic conditions. The cleanup was based on a liquid–liquid partition, and the determination was by LC with UV detection. Recoveries of ER from meat were from 54.0–62.4%, and the detection limit was 0.2 ppm. Although satisfactory separation of ER-A and -B without interfering peaks was observed on the chromatogram, the recoveries and detection limit are not entirely satisfactory for residue analysis.

Conclusion

Antibiotics are considered necessary in current farming practices as veterinary medicines and feed additives to provide economical livestock production. To ensure the safety of foods by monitoring for residues of these antibiotics, chemical analysis methods are required to provide quantification and identification of the residues. However, very few reports have been published concerning the residue analysis of peptide antibiotics. In particular, no analytical method for residual BC and CL has been reported, even though they are the most frequently used of the peptide antibiotics. Therefore, simple, rapid, and reliable chemical analysis methods are required for all peptide antibiotics used as veterinary medicines and feed additives.

In this chapter, the author has proposed several approaches to the establishment of chemical analysis methods for the peptide antibiotics. The florescence detection system with fluorescamine on TLC and the postcolumn fluorescence derivatization LC system with OPA were sensitive enough to determine BC and CL at the same level as bioassays, and they are potentially applicable to other peptide antibiotics. Prepacked C18 cartridges effectively simplified the cleanup procedure for BC and CL. Further work using prepacked cartridges, including the reversed-phase, normal-phase, and ion-exchange types that are commercially available, should lead to more excellent cleanup effects, particularly when different types of cartridges are used in combination.

Capillary electrophoresis, liquid chromatography–mass spectrometry, photodiode array detection, matrix solid-phase dispersion extraction, and on-line column-switching for trace enrichment are all promising, recently developed analytical techniques that have been successfully applied to the analysis of other antimicrobial agents. Therefore, the author believes that it is possible to establish simple, rapid, and reliable analytical methods required for all peptide antibiotics used in animals by investigating these newer techniques, as well as those described here.

Acknowledgment

The author appreciates the valuable discussions with Professor M. Suzuki, Meijo University, on several points in this chapter.

References

1. B. A. Johonson, H. Anker and F. L. Meleney, *Science*, **102**: 376, 1945.
2. Y. Koyama, A. Kurosawa, A. Tsuchiya, K. Takakuda, *J. Antibiot.*, **3**: 457, 1950.
3. P. De Somer and P. Van Dijck, *Antibiot. Chemother.*, **7**: 606, 1957.

4. E. Higashiide, K. Hatano, M. Shibata, and K. Nakazawa, *J. Antibiot.*, **21**: 126, 1968.
5. A. Aszalos, S. Davis, and D. Frost, *J.Chromatogr.*, **37**: 487, 1968.
6. R. J. Stretton, J. P. Varr, and J. Watson-Walker, *J. Chromatogr.*, **45**: 158, 1969.
7. R. Bossuyt, R. V. Renterghem, and G. Waes, *J. Chromatogr.*, **124**: 37, 1976.
8. J. K. Pauncz, *J. Antibiot.*, **25**: 677, 1972.
9. A. Aszalos and A. Aquilar, *J. Chromatogr.*, **290**: 83, 1984.
10. H. Oka, Y. Ikai, N. Kawamura, M. Yamada, K. -I. Harada, Y. Yamazaki, and M. Suzuki, *J. Chromatogr.*, **449**: 448, 1988.
11. M. Iglóy and A. Mizsei, *J. Chromatogr.*, **28**: 456, 1967.
12. M. Iglóy and A. Mizsei, *J. Chromatogr.*, **34**: 546, 1968.
13. A. H. Thomas and I. Holloway, *J. Chromatogr.*, **161**: 417, 1978.
14. Y.Ikai, H. Oka, N. Kawamura, M. Yamada, K. I. Harada, M. Suzuki, K. Komori and Y. Ito, *The 110th Annual Meeting of the Pharmaceutical Society of Japan, abstract,* Vol. 3 (The Pharmaceutical Society of Japan, 1990), 230.
15. K. Tuji, J. H. Robertson, and J. A. Bach, *J. Chromatogr.*, **99**: 597, 1974.
16. K. Tuji and J. H. Robertson, *J. Chromatogr.*, **112**: 663, 1985.
17. J. B. Gallagher, P. W. Love, and L. L. Knotts, *J. Assoc. Off. Anal. Chem.*, **65**: 1178, 1982.
18. Y. Ikai, H. Oka, J. Hayakawa, K. -I. Harada, and M. Suzuki, *J. Antibiot.*, **45**: 1325, 1992.
19. R. G. Bell, *J. Chromatogr.*, **590:** 163, 1992.
20. H. Oka, Y. Ikai, N. Kawamura, M. Yamada, K. -I. Harada, Y. Yamazaki, and M. Suzuki, *J. Chromatogr.*, **462**: 315, 1989.
21. S. Terabe, R. Konaka, and J. Shoji, *J. Chromatogr.*, **173**: 313, 1979.
22. Y. Kimura, H. Kitamura, T. Araki, K. Noguchi, M. Baba, and M. Hori, *J. Chromatogr.*, **206**: 563, 1981.
23. T. J. Whall, *J. Chromatogr.*, **208:** 118, 1981.
24. I. Elverdam, P. Larsen, and E. Lund, *J. Chromatogr.*, **218**: 653, 1981.
25. E. Grell, E. Lewitzki, S. Dehal, I. Oberbäumer, F. Raschdorf, and W. J. Richter, *J. Chromatogr.*, **290**: 57, 1984.
26. M. Horie, Y. Hoshino, N. Nose, H. Nakazawa, M. Fujita, and E. Takabatake, *J. Food Hyg. Soc. Jpn.*, **26**: 337, 1985.
27. M. Nagase and K. Fukamachi, *Bunseki Kagaku*, **36**: 297, 1987.
28. D. W. Gottschall, C. Gombatz, and R. Wang, *J. Agric. Food Chem.*, **35**: 900, 1987.
29. K. Saito, M. Horie, Y. Hoshino, N. Nose, Y. Shida, H. Nakazawa, and M. Fujita, *J. Chromatogr.*, **454**: 387, 1988.
30. W. A. Moats, *J. Agric. Food Chem.*, **36**: 1297, 1988.

31. K. Saito, M. Horie, Y. Hoshino, N. Nose, H. Nakazawa, and M. Fujita, *Eisei Kagaku,* **35**: 63, 1989.
32. Y. Ito, H. Oka, and J. L. Slemp, *J. Chromatogr.*, **463**: 305, 1989.
33. M. Bhatnagar, H. Oka, and Y. Ito, *J. Chromatogr.*, **463**: 317, 1989.
34. K.-I. Harada, Y.Ikai, Y. Yamazaki, H. Oka, M. Suzuki, H. Nakazawa, and Y. Ito, *J. Chromatogr.*, **45**: 1325, 1992.

INDEX